Graeme Marett C Eng MITE

A History of Electronic Communications in the Channel Islands

2017

Copyright © 2017 Graeme Marett

All rights reserved. No part of this publication may be reproduced or transmitted in any form or by any means, electronic or mechanical including photocopying, recording or any information storage or retrieval system, without prior permission in writing from the publishers.

The right of Graeme Marett to be identified as the author of this work has been asserted by him in accordance with the Copyright, Designs and Patents Act 1988

First published in the United Kingdom in 2017 by
The Cloister House Press

ISBN 978-1-909465-63-3

Foreword

When the subject of electronic communications is raised most people tend to immediately think of the telephone, which in today's world is quite natural. However, electrical telecommunications began half a century before the invention of the telephone. The first means of reliable and speedy communication over long distances was the electric telegraph. It was one of many Victorian inventions that had huge repercussions on the day to day life of people everywhere.

The telegraph engineers developed a means of communication that first enabled almost instant communications between tow points and then between many towns. Soon the telegraph was sending messages internationally and within a few years to America and other continents. The telegraph remained the main means of international commerce until after the Second World War. The telegraph arrived in the Channel Islands very soon after the development of submarine cables and was the only means of direct communication with the outside world for over 70 years. Even before the First World War wireless telegraphy arrived in Jersey to enhance shipping communications, which had hitherto relied on visual signalling, and it was used for telegraph traffic when the submarine cables to the Channel Islands failed.

Electronic communications is of course not confined to person to person messaging as in telegraphy or telephony. Broadcasting is also an electronic communication medium and is so widely used that it merges into the backfround of our everyday life. After the First World War wireless was developed into an entertainment medium for both radio and television. The island was well placed to listen to stations from all over Europe including the BBC, Radio Luxembourg and on the neighbouring French coast the English service of Radio Normandy. Early television signals from the BBC transmitter at Crystal Palace were received experimentally locally by keen wireless enthusiasts in the 1930s. After the Second World War television was brought to Jersey in the mid-1950s then later in the 1980s and 1990s local radio was developed. Mobile wireless telephony followed in the 1980s and 1990s.

Nowadays, telecommunications is an essential part of our daily life; we increasingly rely on the technology for communication both business and private. Technological advance has been it strength, it is perhaps the only industry in the history of mankind in which prices have continuously fallen in real terms since the day it first began. We expect more and more for less and less and the engineers that develop and manage the telecommunications systems consistently deliver that promise.

Electronic communications have become so embedded in our lives that most people never give them a passing thought, that is, until they fail. Through this history I hope to make the science and technology more accessible to everyone and enable it to be shown as a social influence as well as a technological miracle.

Table of Contents

Chapter 1 - The Telegraph System..9

Chapter 2 - The Channel Islands Telegraph Company..................13

Chapter 3 - The Submarine Telegraph Company..........................33

Chapter 4 - The Jersey and Guernsey Telegraph Company Ltd.....43

Chapter 5 - The Telegraph Under the General Post Office 1872 - 1914..59

Chapter 6 - The Arrival of the Telephone.......................................71

Chapter 7 - The National Telephone Company 1895 – 1911.........77

Chapter 8 - Telephones and Telegraphs under the Post Office 1912 - 1923..91

Chapter 9 - The General Post Office Telegraphs and Submarine Cables 1923 – 1940...107

Chapter 10 - The 1914 Telegraph Cable and Military Telegraph Circuit..115

Chapter 11 - The first submarine telephone circuit.....................119

Chapter 12 - Wireless Telegraphy...135

Chapter 13 - The War Office Cables..141

Chapter 14 - Telephones under the States of Jersey 1923 – 1940 147

Chapter 15 - Telecommunications in the Occupation 1940 – 1945 181

Chapter 16 - The Post War Years 1945 – 1959..............................189

Chapter 17 - The Automatic Era 1960 – 1972............................221

Chapter 18 - The States of Jersey Telecommunications Board 1973 – 1985..255

Chapter 19 - From Analogue to Digital 1976 – 1993...................275

Chapter 20 - Towards Competition 1995 – 2002........................323

Chapter 21 - The Telegraph the Telephone, the Railway Companies and Way Leaves..351

Chapter 22 - The Post Office Telegraphs and Cables 1945 – 1972 365

Chapter 23 - Civil Defence Communications.............................385

Chapter 24 - Cable Radio and Television Relay Services.............393

Chapter 25 - Local Broadcasting Services.................................421

Chapter 26 - Local Sound Radio..439

Chapter 1 - The Telegraph System

Before the arrival of the electric telegraph much work had been done on the development of optical systems. The optical telegraph had first been proposed by the French inventor who first demonstrated a practical system in 1792. The system was soon adopted by the French Government which developed a comprehensive network spanning strategic towns and cities across the country. The system used a system of semaphore signals by arranging the position of movable arms on a high mast. With the addition of lamps at the end of the arms the system was able to be used both day and night. Its main weaknesses were its reliance on line-of-sight for the sending and receiving of messages and thus was at the mercy of the weather; furthermore each station had to relay the message to the next and thus there was still a considerable delay in the transmission of messages, albeit still much faster than pony express. The system was, nevertheless, quickly copied in other European countries that quickly realized its military potential.

The great leap forward came with the discovery of electricity and the practical application of electrical circuits. The invention of reliable batteries hastened the development of all all aspects of electrical experimentation and this quickly led to the invention of practical systems.

Jersey, being only a relatively small outpost of the British Empire, was fortunate in having one of the earliest submarine telegraph systems. In fact the installation of the first UK-Channel Islands link was made concurrently with the first attempted (but abortive) trans-Atlantic cable in 1858[1].

There was some British Government interest in the installation of such a cable, since the uncertain relationship with the French over the past century had led to the fortification of the Channel Islands as a measure to protect British interests in the English Channel shipping lanes. The islands were substantially fortified and garrisons were maintained well into the early part of the twentieth century. Indeed, the Admiralty had installed an optical telegraph between the islands during the Napoleonic wars using a bespoke system developed by Mulgrave[2] who had been appointed superintendent of telegraphs (1809 – 1815). The system used a two arm semaphore and was carried out between Alderney and Sark and Sark to Jersey and Guernsey. The main islands of Jersey and Guernsey had a network of coastal stations which were used to direct shipping. This system was abandoned by the military shortly after the end of the conflict in 1815, but the States of Jersey were loaned the stations and continued to use the system for commercial shipping for several years thereafter. The optical semaphore links between La Moye, Noimont and St Helier continued until a telegraph line was installed in April 1887 between La Moye and St Helier. There is still some evidence of this telegraph network at Telegraph Bay in Alderney, where a fine granite tower is preserved, and the Signalling Point at La Moye, Jersey which survives as a private residence. Clearly, then, an electric telegraph would be in the best interest of the defence of the realm and the government looked favourably on the enterprise.

The technological advance was remarkable. It had not been long since the first electric telegraph had been demonstrated by Wheatstone and Cooke[3] in 1937. The first commercial telegraph was installed along the Great Western Railway's line from Paddington to West Drayton in 1839. This used a system of 6 wires and demonstrated the feasibility of telegraphy. But

---

1   Atlantic Bridgehead: Story of Transatlantic Communications, Howard Clayton, Garnstone 1968
2   L. L. Robson, 'Mulgrave, Peter Archer (1778? - 1847)', *Australian Dictionary of Biography*, Volume 2, Melbourne University Press, 1967, pp 267-268
3   Steven Roberts, Distant Writing – A History of Telegraph Companies in Britain between 1838 and 1868

it was Samuel Morse[4] who revolutionised communications with the invention of Morse code patented in 1838. The first commercial link using the Morse code opened from Baltimore to Washington in 1844. After this telegraphs blossomed everywhere. Most of the first telegraph circuits were constructed across land using poles and open wires, however the first successful submarine cable was laid in 1851 from Dover to Calais for Thomas Crampton[5] and his Submarine Telegraph Company.

By 1858 there were telegraph lines everywhere, including several submarine systems, an expansion greatly promoted by the railway companies who used telegraphy extensively themselves and also provided the routes down which telegraph systems could be erected between major towns and cities. It was under these circumstances that the desire for a telegraph connection to the UK grew. Jersey businessmen, always keen to make use of every advantage, were greatly enthusiastic about the possibility of almost instantaneous connection with the London stock exchange or their commercial partners.

The coming of the telegraph also changed peoples lives in other ways. It soon became the medium for the dissemination of news as the local press seized upon its ability to provide almost instantaneous intelligence. It was also of strategic advantage since in the 19th Century the British Government still had a substantial military presence in the islands. This ensured that Parliament had a continuing interest in maintaining and improving communications. Consequently the telegraph was underwritten to some extent from British government funds. This interest attracted speculators that saw opportunities in exploiting government subsidies for their own ends.

---

4   Samuel Finley Breese Morse (1901), Trowbridge, John (ISBN: 0548623457)
5   Steven Roberts, Distant Writing – A History of Telegraph Companies in Britain between 1838 and 1868

Chapter 2 - The Channel Islands Telegraph Company

Businessmen agreed that there would be considerable commercial advantages from having a telegraph at their disposal as had already been demonstrated in the United Kingdom. A report in the local newspaper the *British Press and Jersey Times* for 12 February said: '...a submarine electric telegraph cable seems to be a serious proposition...' and a meeting was held, reported the paper on the 16 February, to propose the formation of a company to undertake the provision of a cable. In order to establish a telegraph company it would be necessary to have the expertise and to establish links with an existing carrier. For these reasons the local parties turned to the Electric and International Telegraph Company[6] which had been formed in 1854 as a result of the amalgamation of the Electric Telegraph Company, established in 1845 and the International Telegraph Company formed in 1853 to promote cables to Holland. The Electric had a virtual monopoly on telegraph lines between UK major towns and cities, holding most of the railway company way leaves.

Negotiations opened and a representation was made to Parliament to obtain cable landing rights. On 9 April the local Chamber of Commerce reported that Her Majesty's Government was prepared to contribute toward the maintenance of such a cable provided that it served all of the Channel Islands.

---

6 Steven Roberts in Distant Writing – A History of Telegraph Companies in Britain between 1838 and 1868

The idea of a telegraph link with the UK certainly fired the imagination of the editor of the *British Press and Jersey Times* as he reported on the 20 February that if such a cable was in place news held over from Portugal would certainly have been published much earlier! By 14 May the news had reached London as the *Daily News* reported a proposed cable from Portland to Alderney could certainly be extended to Guernsey, Jersey and even Cherbourg which lay only a matter of 8 miles off Alderney.

By 31 May the Articles of the proposed Channel Islands Electric Telegraph Company had been drawn up at a meeting held in the British Hotel (now Barclays Bank), Broad Street, St Helier. Those present included:

Robert Grimstone[7]	International Telegraph Co. (Chairman)
Douglas Pitt Gamble	International Telegraph Co.
Mr L W Robins	International Telegraph Co.
Mr E M Gordon	R S Newall and Co.
William Penninger	Solicitor
Thomas Colling Bennett	Company Accountant
Mr Le Breton	Secretary Channel Islands Telegraph Co.
Jurat David De Quetteville	Channel Islands Telegraph Co. (Deputy Chairman)
Philip Gossett	Channel Islands Telegraph Co.
F Carrel	Channel Islands Telegraph Co.

R S Newall[8] and Company of Gateshead was, at that time, virtually the only manufacturer of submarine cables. Newall's core business was the manufacture of rope. They had become cable makers by virtue of holding

---

7   Gerald M. D. Howat, 'Grimston, Robert (1816–1884)', Oxford Dictionary of National Biography first published Sept 2004
8   Facts and Observations Relating to the Invention of the Submarine Cable by R.S. Newall (1882)

certain patents on the enclosure of soft cores with iron wire. They had issued an injunction in 1851 against Wilkins and Wetherly who had infringed their patents during the manufacture of the first Dover Calais cable for the Submarine Telegraph Company. The outcome of this legal battle resulted in Newall's completing the cable and thus moving into the world of cable making. The company was, however, never altogether happy with this diversion and following the failure of the Red Sea cable abandoned cable making in 1858 returning only briefly during 1870. R S Newall's representative at the meeting suggested that a return of around 6% per annum could be realized from such a venture. The formation of the company was therefore agreed. A vote approved a yearly retainer of £100 for the Company Accountant and the Company Secretary together with a grant of £50 each to employ a clerk.

The estimated capital required for the project was £25,000.0.0d, in today's terms about £2M, and estimated charges were 3/- for a telegram to Weymouth and 5/- for a telegram forwarded to London. Such prices were well out of the reach of the working man at that time. As a result of this meeting a flotation was made and £30,000 of working capital was raised from a number of shareholders including several local businessmen. The Channel Islands Telegraph Company was effectively a subsidiary of the Electric and International Telegraph Company which already owned and operated several other submarine cables to Holland and Ireland. The company was incorporated in London under the recently introduced Limited Liability Act[9] and its registered office was at the International Telegraph Company building in Great Bell Alley, Morgate, London[10]. The Company Chairman was Robert Grimstone, who was also the Chairman of the Electric and International Telegraph Company.

On 15 June the *British Press* reported, somewhat optimistically:

'Two of the directors of the Channel Islands Electric Telegraph Company

---

9   Limited Liability Act 1855 (18 & 19 Vict. c133)
10 According to Steven Roberts in Distant Writing – A History of Telegraph Companies in Britain between 1838 and 1868, the CITCo was also stated to be at Founders Court, Lothbury (1860) and Telegraph Street City EC (1861). Note that London post codes were introduced in 1857.

returned to Jersey on Thursday last, after having terminated with the Government the arrangements for the completion of the submarine line, which it is said, will be in operation by the ensuing month - we hear the 15th.'

The company did, however, manage to get a grant from HM Government towards the installation and continuing maintenance of the cable. This grant was guaranteed at £1,800 for military and civil traffic or as much as would deliver a dividend of 6% per annum and would be paid for 25 years so long as the cable carried telegraph traffic.

At the end of June 1858 the cable manufacturer and contractor, W T Henley[11] of East Greenwich, arrived in Guernsey ready to prepare the trenches for the land part of the cable and he was expected in Jersey soon after. On 6 July the London *Shipping Gazette* reported: 'Originally intended to go from Weymouth via Alderney to Jersey then Guernsey, the cable will now go Alderney-Guernsey-Jersey, landing at Lancress bay Alderney and St Martins Point Jersey. Expected charge 5/- per telegram.' The editor of the *British Press,* which reproduced this article, commented that: 'the correspondent is rather wild with insular geography.'!

The final sea route taken was from the Island of Portland across the portion of sea known as *The Shambles* and onward to Alderney. From Alderney it was taken to Fermaine Bay in Guernsey and then from St Martins Point to Greve au Lençon (now called Plémont Bay). The cable was laid by the cable ship *Elba*[12] which was owned and operated by the cable manufacturer R S Newall and Company[13] and possibly the first ship properly fitted for submarine cable laying.

On 27 July the iron tubes which would cover the cable in its passage through the town streets, *'for greater security from accident*', arrived in Jersey. On 3 August the *Elba* arrived from Birkenhead to lay the cable, which was 'covered in Gutta Percha and rolled off a large drum into the sea and

---

11 'William Henley, pioneer electrical instrument maker and cable manufacturer 1813-1882' by A F Anderson , 1985-07 UK0108 NAEST 045/118
12 Cable Ships and Submarine Cables, K R Haigh, Adlard Coles 1968
13 The Atlantic Cable http://www.atlantic-cable.com/Books/Newall/index.htm

onto the shore'. Gutta Percha is a natural substance obtained chiefly from the latex of the Malaysian *Sapotaceae* genus of rubber trees. It is harder than normal rubber and much less flexible. It is, however, waterproof, highly resistive to electric currents and very hard wearing and was used extensively as an insulating material in the early days of electrical equipment and continued to be used for submarine cable into the twentieth century.

The cable was manufactured in Gateshead[14] by Messrs R S Newall and Co. who had been involved in the first abortive trans-Atlantic cable earlier that year. The cable used for the deep sea part was constructed of a No. 1 Gauge (0.3" or 7.62mm) copper conductor covered in gutta percha then served in tarred yarn. This part of the cable would have been supplied by the Gutta Percha Company[15] of West Ham to Newall's as they were the only company at that time with the expertise to produce good quality insulated wire. The outer part of the cable was then lapped by 10 No 6 Gauge (0.192" or 4.877mm) iron wires resulting in a cable that weighed 2½ Tons per mile. The shore ends of the cable, which are subject to more wear and tear because of the tidal flows and wave motion, were lapped with 10 No 2 Gauge iron wires which resulted in a thicker cable weighing 6 Tons per mile. It is likely that the cable used in the Channel Islands link was similar, if not the very same, to that used in the Red Sea cable which had been laid earlier that summer as the supply contract had allowed the company to retain the unused cable[16].

The land part of the cable was constructed in a similar manner except that the armouring iron wires were not necessary and the cable was left finished at the tarred yarn stage. The gangers installed the cable in a 20" (50cm) deep trench, in busy town areas cased in a cast iron tube and in rural areas into a prepared creosoted wooden trough laid in the bottom. The wooden troughing was made of two hollowed out square sections of timber treated with creosote. The upper and lower sections were identical and

---
14 R S Newall rope patent   http://www.afundit.co.uk/washington1.htm
15 John Longman http://www.bouncing-balls.com/index2.htm
16 The Invisible Weapon: Telecommunications and International Politics, 1851-1945, Daniel R. Headrick, 1991 ISBN13: 9780195062731

when laid on top of each other formed a circular duct for the cable and then secured by nails or straps. It was not intended to permit cable being drawn through but rather as a form of protection from earth movements caused by passing traffic and to protect the cable from being damaged during subsequent digging. This method of cable laying had been developed by William Henley and Charles Bright[17] during the laying of the underground sections of the English and Irish Telegraph Company[18] line from Liverpool to Manchester in 1852 and subsequently on the Manchester to London section where, despite it's high initial cost, had proved to be extremely reliable in service. The route taken from the telegraph office was from St Helier via Half Way House, Millbrook, St Lawrence valley, (now commonly known as Waterworks valley) and past St Ouen's manor to Greve au Lechçon or Sand Eel bay (now commonly called Plémont Bay). William Henley himself supervised the installation using a workforce of some 70 men. During construction it was also reported one day that 'some mischievous or malicious person had cut the cable near Salérie, Guernsey but that this would in no way impede the rapid progress', indeed by 10 August the cable was safely in place on both islands and on 17 August it was terminated in the town office at a building on the corner of Church Street and Library Place. The Guernsey telegraph office was sited at the Guard House, South Pier, St Peter Port. On the UK side the offices at Weymouth were connected along the London and South Western Railway to the Electric's office in Southampton.

The cable installation was a turn-key contract supervised by the manufacturers Messrs Newall and Co. After testing it was handed over to the Channel Islands Telegraph Company ready for service. The cable was accepted on behalf of the company by Mr James Graves, who had been appointed the Chief Electrician to the Channel Islands Telegraph Company and was stationed at the Jersey office. The manufacturers only offered a 30 day warranty on the cable.

---

17 Charles Bright, Submarine Cables: their history, construction and working (Arno, New York, 1974 [1898])
18 The Worldwide History of Telecommunications, Anton A. Huurdeman, ISBN: 9780471205050 John Wiley & Sons, Inc.

On 24 August the *British Press* reported: 'An Electric Ball - mounted at Fort Regent- will be dropped to signify noon synchronised by Electric Telegraph, as is the custom in all other important ports in England, regulated to Greenwich.' It is not known if this actually came into effect since Jersey did not officially adopt Greenwich Time until 1896.

The *Jersey Times* reporter was privileged to see on 27 August the newly installed Electric Telegraph equipment, manufactured by Siemens and Halske of Berlin[19]. The equipment was of the Relief Recorder[20] design which had been proven in service for some time. It was driven by a system of springs and escapements similar to that used in clocks and reproduced the incoming Morse code in relief onto thin paper tape so that it could easily be interpreted and written on a telegraph form by the operator. At this stage of development of the telegraph, there was no automatic working and all messages had to be sent by hand. The incoming messages could be read later from the tape, but when forwarding messages, as in the case of a telegram from Jersey to London via Weymouth, the intermediate operator had to re transmit the incoming message, inevitably there was some delay. Each 'dash' was ideally three times the length of a 'dot' and the time between 'dots' and 'dashes' should be equal to the length of a dot. The space between letters was equal to a 'dash' and the space between words equal to seven 'dots'. A skilled operator could reach speeds of up to 70 words per minute for short periods, although 30 to 40 words on average was considered very good. It was a happy, if accidental discovery, that operators could interpret the incoming messages by ear, distinguishing the 'dots' from the 'dashes', and experienced operators could even determine who was sending the code! Consequently, equipment was designed to give a good audio signal as well as recording the message on tape. This speeded up the forwarding of messages.

On 1 September the States of Jersey debated a motion proposing a celebration to commemorate the opening of the town Telegraph Office. All schools were to be given a public holiday on the day of the official opening and orders for the decoration of public places was given. Such was the local

---
19 Siemens: From Workshop to Global Player - Wilfried Feldenkirchen. Piper.
20 Perhaps like this example: http://www.telegraphsofeurope.net/page28.html

Image permission of the Société Jersiaise

This engraving shows the public interest that was generated by the opening of the telegraph in Jersey. The Channel Island Telegraph Company offices were at the junction of Library Place and Church Street in St Helier. This office was later shared by the Submarine Telegraph Company's French telegraph line. The office was closed on 29 July 1889 when the Submarine Telegraph Company Jersey operations were ceeded to the GPO. The original building was demolished in the 1960s to make way for a new office block development.

impact of the event.

The office[21] was officially opened on 7 September and a long parade was held which wound from the Royal Square through Mourier Lane (now the upper part of Halkett Place), Queen Street, Hilary Street, Beresford Street, Halkett Place, King Street, Charing Cross and Broad Street eventually ending up outside the telegraph office in Library Place . The route was decorated with bunting and such was the public excitement 'multitudes of holiday proportions' had started gathering in King Street as early as 7 o'clock. A double archway decorated with evergreens had been erected in Charing Cross, the signal mast at Fort Regent had been decorated with flags of all nations and the ships in the harbour were all bedecked. At 9 o'clock a salute was fired and the crowds were thronging through the streets. A special Morning Service was held in the Town Church at 10.30, with the Lieutenant Governor in attendance. The lesson, read by the Dean, was taken from the Second book of Exodus Chapter 12 Verse 26: "What mean ye by this service?" followed by a sermon extolling the greatness of this achievement. At 12:30 the band of the Royal Artillery played in the Royal Square. At 2 o'clock all the dignitaries gathered ready for the parade. The procession set off headed by the band of the Royal Artillery followed by the assembled Civic leaders and a huge crowd. On reaching the telegraph office the Bailiff, the Crown Officers, the Constable of St Helier, George Philippe Benest, and the Directors of the Channel Islands Telegraph Company entered the office and handed the clerk a telegram for Her Majesty Queen Victoria. Three minutes later the Weymouth office confirmed the onward transmission of the 145 word message to London. The inaugural telegram was as follows:

To the Honourable S H Walpole, Her Majesty's Principle Secretary of State for the Home Department.

The Directors of the Channel-Islands' Telegraph Company, on behalf of the people of the Islands, Solicit that you may be pleased to lay before Her Most Gracious Majesty, this the first message conveyed by their telegraph.

---

21 Société Jersiaise Photographic Collection http://www.societe-jersiaise.org/adlib/009303.jpg

Though the establishment of this rapid means of communication with the Mother Country is an event of minor importance to the Empire at large, it is one of heartfelt satisfaction to Her Majesty's Loyal and Devoted Subjects here as tending to draw still closer the bonds which for nearly one thousand years have linked these Islands to the Crown of England and more firmly to secure that connection, the foundation of their liberties and their prosperity, and which, like their forefathers, they would deem no sacrifice too great to preserve.

Jersey September 7th.

The reply from the Queen was received early on the following morning and read:

Sept 8th 1858

Earl of Derby to the Directors of the Channel-Islands Telegraph Company, Jersey.

Holyrood Palace, Tuesday night, 7th September, 1858.

The Queen has received with the highest satisfaction, the announcement of the successful completion of a Telegraphic Communication with the Channel-Islands, and while Her Majesty congratulates the Channel-Islands Telegraph Company upon their success she rejoices in the more rapid means of communication and the closer connexion thus happily established with a portion of her dominions hitherto locally separated, but always united to her Crown by a spirit of unswerving loyalty unsurpassed in any part of them, and of which the Message just transmitted on behalf of the people of the Islands contains a very gratifying expression.

Although the line was complete the workmen did not finish for some time as it was reported in the paper on the 14 September that '*painting and papering etc. was still being done in the Guernsey office although a fair amount of messages were being sent.*'

Such was the public interest in the new telegraph that a lecture was arranged at the Queens Meeting Rooms, Belmont Road, St Helier. The lecturer, Mr Martin a representative of the company, explained the technical details of the system and noted that 'upward of 2 millions of miles of telegraph cables are laid in the UK alone.'

From the outset the new cable was beset with problems. From opening on the 7 September the first fault resulting in a breakdown of communication occurred on the 26 January 1859. The fault was diagnosed as being in the

Jersey shore end of the cable and the local representative, James Graves[22], reported that serious chaffing of the cable had resulted in the breakdown. A new shore end was laid and the cable was fixed to rocks and passed through iron tubes at the worst points to protect it further. Service was restored on 22 February.

The cable again failed on the 22 April. This time the fault was diagnosed as being in the Portland to Alderney section off Portland. The Electric and International Telegraph Company chief engineer William Preece[23] was despatched to oversee repairs. He had some difficulty in locating a suitable repair ship but eventually secured the grappling and jointing equipment and set about repairs on board the South-Western Company steamer the *Prince*[24]. The year 1859 proved to be one prone to very stormy weather and repairs were often held up for several days. The fault was eventually found 4 miles south of Church Hope Cove, Portland and the cable returned to service on the 15 May.

On the 20 May a contract for the supply of telegraphic news was announced between the Channel Islands Telegraph Company and the British Press. This resulted in a special section in the paper being devoted to the latest telegraphic intelligence.

On 7 June a lightning storm resulted in another break in service, this time the fault was found to be in the receiving equipment at the St Helier office, a coil having burnt out.

At the first half yearly meeting held in June the books showed that out of the original capital of £30,000 raised through the share issue some £25,495-14-6d had so far been spent on the initial installation and repairs. Despite the troubles so far experienced on the Company's cable, the directors felt confident enough to give a 5/6d dividend to shareholders, this representing a return of 9%. However, a vote was also passed to request further funds for repairs from the British Government.

---

22 The writings of James Graves and their historical significance D. de Cogan IEE Digest 2003, 11 (2003)
23 Sir William Preece, F.R.S. E.C. Baker, ISBN-10: 0091266106
24 Cable Ships and Submarine Cables, K R Haigh, Adlard Coles 1968

On 20 September the cable again failed and the fault was found to be 3 miles off the Jersey shore end. A steam tug, the *True Briton*[25] under Captain Head, was chartered and a new section of cable spliced in. This fault was due to two kinks in the cable obviously there since the cable was laid. The new section of cable was of a later type manufactured by Messers Newall and Co that had recently been laid in the Red Sea. (The terms of the agreement with the British Government allowed R S Newall to retain any unused cable, a contractual clause which may have contributed to the failure of the Red Sea cable, inasmuch as the cable was laid too tightly thus contributing to it's early demise.) Service was restored on 18 October.

During this down-time the *British Times* continued to report telegraphic news items with the additional note that they were forwarded by mail packet from Guernsey.

The cable again failed on 4 November. The EITCo's own cable ship the *CS Monarch*[26] was despatched with Chief Electrician William Preece on board. Two faults were found at 7 and 12 miles south of Portland. The sea bed was found to be rocky and so the repaired section was shifted eastward resulting in the laying of an extra 3 miles of cable. Service was again restored on 25 November.

While the cable was out of service again, the Company half yearly meeting was held at the offices of the Electric and International Telegraph Company in Morgate, London. Despite continuing cable faults, the directors issued a slightly reduced dividend of 5/- to shareholders.

On 7 January 1860 another fault was found on the Alderney to Portland section. William Preece was again on station to do repairs aboard the cable ship *CS Resolute*[27] on hire from the recently formed submarine cable manufacturer Glass, Elliot and Company. Again, as in 1859, stormy weather held up repairs. The cable was not returned to service until 18 February, a delay of some 6 weeks. Only 9 days later it was down again between Jersey and Guernsey. A ship was chartered from the Submarine Telegraph

---
25 Ibid.
26 Ibid
27 Ibid.

Company and William Preece reported the cable repaired on 10 March.

On 8 June the cable again failed between Jersey and Guernsey 2½ miles off Jersey. This time the tug *Dumfries*[28] was employed. James Graves joined it on 20 June when it arrived off Jersey and the cable was restored to service that evening. It was reported that this was the first time that a cable had been grappled and repaired in one day.

The Company half yearly meeting held the same week was unable to offer any dividend to shareholders because of the expenses incurred in cable repairs. The Company reported that it owned some 123 miles of telegraph cable and employed 13 people.

The cable failed again, this time 6 miles off Guernsey on 20 July. The tug *Dumfries* was available and, because of the clement weather, service was restored on 3 August.

On 24 August Mr Ayrton MP (Conservative) raised a question in the house on the award of £1800 for the repair of the Channel Islands Telegraph Company cable. Sir G Cornwell for the Liberal Government replied, to back bench cheers, 'that it was important to keep telegraphic links to all parts of the Empire open'. This does, however, highlight the concerns felt about the reliability of submarine cables at this time. Indeed, the British Government had suffered severe financial setbacks over the failure of the Atlantic cable in 1858, after only 10 weeks of operation, and the Red Sea cable to India which had failed in March without ever carrying a single telegram[29]. An article in the influential industry magazine *The Builder* on 27 August raised questions on the construction of telegraphic cables and whether Gutta Percha should be replaced by India rubber for such ventures. These expensive failures prompted a change of attitude within British Government circles and as a consequence no further funding or underwriting of telegraph cables was made for over 20 years.

On 17 September a fault occurred on the Alderney to Portland section again and on 26 September the Guernsey to Jersey section failed. Both

---
28   Ibid.
29   The Invisible Weapon: Telecommunications and International Politics, 1851-1945, Daniel R. Headrick, 1991 ISBN13: 9780195062731

faults were repaired by the EITCo's cable ship *CS Monarch*, restoring service on 1 October.

The November half yearly meeting again was unable to issue a dividend to shareholders, reporting that 8 faults had occurred since the cable was opened. Shareholders voted to explore the possibility of a replacement cable between the Isle of Wight and Alderney. A further vote passed a motion to make representations to HM commissioners for an extension to the grant.

On 27 November William Preece, the then Chief Engineer of the Electric and International Telegraph Company, presented a paper on *The Maintenance and Durability of Submarine Cables in Shallow Waters*[30] to the Institute of Civil Engineers in London. His report specifically mentioned the Channel Islands Telegraph Company cable from Portland and he described in detail the construction of the cable and the route it took. The route consisted of 93½ miles of submerged cable and 23 miles of land section. The submarine cable was constructed to two standards; that required for deep water and that for shore ends. Shore ends are subjected to more wear and tear and are therefore thicker and stronger. The Channel Islands Telegraph Company had suffered some 11 faults since they had received the cable from the contractors in August 1858 and they could be classified as follows:

> Two due to the careless laying of shore ends (2 kinks found in the cable off Jersey)
>
> Four due to the dragging of ships anchors in the vicinity of the cable, these all being in the Jersey to Guernsey section
>
> Five due to abrasion on rocks, these all being in the Portland to Alderney section.

Preece determined that the laying of the Portland to Alderney section was a mistake. The cable traversed some particularly rocky areas, although the Admiralty charts from which the course of the cable had been decided had indicated a sandy bottom. A sandy bottom is the best surface for a submarine cable as there is less likelihood of abrasion and of ships dragging an anchor. He considered that, in future when cable routes were selected, a

---

30 Minutes of the Proceedings of the Institution of Civil Engineers Volume 20 Issue 1861, 1861, pp. 26-48 E-ISSN 1753-7843

thorough survey of the area should be carried out before laying, as it was clear that the Admiralty charts were not always correct. He also voiced concerns about the quality of construction of the cables as he had found severe corrosion on cables in areas of tidal runs off Portland. This was attributable to the high alkaline content of Portland cement stone.

A description of the methods used in finding the position of faults on the submerged cables was also given. This was done in conjunction with another telegraph engineer Cromwell Fleetwood Varley[31] and outlined the processes in determining wire lengths from their electrical resistance. This method is still used today by electricians for fault finding on cables and is better known as the Varley Megger test.

The Channel Islands Telegraph Company troubles went on and on. On 1 January 1861 the cable between Alderney and Guernsey again failed. The tug *Dumfries* was once more called to assist and Preece joined the ship on 10 January. The bad winter weather again caused delay and the *Dumfries* was considered unsuitable to continue with the repairs after the extent of the problem was realized. The Electric and International Telegraph Co. cable ship the *CS Monarch* was despatched from Greenwich on 30 January but bad weather meant that she had to shelter in Southampton until 12 February. When the ship arrived in Guernsey the local engineer James Graves joined her and work began. Some 8 miles of cable between the islands had to be replaced. The cable was badly corroded with copper and investigations indicated that a ship laden with copper ore had sunk in the vicinity some years previously. Mr A C le Bois of the Jersey telegraph office announced that the cable had been returned to service at 2:20PM on 26 February.

The problems continued and on 27 March James Graves reported a failure yet again on the Alderney to Portland section some 18 or 20 miles south of Weymouth. Once more bad weather held up repairs and not until 23 April did the *CS Monarch* manage to buoy the broken ends. Preece boarded the *CS Monarch* and set sail for Jersey to collect spare cable and left for the

---

[31] Lee, A.G (1932) "The Varley brothers: Cromwell Fleetwood Varley and Samuel Alfred Varley", *Journal of the Institution of Electrical Engineers*, **71**, 958-64

repair on 29 April. The cable was returned to service the following day at 9:30PM. The same day a letter appeared in the *British Press* from James Graves announcing that he had been appointed the Chief Electrician aboard the Electric and International Telegraph Co. cable ship *CS Monarch*. He expressed his regret at having to leave the island after nearly 3 years but was confident that his replacement, Mr A Fields, would continue in his footsteps. James Graves replaced William Preece who was later to become Sir William Preece, the Chief Engineer of the General Post Office. James Graves himself was a very able engineer who also became famous for his invention of the 'sea earth[32]' method of telegraph transmission in submarine cables while working for the Anglo-American Telegraph Company[33] in the late 1860's.

The final straw came on 17 June when the *British Press* announced the cessation of telegraphic communication with England. The cable had again failed between Alderney and Portland but this time the Channel Islands Telegraph Company had run out of funds. They had spent all their liquid capital, some £4,000, on the previous 13 faults and had no assets left for repair. With the cable down they also had little or no hope of redeeming the position as the major source of their income had gone. Shareholders and the British Government were unwilling to provide further funding. The Board of Trade had recently completed an enquiry into submarine cables and noted the disquieting fact that out of 11,364 miles of cable laid to date, a little more than 3,000 were working[34]. The Government realised that the new cable operated by the Submarine Telegraph Co from Jersey to France, provided a service and it had so far proved more reliable and at that time relations with the French had greatly improved. In addition the STC had, on the occasion of the fault in March, reduced their price for a telegram to London to that charged by the Channel Islands Telegraph Company, 5/-. The writing was on the wall as the *British Press* in its edition of the 18 June carried an advert from the STC reaffirming the reduced rates for calls to the UK.

---

32 Submarine Telegraphy – a Practical Manual, Italo De Guili, Sir Isaac Pitman & Sons Ltd 1932
33 Baglehole,K.C., A Century of Service. Cable & Wireless 1868-1968.
34 The History of the Institution of Electrical Engineers (1871 – 1931), Appleyard R, 1939

In all, the link to England had been in place some 34 months and in that time it had been out of service for a total of 10 months; a sorry record. In fairness to the company, the paths chosen for the cable were not the best; the charts provided by the Admiralty being defective. The route taken to Alderney was also questionable in the light of experience and the original cable had not been of the best quality, a problem which should have been addressed to the contractors. Many of the faults, especially those in the Guernsey to Jersey section, were caused by the dragging of ships anchors and it is possible that some steps, possibly by the introduction of local laws, could have been taken to prevent or reduce the incidence of such faults.

The Channel Islands Telegraph Company continued, however, with its links to Guernsey and Alderney still intact. While these cables still worked there was still some hope for the company as revenue could still be generated on messages passed on to the Submarine Telegraph Company via the Jersey office. The final blow came on 24 February 1862 when the cable between Guernsey and Jersey failed. Although the Alderney to Guernsey section remained intact the Guernsey *Star,* on 21 May, announced with regret that the office in Guernsey was to close on Friday 30 May, '...*the shareholders being well advised not to throw good money after bad'.*

An attempt to revive the company was made on 19 July 1862 when the directors of the Channel Islands Telegraph Company met with William Preece of the Electric and International Telegraph Company together with Messrs Silver and Co. (later the India Rubber, Gutta Percha and Telegraph Works Company) of Woolwich, to sound out the possibility of laying a new cable from Jersey via Alderney to the UK, utilising the remaining link between Alderney and Guernsey, which was practically a new cable. William Preece said at the meeting that a suitable course over a sandy bottom would have to be sounded before proceeding further. He also said that the Channel Islands Telegraph Company shareholders should not feel too aggrieved as many cables had foundered in the short history of submarine telegraphy. However, nothing further resulted from that meeting.

A further attempt a resurrecting the company was made in February 1863. Jurat De Quetteville made a proposal in the States that an advance of

£12,000 should be made out of public funds as an interest free loan to the company. This would be repaid over the next 25 years out of the 6% annual grant on the original installation costs of £25,000 allowed by HM commissioners for the maintenance of the cable. This offer was well received by the directors and shareholders of the CITCo at a meeting held on the 4 February in the newly formed Mercantile and Commercial Club at the Union Hotel, Royal Square, St Helier. A meeting of the local Chamber of Commerce held the following day also endorsed the proposal. However, there was considerable disquiet about the proposal from the public who were suspicious that the proposed loan would be used to offset the losses made by the shareholders, especially as those endorsing the proposal were shareholders in the company, including the originator of the idea Jurat De Quetteville, who was in any case an unpopular politician. There was also concerns that unless a new cable were laid, the existing cable would continue to be just as fault prone as it had already proved to be. The loan of £12,000 was not sufficient to replace the cable and HM commissioners were not forthcoming in offering to fund the difference and, as the 6% grant depended upon the cable remaining serviceable, there were grave concerns that the public funds would indeed ever be repaid. The company was, however, not wound up officially and remained on the companies register for a further 8 years.

CITCo 1858 – 1861 Cable Outages

Date out	Date repaired	Days	Reason	Repair Ship
26 January 1859	22 February	27	Jersey Shore End	?
22 April	15 May	23	Off Portland	The Prince
20 September	18 October	28	3 miles off Jersey	True Britton
4 November	25 November	21	Off Portland	Monarch
7 January 1860	18 February	42	Off Portland	Resolute
27 February	10 March	12	Guernsey – Jersey	Contractor's ship
8 June	20 June	12	2½ Miles off Jersey	Dumfries
20 July	3 August	14	6 Miles off Guernsey	Dumfries
17 September	1 October	14	Off Portland	Monarch
26 September	1 October	0	Guernsey – Jersey	Monarch
1 January 1861	26 February	56	Alderney-Guernsey	Dumfries/Monarch
27 March	30 April	34	Off Protland	Monarch
17 June	Abandoned			
Total outage at Jersey		283		

Table 1

Chapter 3 - The Submarine Telegraph Company

As early as 1858 there were rumours that a cable would be laid from Jersey to France. The Submarine Telegraph Company, founded by Thomas Crampton, had laid the first successful telegraph cable across the English Channel in 1851. By 1858 it was already an established international telegraph carrier and had several cables connected to France and held a license from the French Government to carry telegraphs across French territory.

During the summer of 1859 the Submarine Telegraph Company made applications to the UK and French Governments for permission to run a cable from Jersey to France. At first, the Channel Islands Telegraph Company and its parent company the Electric and International Telegraph Company, who were rivals of the Submarine Telegraph Company, raised objections to the laying of a shore end in Jersey. As a consequence, the States were initially advised by the British authorities to prevent any cable being landed in Jersey. After further negotiations, however, the Channel Islands Telegraph Company withdrew its objection and the Submarine Telegraph Company was granted a license by the British Government. During September Her Majesty's Government appointed the Earl of Malmesbury[35] to head negotiations with France on behalf of the Submarine Telegraph Company to renew the license to operate on French soil and for permission for the Jersey

---
35 An Overlooked Entente: Lord Malmesbury, Anglo-French Relations and the Conservatives' Recognition of the Second Empire, 1852

cable. The French Government was at first reluctant to renew what was a virtual monopoly but in the end conceded and renewed the license for 25 years, half the period initially requested. This, in effect, opened the way for the French cable. The Earl of Malmesbury was an intrepid negotiator and the Conservative government made great use of his talents and his friendship with Louis Napoleon Bonaparte to develop Anglo-French commercial relations.

The route to be taken by the new cable was from Fliquet Bay in Jersey to Pirou, on the Normandy coast south of Lessay, and on to Coutanches. On the 10 January 1860 the cable ship *CS Resolute* owned by the independent telegraph cable engineer W France and chartered to the contractors Glass, Elliot and Co, arrived off St Catherine's with the cable and landed the shore end at Fliquet and proceeded to pay out the cable to Pirou. A contemporary description said that the cable consisted of 7 copper strands covered with gutta percha to a diameter of $\frac{3}{8}$", having the same dimensions as the (abortive) Atlantic cable laid in 1858. The outer cover is made up of 12 No.5 gauge iron wires. The resulting cable was slightly more substantial than that of the Channel Islands Telegraph Company. The cable was laid under the supervision of Mr Canning on behalf of the contractors with Captain Bright of the Submarine Telegraph Company in attendance.

The land line in Jersey was laid underground by the cable manufacturer and contractor, W T Henley of Woolwich, from the shore landing at Fliquet, via St Martins Church, Five Oaks, St Saviours Road, St James Street, Colomberie, Hill Street to the Church Street telegraph office. The STC was obviously more parsimonious than the Channel Islands Telegraph Co., as the cable was laid directly into the ground without protection. The friction between the Submarine Telegraph Company and Channel Islands Telegraph Company must have been greatly lubricated as, in the event, they shared the same office. The cable laying was completed by 30 January despite appalling weather, the trenches being continually filled with rain water.

A celebration dinner was held for the contractors and guests at the Royal Yacht Club Hotel, the Weighbridge, St Helier.

The French cable link opened for business on 7 May 1860, the connections at the French end accounting for the delay. An advert in the *British Press* announced the call charges:

The Submarine Telegraph Company

Jersey to Coutances	2/6
Avrances, St Malo, Caen, Grandville & Cherbourg	3/6
Harvre	5/-
Paris, Bolougne	6/-
Bordeaux	7/3
Marseilles	8/6
To Great Britain (via Paris) 20 words	11/6

In conjunction with the British and Irish Magnetic Telegraph Company.

As can be seen, these were substantial charges; the cost to London being more than twice the Channel Islands Telegraph Company charge. This being in the light of the direct connection provided by the Channel Islands Telegraph Company, however, even at this time the directors of the Submarine Telegraph Company must have had their suspicions about the long term integrity of the northern cable. They were right, as on 20 July the Channel Islands Telegraph Company cable failed again. The Submarine Telegraph Company entered the following advert in the British Press on 22 July:

The Submarine Telegraph Company

Telegrams to England 11/- per 20 Words

The opportunity to take business from their rival had prompted a reduction of 6d in their rate!

An earlier failure (7 June) prompted a letter from Charles Alexander Gerhardi, the local Superintendent of the Submarine Telegraph Company, to be published in the British Press dated 22 June. This referred to a telegram received by the famous French writer Victor Hugo. It had been claimed that while it had arrived too late to be of use, the letter explained that it only took 4½ hours for the telegram to be delivered having been sent from

London via Paris.

The Submarine Telegraph Company cable proved to be more robust than that of the Channel Islands Telegraph Company. The area of sea that it crossed was shallower, the bottom sandy and it was less susceptible to the stormy seas and tidal flows of the Channel. It also had the advantage of being newer and using more up-to-date technology. The Submarine Telegraph Company also took more care of their investment, regularly warning fishing vessels of its presence through notices in the press. They were thus able to take financial advantage of the periodic failures of the Channel Islands Telegraph Company cable and by March of 1861 had reduced the cost of 20 word telegrams to 5/- in direct competition with the Channel Islands Telegraph Company. Throughout the troubles of its rival, the Submarine Telegraph Company cable held firm. The final failure of the Alderney to Portland section must have come as an unexpected bonus to them as they now had unrivalled access to all telegraphic traffic leaving the islands. Following the break in the remaining Jersey to Guernsey section the Submarine Telegraph Company signed a contract with the *British Press* to provide telegraphic news services commencing on 23 April 1862. On 17 June they opened a sub-office in Guernsey, appointing Mr S Barbet of the High Street, St Peter Port as their agent. Telegrams were passed via steam packet for onward transmission from the Jersey Office.

On 19 January 1863 the STC set up an experimental link between its office in Jersey and the London office in Threadneedle Street by connecting its lines through France via Coutances, Caen, La Harvre, Dieppe and Beachy Head in a continuous metallic circuit of 380 miles. Those present in the Jersey office included the Manager, Charles Gerhardi, Mr W H Le Feuvre, who was also a director of the CITCo and Mr M V Wardley of the *British Press*. The circuit worked perfectly and a call was set up between officers at the London office and Jersey. It is interesting to note that the times recorded for the connexion were 6:33PM in London and 6:27½PM in Jersey, synchronisation to Greenwich still not being in place. By coincidence two of the clerks at the London office were young Jerseymen, Messrs Gavey and Prichard, who had recently completed their training. The conversation consisted of general chit-chat about the weather and a remark about the recently opened Metropolitan underground railway.

In December of 1863, Gerhardi announced that a telegraph line would be laid from St Helier to Gorey pier. At this time Gorey was used as a commercial port and there were also a large number of boatyards in the area. In variance to the lines so far installed on the island, this new line was constructed using poles and open copper wire. The probable route taken by the new line which began from a pole at the bottom of Belvedere Hill, Georgetown was through Longueville, past Grouville mill to Verclut, across Fauvic Common and onward to Tower Number 5 where a reference was made to a very long span of wire, some 430 yards, from the tower to a pole on a small hill on Gorey common. This route is unsubstantiated but is based on the most direct route between the known points. The poles then continued along the common to an office on Gorey pier. The pole at Belvedere Hill was connected to the office in Library Place by underground cable. The project was overseen by Charles Gerhardi and construction took place during December 1863 and January 1864. The overhead construction seems to have been carried out on a strict budget as the poles were placed on average 220 yards apart. This is considerably longer than modern construction where 55 yard to 80 yard wire spans are more normally used, but given that only one wire needed to be suspended, this was probably entirely satisfactory. The work was also punctuated by incidents of stone throwing damaging the insulators, which prompted a notice to be posted in the local press. The project was completed on 20 January and the line was opened for service on Saturday 23 January with this inaugural telegram sent to HM Lieutenant Governor:

I take this liberty to address to your Excellency this, the first telegram to be transmitted by the new line between St Helier and Gorey, to inform his Excellency that the line is from this moment open to the public.

The company charged 6d for a telegram from Gorey to St Helier.

The integrity of the STC cable was such that the submarine section only failed 7 times in the period 1860 to 1870 on each occasion due to dragging anchors or oyster trawling. These failures were often followed by a considerable delay in repair as the STC usually placed the Jersey link at the bottom of its priorities, having several more commercially lucrative cables between England and various continental countries. There were occasions where the cable remained out of service for months at a time, September to November 1864, December 1865 to February 1866 and April to July 1869 were notable occasions. These protracted delays caused much inconvenience to both business traffic and the dissemination of news in the local press. The company always used its own two cable ships the *Retriever* and the *CS Resolute*[36].

In September of 1865 came what must have been the first local dispute between public utilities. On 22 September the cable failed. At first it was thought to be a submarine fault but after investigation it was discovered to be in the land line between Fliquet and St Helier. Because the fault was a clean break it was not possible to discover the exact location of the fault with the testing equipment to hand. Consequently, Charles Gerhardi relocated the office temporarily to the shore station at Fliquet and arranged a pony express service from the town office while further testing took place. On 28 September after further investigation the temporary office was relocated to the St Saviour's Inn near St Saviour's church, a more convenient location. Four days later the cable fault was located in Colomberie outside the premises of a certain Miss Hemery. Apparently she had just had a gas pipe installed and during the works the trenchers had severed the cable. Charles Gerhardi called Mr Morris of the Gas Company to the scene to show him the damage and explaining that the STC would expect recompense for the damage. It was estimated that the total cost would be in the order of £30 to £40 which included the cost of getting a Gutta Percha cable jointer over from England. Meanwhile a further temporary station was set up in Westaway's Yard in La Motte Street.

Following this incident, a close watch was kept on the Gas Company's activities and during the laying of a new main down St Saviour's Road, Charles Gerhardi arranged for the cable to be encased in bitumen covered wood ducting.

---

36 Cable Ships and Submarine Cables, K R Haigh, Adlard Coles 1968

On the 30 July 1866, the STC arranged a special illuminated star, lit by gas, outside its offices to celebrate the connection of the new Atlantic cable. The attraction drew a crowd of some 500 who gazed in wonder until it was extinguished when the office closed at 11:00PM. The following day the Stars and Stripes and the Union Jack were raised outside the office. Charles Gerhardi had more reason than most to celebrate this occasion as he had been involved on the first attempt at laying an Atlantic cable in 1858 having been with the Newfoundland party on board the cable ship *Niagara* which had laid the western half of the cable.

In March 1867 the UK government proposed the Telegraph Bill which would bring all the telegraph companies into the ownership of the Post Office. This was largely based on a paper by Edwin Chadwick, a noted social reformer and sponsored by the eminent civil servant Frank Ives Scudamore, which extolled the virtues of the Post Office running the telegraph system noting that it had over 10,000 offices compared with the private telegraph companies 1,900. There was also an underlying tone of national security in that the telegraph network could be used for military purposes.

In August the STCo company secretary announced increases in telegram charges from Jersey, 6/8d to London and 7/8d to elsewhere in the UK. Charges to France remained unchanged.

In June 1868 the Electric Telegraph Bill was presented to parliament. It was estimated that the nationalization of the telegraph companies would cost between £3 million and £4 million. The STCo cable from Jersey to France, however, did not come into the remit of the bill since it had no landing on the UK mainland.

In October STCo adjusted their pricing again lowering international rates to Italy from 13/6d to 7/6d, Constantinople 20/6d to 10/10d and to Malta from 22/- to 10/-. These reductions were due to a new agreement with the French authorities and reflected the continuing development of the European telegraph network.

During May 1870 the STCo decide to replace their French cable from Pirou to Fliquet. This replacement was carried out under the agreement signed with the French authorities when the original cable was laid in 1859. The cable was manufactured by Bullivant and Allen and laid by Stoffel and Co. of London. At the same time, the landings at Fliquet and Pirou were improved with the construction of cable huts which facilitated convenient test points in event of faults.

Business was good for the STCo as for some time they had had the monopoly on telegraph traffic from the island. Reductions in international prices had stimulated growth and their strangle-hold on UK traffic had maintained revenues. But during 1870 the Jersey and Guernsey Telegraph Co was formed to take advantage of the generous terms offered by the Post Office under the extension bill to the Telegraph Act which incorporated the provision for purchasing the assets of the Isle of Man and Channel Island telegraph companies. The likelihood of a rival would mean that UK telegram prices would drop. The Post Office had a flat rate charge of just 1/- for 20 words anywhere in the UK and it was likely to be extended to the Channel Islands following the takeover.

Despite the opening of the new UK cable, the STCo continued to fare well. The company announced a 15% dividend at their half year AGM in August 1871. The company had also commissioned a new purpose built cableship to be named the *CS Lady Carmichael* in honour of the wife of the company chairman, Sir John Carmichael. The reduced price to the UK impacted on that traffic but new business was obtained by onward transmission of telegrams through the Post Office. In order to protect their investment in the new cable the company commissioned a cable tower to be constructed at Pirou to mark the course of the cable for the navigation of fishing boats.

The company continued to pay 15% dividends for the next few years as international business continued to grow and as they secured a better business relationship with the Post Office. In August 1872 the company opened offices in the Post Office Threadneedle Street telegraph office. Business boomed to such an extent that the company was able to increase dividends in the year 1876 to 16½% and the following year to 17½%. The new Jersey-France cable proved to be a worthwhile investment as the number of cable faults decreased. Between 1870 and 1889 (when the company was taken into the ownership of the Post Office) only 9 submarine faults occurred. On one of those occasions, in December 1877, a telephone was used during testing between the *CS Lady Carmichael* and the Fliquet shore end. This was only one year after the Bell patent had been filed.

During some of the prolonged cable failures on the UK link, the company provided services for the Post Office as well as news reports for the local press. No doubt they used these opportunities to their financial advantage. Messages to the continent were gradually reducing but it seems that telegrams to the UK via France remained high since in February 1881 they reduce charges to the UK via France by 2/6d to 8/6d to London and 9/6d elsewhere.

With the consolidation of the telegraph network and the growth of the telephone, the Post Office decided to incorporate many of the Anglo-Continental international cable circuits into its network. Thus the European operations of the STCo were to be absorbed into the Post Office. Terms for the sale and transfer of the STCo infrastructure and employees were agreed during 1888 and the sale was finalized in 1889 at a cost of £67,163. The Jersey operation had its own licence extended until 31 March that year so as the final arrangements could be made locally. On 29 July, the STCo office equipment installed on the corner of Church Street and Library Place was transferred to the main Post Office building at Grove Place (now the home of the Mechanics Institute, Halkett Place). The company cable ship the *CS Lady Carmichael* was renamed *CS Alert* by the Post Office in 1894 and sold out of cable laying in 1915.

The STCo's overseas operations continued and the company flourished abroad, eventually comprising a substantial part of Cable and Wireless which was formed on the 1 April 1929 from a number of British overseas cable companies.

Submarine Telegraph Company 1860 – 1889 Cable Outages

Date out	Date repaired	Days	Reason	Repair Ship
23 Jan 1863	3 Feb 1863	11	Fault off Pirou	Retreiver
15 Feb 1863	24 Feb 1863	9	Fault off Fliquet	Retreiver
30 Sept 1863	1 Dec 1863	61	Fault off Fliquet	Retreiver
13 Sept 1864	11 Nov 1863	59	Fault mid channel	Retreiver
1Dec 1865	19 Feb 1866	80	Fault off Pirou	Resolute
28 Apr 1866	15 May 1866	17	Fault off Pirou	Resolute
12 May 1868	27 May 1868	15	Fault off Pirou	Resolute
15 Sept 1868	15 Oct 1868	30	Fault off Pirou	Retreiver
24 May 1869	5 July 1869	42	Fault mid channel	Resolute
			Cable replaced	Resolute
11 Nov 1870	28 Nov 1870	17	Fault off Fliquet	Resolute
7 Feb 1871	14 Feb 1871	7	Fault off Fliquet	Resolute
21 Oct 1871	23 Jan 1832	94	Off Fliquet	Lady Carmichael
23 Mar 1872	12 April 1872	20	Off Fliquet	Lady Carmichael
	8 July 1873	1	Fliquet and Pirou ends	Lady Carmichael
4 Nov 1877	5 Dec 1877	32	Off Fliquet	Lady Carmichael
16 Mar 1878	7 April 1878	22	Off Fliquet	Lady Carmichael
6 Feb 1884	28 Feb 1884	22	Off Fliquet	Lady Carmichael
1 Jan 1885	23 Jan 1885	22	Off Fliquet	Lady Carmichael
Total Outage		**561**		

Table 2

Chapter 4 - The Jersey and Guernsey Telegraph Company Ltd.

The history of this company is as brief as it is controversial.

Before proceeding with the description of the company it will first be necessary to describe the state of the telegraph industry at the time. In 1867 the British Government decided that the private telegraph businesses in the Kingdom should be nationalized. The driving force behind this policy was the eminent civil servant Frank Ives Scudamore who sponsored the ideas of the great social reformer Edwin Chadwick[37], who had also been responsible for the development of public health policies. It was claimed that the telegraph companies often worked in cartels and the larger companies had monopolised the most profitable routes with way-leave agreements which were often not implemented. As a result of telegrams between smaller towns having to pass through many hands the cost of such messages was often prohibitive. Businessmen, who were fully aware of the trading advantages of the telegraph, were keen to establish a more uniform tariff system throughout the country. The government appointed a commission to examine the issue and at the end of that year the resulting report suggested that the telegraph system should be consolidated under the direction of the Post Office, who was the officially appointed government messenger.

---

37 Finer, Samuel Edward (1952). *The Life and Times of Sir Edwin Chadwick* (Reprint ed.). Taylor & Francis. p. 6. ISBN 9780416173505

The government report resulted in the Telegraph Act of 1868 which provided for the Post Office to purchase, at its discretion, any telegraph company operating on the UK mainland. The terms for compensation were extremely generous. They included the capital costs of installed lines and equipment and also a formula which allowed for loss of profits over a 20 year period. The announcement of the bill resulted in a rush by speculators to install telegraph lines wherever none existed, for example, to the Scilly Isles which had long been considered an unprofitable venture. The 1868 Act, however, excluded foreign cables and thus the Act was amended in 1870 so that it included the Channel Islands and the Isle of Man.

Since the Electric and International Telegraph Company was mainly concerned with routes within the British Isles, the directors protested to the government that they would be marginalized by the bill leaving them with only one route, to Holland. The government therefore amended the bill the following year to include the special case of the Dutch cable. The Post Office subsequently sold this cable to the Submarine Telegraph Company. This enabled the Electric to be fully wound up and thus released its capital for speculation in overseas ventures.

With the door thus opened, the directors of the Isle of Man Telegraph Company, which had been effectively a subsidiary of the Electric and the Channel Islands Telegraph Company, could see a way of making capital out of the sale of their companies to the Post Office. Therefore representations were made to the government to amend the Act to include the Isle of Man. The CITCo was, of course, a moribund company but nevertheless still registered. After talks with the Post Office during July 1869, the local directors approached the parent company with a view to reviving the company and thus profiting from a subsequent sale to the Post Office. However the chairman Robert Grimstone[38] was now more interested in overseas development and refused to fund any new cable out of the Electric sale. This left the local directors with two choices; to fund the venture themselves or to seek funding elsewhere.

---

38 Robert Grimston prepared and published "The Statement of the Case of the Electric & International Telegraph Company against the Government Bill for Acquiring the Telegraphs" in 1868

Two of the directors Phillip Gossett and F Carrel decided to try and revive the company and approached the Globe Telegraph Trust, a venture capital company with offices in Nicholas Lane off Lombard Street, London, which provided funds for speculative cable projects. The Globe at first seemed willing to pursue the venture and on behalf of the CITCo opened negotiations with the Board of Trade for landing rights. A front company was formed calling itself the Channel Islands Telegraph Association based at 7 Great Winchester Street, London. The new association opened talks with William Henley Telegraph Works for the provision and laying of a new cable.

At the same time two other directors decided to form a new company. This company, which was called the Jersey and Guernsey Telegraph Company, was founded by William Henry Le Feuvre, a flamboyant locally born civil engineer and entrepreneur who also had interests in the Jersey Waterworks Company and the Jersey Railway and Tramways Company, as well as several ventures in the UK, and the infamous Jurat David De Quetteville. The company issued its Articles of Establishment on 26 January 1870 but not before opening tentative negotiations with the Board of Trade during the latter part of 1869. Among the shareholders was William Preece, a personal friend of William Le Feuvre, who, following the demise of the Electric, had been appointed Chief Engineer (Southern District) of the Post Office telegraphs. The company was incorporated on 4 February 1870 with an issue of 15,000 £2 shares, just 2 days before the expiry of the lodging of a purchasing option deadline imposed by the Post Office and 2 days after receiving a cable landing license from the Board of Trade. The new company board consisted of:

Mr W Le Feuvre	Chairman
Mr C E Philips	
Mr Hemery Le Breton	
The Rt Hon E Haviland-Burke MP	
Captain Thomas Carr	
Mr ? Gauntlet	Secretary

William Preece was appointed the company's chief engineer.

Meanwhile, due to a clerical error by a junior clerk at the Board of Trade, the Globe received a letter which implied that a landing license would be granted exclusively to them. This led Nathaniel Holmes and Louis Stoffel, two of the principles of the company, to conclude negotiations during February with William Henley on the laying of a new cable. However, during the following few weeks the Globe apparently got cold feet and failed to come up with the expected funds. Nevertheless, William Henley started manufacture of the necessary cable and made plans for the laying of a cable from Start Point, near Salcombe, Devon to Guernsey.

At the same time the JGTCo appointed Bullivant and Allen Ltd of Millwall as prime contractors for the laying of the cable. This was a strange choice as they were mainly rope manufacturers not cable makers. In fact Bullivant and Allen approached William Henley for a quotation on the supply of a cable. Henley offered them the Globe cable after adding a percentage to the £25,000 that he had already quoted. Eventually, Bullivant and Allen purchased the cable from Henley's rivals Glass, Elliot and Company of Greenwich[39].

The Bullivant and Allen quotation to the JGTCo stated that:

The cable shall be made of three copper wires of No 16 BWG covered in three layers of gutta percha No 4 BWG served with well solutioned yarn.

The deep-sea section shall be covered in 9 iron wires of No 4 BWG and the shore ends in No 0 BWG[40].

The terms of supply stated that 90 nautical knots would be provided for the sum of £25,000 - 0 - 0d and that if the final cable should exceed that, Bullivant and Allen could elect to make up the difference in JGTCo shares. They also quoted 5 years maintenance at the sum of £1000 - 0 - 0d per annum.

At the end of April the land-line contractors, Warden and Co of

---

39 Glass, Elliott and Co merged with the Gutta Percha Company in 1864 to form the Telegraph Construction and Maintenance Company.
40 BWG is the Birmingham Wire Gauge. No 16 is 0.0625" diameter, No 4 is 0.2253" and No 0 is 0.3240".

Birmingham arrived in the islands to lay the local cables. The new company acquired offices in Hill Street and a cable was laid in iron pipes down to the site of the new railway on the St Aubin's Bay seafront. The JGTCo and the Jersey Railway Co[41] shared a number of directors and no doubt this fortunate relationship led to the ready usage of the Railway Company's land for telegraph poles for the routing of the overhead wires. From this point the line was taken overhead along the course of the railway to Beaumont. From here a new route of telegraph poles took the line on to Plémont via St Peter and St Ouen. The railway also had its own private telegraph lines along this route which continued on to St Aubin. As a result the St Aubin's post office was also linked to the telegraph circuit from the beginning of operations.

While the JGTCo were busy laying land lines, William Henley's cable ship the *Caroline*, a paddle steamer which had been used in the laying of the shore end of the first of the two new Atlantic cables laid in 1865/6[42], arrived in St Helier Harbour with the new cable on board. Due to a combination of bad weather and legal problems the ship laid up until the end of the following month.

On the 20 May 1870 the JGTCo held its first ordinary shareholder meeting. Here, under questioning from the shareholders, the directors refuted the claim of the Globe to landing rights claiming that they had applied two years previously for a license. This is strange since as at that time the CITCo was still moribund and presumably, therefore, still entitled to its original license. William Le Feuvre admitted that technically he was still a director of the old company, an admission which must have put him into a tricky position; nevertheless, the meeting produced the necessary two-thirds majority to carry the proposal to sell out to the Post Office.

On the 6 June, the *Caroline* left St Helier harbour to the sound of McKey's band and made way to Greve au Leçon (Plémont) to attempt to lay the cable. However, the British government had sent a small warship, *HMS Dasher*, to the spot and together with the men of the 17th Regiment on shore prevented the landing of the shore end. All these events were

---
41 The Jersey Railway (J.R.& T.), N.P.R. Bonsor, Oakwood Press 1986
42 Cable Ships abd Submarine Cables, K R Haigh, Adlard Coles 1968

watched by a small crowd of locals. The *Caroline* eventually backed off and retired in the direction of Guernsey.

On the 8 June yet another strange event occurred. A certain Henry Jones was arrested by the Constable of St Ouen at Plémont, charged with the cutting and taking away of a telegraph cable. He came before the magistrate on the 21 June to answer the charge. His Advocate claimed that he had been carrying out instructions issued by a Mr Bennett of the CITCo. The magistrate was unable to immediately determine this and so called for witnesses, the case being remanded until the 12 July. When he next appeared the witnesses included William Le Feuvre who substantiated the defendants claim that a letter signed by Mr Bennett was genuine and that as company secretary must have had authority to order the cutting. The case was therefore dismissed. There appears to be no reason for this action. Certainly, Mr Bennett was still secretary to the moribund company but what possible advantage could be gained from cutting the cable remains a mystery. It could be that it was a vain attempt to prevent the new company from making use of the land portion of the old cable but as the new overhead line was already in place at that time it makes that proposal unlikely.

On the 1 July the parliamentary Select Committee enquiry into the compensation claims associated with the Channel Islands cable was opened in London. The claimants were the Channel Islands Telegraph Company, the Jersey and Guernsey Telegraph Company, William Henley, the Globe Telegraph Trust, the Jersey Railway and Tramways Company and the Submarine Telegraph Company. The case of the Jersey Railway was dealt with quickly the chairman ruling that as they had no telegraphs connecting directly or indirectly with the Post Office their claim was invalid. The rest of the cases were far more protracted. The Globe claimed to have had exclusive landing rights granted to the Channel Islands Telegraph Association (CITA). The Post Master General, Frank Ives Scudamore, refuted this claim suggesting that this must have been a clerical error. The only licence was held by the JGTCo. They denied any connection between the CITA the CITCo while admitting to sharing directors. William Henley argued that he had gone to considerable expense in producing a cable and keeping a cable ship on station for several weeks while the government procrastinated. The STC argued that the 1868 Telegraph Bill defined the Channel Islands as 'islands off Europe' and this identified the islands as foreign under the law thus they were due compensation for loss of traffic due to the new direct cable. The Jersey and Guernsey Telegraph Company sought to establish themselves as the only legitimate company serving the islands.

The commission was to last a long time, indeed very many months. At the time of the first hearing the law extending the terms of the Bill to include the Channel Islands and the Isle of Man was only just passing through the House of Lords. The initial sitting defined the Telegraph Companies to be included in the Telegraph Bill (1870) for consideration. By the end of the month, however, the Bill had passed through its final parliamentary stages and passed into law. There now only remained the matters concerning the terms under which the JGTCo would be transferred to the Post Office. The Globe, and the CITCo were not included.

Incredibly, at the time that these negotiations were going on the telegraph cable connecting the JGTCo had not even been submerged. There was some disquiet about this back in Jersey where the Chamber of Commerce accused certain States members of complicity by excluding the former company from the Bill extending the Telegraph Act to the island[43]. The chairman of the company, W H Le Feuvre, tried to pacify the local business community by publishing a letter in the British Press explaining that the land line portions were already working and that the equipment for operation was installed.

Meanwhile, undeterred by the legal complexities of the compensation question, William Henley proceeded with his planned cable. The *Caroline* once again set sail from Henley's Cable Works at Woolwich on 17 September and headed for the Channel Islands. His intention was to lay the Alderney to Guernsey section after which the ship would go on to complete

---

43 Telegraph Act 1870, 1870 CHAPTER 88 33_and_34_Vict, An Act to extend the Telegraph Acts of 1868, 1869, to the Channel Islands and the Isle of Man.

the link between Guernsey and Jersey which had been interrupted in the spring. The deep sea section had already been laid. The *Caroline* was going to pick up the submerged section to splice and land the shore ends to complete the route. At 7:00pm on 26 September the *Caroline* docked in Guernsey having completed the Alderney to Guernsey section. The boat laid-up until Wednesday 28 September because of fog, but arrived off Greve au Leçon (Plémont) at 12 noon. A cable of 20 tons per knot with a diameter of 3½" was landed. The British Press reported:

'Sixty men swimming and in boats brought the cable ashore. Scrambling like cats they hauled the cable up the cliff in a great feat of engineering and tied the cable to the post of the JGTCo. The chairman W H Le Feuvre was present along with William Preece and a very few spectators. The event was recorded by the camera of Mr Sharpe of King Street. The toast was drunk with the master of the *Caroline*, Captain Galelio representing Mr Henley'

The cable was tested by the local electricians Mr Mayo, Mr Winter with the STC electrician Mr Gerhardi present.

By 18 September the *Caroline* had finished all the shore ends of the Guernsey to Jersey and Alderney to Guernsey sections both cables having been tested ready for connection. It seems, however, that illiam Henley decided not to proceed with the cross channel cable to Start point.

On the 8 October the *International* left the Thames under command of Captain Beesley having collected the 'official' cable from Glass, Elliot and Co of Millwall. On October 11 it picked up Captain Carr, one of the JGTCo directors, from Dover and set sail for Dartmouth. Bad weather intervened and the ship had to shelter in Portland until 22 October. On that day Mr Bullivant of the official contractors, announced that the ship had set sail for Dartmouth but because of more bad weather had to hold over in Plymouth. The *International* was a new ship of 1,381 tons and 240 feet long which had been specially built for the India Rubber, Gutta Percha and Telegraph Works Company[44] entering service at the beginning of 1870. The Dartmouth to Guernsey cable was her first contract. The route had been chosen because of the favourable conditions afforded by the sea bottom across this route.

---

44 Cable Ships and Submarine Cables, K R Haigh, Adlard Coles 1968

The first cable had suffered serious damage because of the rocky bottom at Portland and indeed in the intervening years the technology of cable construction had greatly improved. Nevertheless it is always important to use the best route for any submerged cable to minimise potential failure.

The *International* eventually arrived at Dartmouth on 26 October and dropped anchor in Stonehole Bay between Salcome and Bolt. Again there were delays as the original route for the cable shore end had run into way-leave problems. Further delays resulted when the *International* lost her anchor and had to dock in Dartmouth harbour for repairs. At last, early on 2 November, the deep sea cable laying began. The cable was landed at Lancress bay at 8pm that evening and buoyed.

When the ship arrived at Lancress Bay it was joined by W H Le Feuvre and his electrician Mr Winter. The next section was from St Martin's Point to Plémont. It was decided to lay it from Jersey and so the ship set sail for Plémont the following morning. On arrival a telegram was sent to town to invite Mr Gerhardi of the STC and Mr Edward Le Couteur of the South Western Company to join the vessel. The shore end was landed and hauled up the cliff with the aid of two horses. There was a minor accident witnessed by the Rev P A Le Feuvre (the brother of the chairman) and a few ladies who had come to watch. One of the workmen slipped and fell into the cable tank on board the *International*. The Lt Governor, Major General Guy CBE, also arrived during the morning and the events were recorded by Messrs Asplet and Green, photographers. The cable was eventually landed and secured and the end sealed whereupon the men repaired to the Picnic Room, half way up the hill, for grog. Afterwards there were cheers for the Governor and for Mr Vincent who had loaned the horses. The men rejoined the ship the injured man having hurt his ankle.

The ship set sail for St Martin's at 3:00pm and secured the sea section to a buoy. A new shore end was laid alongside the Henley cable. The ends were spliced; 'first the copper ends were brazed together then covered using gutta percha and Chatterton's compound (which is a mixture of tarry resins) to complete the joint.' The protective wires were spliced over the joint and the cable buried.

The first telegram was sent on the 8 November from Compass Bay to Lancress Bay, it read:

'Preece Exeter to Winter Guernsey. It will be at least a fortnight before the land line will be ready. I should like to see you in Southampton as soon as Clarke (*the local linesman*) has started the Alderney line and you can get away.'

Shortly afterwards, a dispute arose between a landowner at the Dartmouth end who objected to the cable passing over his property. It was decided to move the shore end to Dartmouth harbour but during the move the coastguard arrived and drove the workmen away with his cutlass. Apparently, the Government order issued in April to prevent the *Caroline* (Henley's ship) from landing a cable had not been rescinded!

On the 16 December the new telegraph was installed at the JGTCo offices in Hill Street. The official opening was planned for the 21 December but meanwhile a celebratory dinner was held at the Royal Yacht Club Hotel (now called the Royal Yacht Hotel) at the Weighbridge. The 120 guests included the officers of the company, the Lt Governor Major General Guy CB, the Dean Rev W C Le Breton, the German Count von Maltke, Jurat Le Bailey and the Bailiff. In the evening an exhibition of the electric light was presented by a Mr Browning. Inurgural telegrams were despatched to Queen Victoria at Osbourne, the Marquis of Hartington, the PMG and to Mr F Scudamore.

At the dinner the well known Jersey lawyer Robert Pipon Marett, a shareholder, made a witty speech mentioning the cost of installing the cable which was estimated at £30,000. He also referred to the recent incident of cutting the old CITCo cable and looked forward to the success of the new venture. Also mentioned was the fact that since the GPO takeover of the telegraph service over nine million messages had been sent netting £250,000 for the government, this being more than that raised through the postal service. The telegraph office would open with a uniform message rate of 1/- for up to 20 words, except for messages to Sark. The GPO had also agreed to purchase the Henley cable from Guernsey to Alderney.

On the 23 December the new office was decorated with a gaslight display of the Prince of Wales' Feathers and the words 'Ich Dien'. Mr Waterman was appointed the first telegraph clerk.

On the 11 March 1871 the following statistics were published in the *Jersey Times and British Press* showing the traffic from the opening to 1 March:

Station	Date of Opening	Receipts	Messages Sent			Received Messages
			Through	Local	Total	
Guernsey	21/12/70	£126-18-6	1716	359	2075	2452
Jersey	21/12/70	£298-12-7	3947	308	4255	5264
St Aubin	21/12/70	£3-6-0	49	14	63	80
St Ouen's	21/12/70	£1-1-0	7	14	21	14
St Peter's	21/12/70	£3-12-3	49	18	67	62
St Saviour's	21/12/70	£13-13-1	166	48	214	251
Alderney	18/1/71	£11-7-1	85	113	198	151
		£458-10-9	**6019**	**874**	**6893**	**8274**

Table 3

This illustrates the popularity of the lower rate, which enabled more people to take advantage of the service. It also showed the considerably more buoyant Jersey market which accounted for almost 3 times more traffic than the other two islands together.

On 12 May a further statistical analysis of the traffic was published for the four weeks ending 18 March:

Government Messages	20
Shipping	558
Banking and Stock	154
Butter	22
Fishmongers	9
Greengrocers & Fruiterers	28
Cattle	32
General Trade	1101
Domestic	1031
Continental	85
Prepaid	31
Non-classified	51
Racing and sport	2
Total	**3124**

Table 4

This further shows the importance of the telegraph to business but also shows that it was becoming popular for domestic messages too, with over one third of all messages.

This increase in traffic was about to lead to changes in the technology used. A good Morse key operator could average about 35 words per minute but the current rate of traffic was beginning to strain the operators. In addition there was only a single line to the UK which was shared with Guernsey; therefore it was decided to update the equipment to the Wheatstone automatic telegraph which could send over 100 words per minute by use of punched tape. This enabled the messages to be prepared 'off-line' and then sent in batches when the line was free. Preparation was carried out on a Wheatstone 'stick' punch[45]. This was a perforating instrument which was operated by the telegraphist using rubber hammers, rather like playing a xylophone with only three notes. Each punch represented either a 'dot', 'dash' or space. This would be installed only at the main office. Incoming messages are recorded onto tape in Morse code

---

[45] Post Office Electrical Engineering Journal, Telegraphy Vol 49 p166 - 171

and then transcribed by hand onto the telegraph forms for delivery by the telegraph boys.

On the 16 June the refurbishment to the main Post Office in Queen Street was complete and a new entrance was also provided from Hill Street. The following day the J&GTCo office was relocated to the post office itself.

In the meantime, the J&GTCo were in intensive arbitrations with the Post Office commissioners regarding the purchase of the company equity under the Telegraph Act 1868 which would be soon extended by the 1870 Order to include the Channel Islands and Isle of Man[46].

The arbitrations with the Post Office were held before the Rt Hon Russell-Gurney, the Recorder of the City of London and were completed by 24 May 1871. This enabled the board of the J&GTCo to call an Extraordinary General Meeting on 27 July at St Antholin's Chambers, 26 Bridge Street, London. The chairman, Mr F W Le Feuvre, read the board report, minutes were taken by the company secretary Mr Gauntlet.

The company had been formed 18 months earlier and had progressed to the situation where it was viable and thus available for purchase by the Post Office. The negotiations with the GPO had resulted in a settlement of £54,920 plus any accumulated interest. The expenses of the arbitration were to be paid by the Post Office, this amounted to £2,500. In addition there was expected to be tax due to the sum of £1,000 to £1,500, the company solicitor Mr G Bristow would resolve this issue. Once the taxes were paid, the company could be dissolved and a final dividend paid to shareholders. Mr E Haviland-Burke MP seconded the chairman's proposal and the resolution was carried. The meeting complained that the actions of the 'buccaneering' Mr Henley may have affected the final payment, which, the meeting felt, was not as generous as earlier settlements. It was also noted that other companies had experienced 'some difficulties' in receiving the money from the Post Office.

Mr Bennett was appointed auditor and Mr Buchan announced that the

---

[46] Telegraph Act 1870 CHAPTER 88 33_and_34_Vict An Act to extend the Telegraph Acts of 1868, 1869, to the Channel Islands and the Isle of Man. 9th August 1870

final return could be as much as £1-7-6d per share.

That might have been an end to it as the company had been formed as a speculative venture to capitalize on the generous terms offered by the Post Office in the nationalization process. The original Bill did not include the offshore islands and it was only by vigorous campaigning that the law was extended. The company chief engineer was William Preece, who was also a substantial shareholder. At the time the company was formed Preece was also employed by the Post Office and was chief engineer of the southern region. In these days of closely regulated financial dealings that would almost certainly be classed as insider dealing. It is extraordinary that even in Victorian times that Preece's dual role was not scandalized in the press. The public enquiry had highlighted this issue but when Preece was questioned in the witness box, he refused to divulge his financial associations with the company, merely stating that he had been employed as an 'Engineer Consultant'.

However, the company was not wound up following the transfer of assets to the Post Office. There remained a continuing dispute between the board and the Post Office over the allowance made by the Post Office for the 'Good Will' on takeover and the forecast in traffic growth. The company claimed that the GPO had undervalued this and they continued to press for a further settlement. The sum settled on by the principals was £11,118.10.5d, and, in addition, a further £300.0 .0d in legal fees. This dispute continued for many years. The Post Office made an offer of £6,350.17.4d on the 15 April 1874. The company refused to settle for this sum and continued pressing for the full claim. Eventually, in January 1879, the board accepted the Post Office offer on condition that the legal fees were paid. The company was finally liquidated on 29 January 1879.

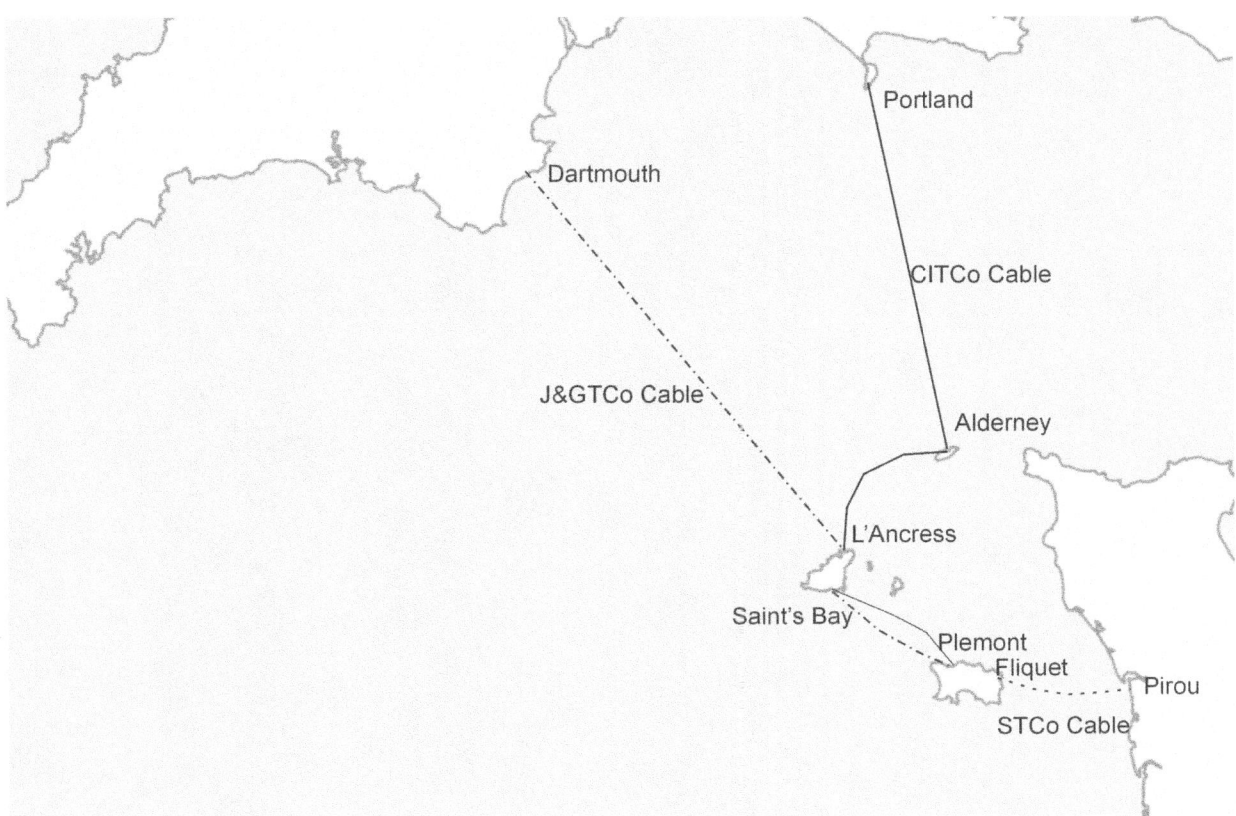

The figure above shows the approximate paths of the telegraph cables installed up to the takeover of the Jersey and Guernsey Telegraph Company by the GPO in 1872. The Channel Islands Telegraph company cable from portland had been abandonned at this time but had remained submerged. Its path was ill advised having passed over the Shambles, a notioriously rough passage south of Portland. The later cables of both the Submarine Telegraph Company and the Jersey and Guernsey Telegraph Company were better surveyed and were laid over more favourable sea bottom.

Chapter 5 - The Telegraph Under the General Post Office 1872 - 1914

The Jersey and Guernsey Telegraph Company finally ceded operations to the Post Office in August 1872. At the half-yearly meeting of the J&GTCo the Chairman, W H Le Feuvre stated that a writ against the Post Office claiming approximately £12,000 expenses due to the protracted enquiry was still outstanding, but the GPO had paid £6000. After expenses a sum of £1,000 was announced and a bonus dividend to shareholders of 2/6d per share was proclaimed. The witnesses to the enquiry had to be paid out of the £6,000 and one single witness had claimed £600 of which the J&GTCo had only paid 300 Guineas, the witness had threatened litigation. After further discussion the final dividend was settled at 4/- per share.

Since the adoption of the 1/- rate, over 600 messages per day were now being passed and, as a result, the profit margin achieved by the company was greater than under the old 2/6d rate. Shareholders carried the Directors recommended motion, decided at the board meeting 16 July, to wind up the company in favour of the GPO takeover. The meeting was closed with that resolution and the company's operations be passed into GPO control, however the company remained active while outstanding settlement matters continued (see above).

Despite the new cables, there were still faults, although rather less frequent than under the CI Telegraph Co. Interruptions were frequently caused by faults on the UK mainland, the first such failure being reported on

9 August 1873.

The first submarine fault under GPO ownership occurred on the 27 February 1874 when the cables between Guernsey and Jersey failed. The *Jersey Times and British Press* announced that telegraphic news had been despatched via the Steam Packet from Guernsey. On 5 March an announcement in the GPO, Queen Street, St Helier read:

'Cutters will be despatched either from St Helier's or Plémont, about 5pm daily during the interruption of the telegraphic communication between here and Guernsey. They will be sent by telegraph to Plémont (where a temporary office had been established). *The cutters are expected to arrive in Jersey on the return trip at about 6 or 7pm daily'*

And on 11 March a letter from Chas E Winter, Superintendent Post Office Telegraphs, was published in the *Jersey Times and British Press*:

'I have tested the cable submerged by the steamer *Caroline* and it is broken about 2½ miles off Jersey. The other cable submerged by the steamer *International* is broken 4 miles off Jersey. The PO has engaged the services of the *International* which is currently in the Firth of Forth and should arrive within the week.'

Note that the W T Henley cable was now being used by the GPO as a spare. It is not known whether Henley ever received settlement for this cable. Since both cables were damaged, it I likely that this was caused by trawling or a dragging anchor.

The GPO utilised the services of the Steam Packets for forwarding telegraphic messages to Jersey. A newspaper report from Guernsey asked:

'Why do the cutters delivering telegrams have to pay 7/- harbour fees? Surely they could levy such a duty on telegrams too!'

Note the use of the word 'telegram' now becoming common usage rather than telegraphic message. The press was full of alternative suggestions including the use of carrier pigeons!

The *International* arrived in Guernsey on 6 April and the cables were restored on the 19 April.

A further interruption occurred on 31 May 1876 and was restored on 24 June. However, when the cable between Alderney and Guernsey failed on 2 September the GPO engineer, Mr Powers, announced that 'it was unlikely that the GPO would despatch a ship to repair the cable.' The local press called for the War Office to intervene. Nevertheless, the cable remained unrepaired until the 7 May 1877 when the GPO despatched the *International* to make the repair. Unfortunately the GPO engineer, Mr Power, was unable to detect the exact location of the break and the fault remained unfixed. In the same 31 May 1876 edition of the *Jersey Times and British Press* a short note on the invention of the telephone was recorded.

On the 14 May the Guernsey – Dartmouth link failed; telegrams were despatched via the offices of STC. The *International* was again despatched an on the 27 July the cable was returned to service. The urgency for repair seems to be somewhat mitigated by the presence of the STC cable. The French route enabled continuity of service throughout breaks with the mainland. It is not known what financial arrangements were in place between the STC and the Post Office in these cases.

The cable failed again on 23 February 1878 between Dartmouth and Guernsey. Once again, the STC came to the rescue. The GPO again showed no sign of urgency and Guernsey's Lt Governor tried to intervene to hasten the progress. On the 12 March the GPO announced that the *International* would be despatched after repairs to the Scilly Islands link. It duly arrived on 27 March and after extensive testing declared on 2 April that the cable had failed in 3 places. Meanwhile, on the 16 March the STC cable failed, resulting in total loss of communication. The local newspapers now had to rely on despatches received via the Steam Packet service. The STC cable was returned to service on 7 April. The GPO cable was restored soon afterwards.

A further fault occurred on 29 November off Dartmouth. The cable was repaired on 16 December.

In July 1879 the cable ship *Dacia*, on charter from The India-rubber, &c., Telegraph Company's Works, Silvertown, London, was despatched by the War Office and the Guernsey – Alderney cable was finally returned to service, according to the press, by Corporal Bowden, a military engineer. The cable had not been operational for almost three years.

It was some time before the next failure on 2 February 1881. The fault was diagnosed as being near Dartmouth and it was repaired just off Compass Cove on 23 March. A further fault in the river mouth at Dartmouth put the cable out of service from 20 to 27 October.

In the meantime, the main Post Office in Jersey relocated from the Queen Street office to Albert Hall in Grove Place (now known as Halkett Place), nowadays the home of the Mechanics Institute. A new 42' long counter lit by gaslight was installed and special compartments for the writing of telegrams were provided. The refurbishment was carried out by Messrs Fallaize and Tostevin, local builders. Speaking tubes were installed between all floors. The number of overhead telegraph lines to minor post offices was growing and this seemed to provide sport for local youths since in March the Post Master General issued a warning of prosecution for 'The throwing of stones at the Telegraph Lines'. Telegraph insulators are quite fragile, being made from porcelain or glass, and can easily be shattered with an accurately aimed pebble.

The next major submarine cable fault was on 11 February 1884. This time, the GPO decided to upgrade the route by replacing the cable entirely. The *CS Monarch* was despatched on 1 March with new 3-core cables. The entire cable system was to be replaced, the upgrade in anticipation of the new proposed 6d rate. In the meantime, the *CS Monarch* repaired the old cable and service was restored on 21 March.

Between the 12 March and 4 September the *CS Monarch* laid the new cables between Jersey and Guernsey and in August between Guernsey and Dartmouth. This job was done in sections, as on 26 March the Monarch was sent to Scotland for repair duties. The new cables were completed by 4 September and the 3 core system allowed a dedicated direct connection for both Guernsey and Jersey to London and a shared section to Exeter. The *CS Monarch* recovered the old cables before leaving the area, including, presumably, the W T Henley cable. This appears to be the first time that the recovery of old cables was performed in the Channel Islands; formerly, faulty cables had simply been abandoned. This probably reflects the maturing nature of the submarine cable technology industry, since recovered cables could provide valuable scrap metals.

The new cable brought much needed stability to the telegraph service and with the introduction of the new Inland 6d rate (for up to 20 words), traffic had increased 48% by 2 October 1885. After this replacement, submarine cable failures became less frequent. There were no major cable faults on the UK section until 7 January 1894, the *CS Monarch* repairing the cable on 24 January. The long-time failed Alderney-Guernsey cable laid by Henley was repaired at the same time.

A freak accident occurred in the main post office; during building work on 22 March 1886, 5cwt (250Kg) of mortar fell upon the telegraph equipment. Fortunately, no-one was injured and there was no loss of service reported.

From 5 April 1886, the Post Office adopted Greenwich Meantime (GMT) for the timing of all telegraph transmissions. This brought the Channel Islands post offices into line with the UK and removed confusion over the timing of telegrams. This may seem trivial, but could have resulted in legal disputes had the difference in time not been taken into account with telegraphed instructions, for example, purchases of shares on the stock market. The States of Jersey did not officially adopt GMT until 21 November 1896.

On the 29 July 1889 the local STC operations were handed over to the GPO and its offices transferred to the main post office. Meanwhile the local Jersey network was being augmented. The telegraph was extended to more local rural postal sub-offices. Improvements in technology meant that new telegraph equipment required little skill to operate. The Wheatstone ABC telegraph[47], for instance, could be operated with little training and although slow in operation, about 15 words per minute, it was reliable. This made the deployment of equipment easier as staff could be readily trained to operate the sender and receiver. The telegraph was extended from Millbrook sub-office to St John by a new overhead telegraph route opening on 12 March 1894. Havre des Pas sub-office was added on 18 April.

---

47 Sir Charles Wheatstone FRS, 1802-1875 by Brian Bowers IEE Publications ISBN 0852961030

On 29 September 1896 HM Public Building and Works Department accepted a tender for the erection of cable house at Plémont from Messrs Dart and Son of St John, Jersey, who had recently carried out works at the Head PO as noted above. The hut was constructed in dressed granite and completed with a Welsh slate roof. The construction of the cable hut enabled a more convenient place for engineers to effect tests. This cable hut survives today, although it has subsequently been converted into a public convenience.

The military telegraph, which was administered by the GPO, was also extended from Government House to the arsenals in the north and west of the island. There were also extensive telegraphs operated by the local garrison extending from Forth Regent to the signal station at Corbière and serving establishments in between.

St Martin's PO was connected to the telegraph on 12 June 1895 using poles erected along the railway to Gorey, then up Mont du Gouray and along Grande Route de Faldouet to St Martin's Church. This pole route also carried for part of its length the line for the Fliquet-Pirou cable which replaced the earlier underground cable installed by the STCo. This wire was brought down by snow as reported in the *Jersey Times* on 25 January 1897.

On the 21 January 1902, the GPO installed a Wireless Telegraphy transmitter at Fort Regent on behalf of the Admiralty. This was to provide both ship to shore and communication with the UK. The popularity of the telegraph continued, as on the 3 June 1902, the GPO reported that they had handled over 30,000 words in messages following the declaration of peace in the Boer War.

The GPO advertised on 3 October 1903 for a new site for the head main post office. The site at Grove Place had become too small for the growing business. The advert placed in the local press by WS Rushton, the GPO Surveyor for the Portsmouth district, requested responses by 15 October. The notice specified that any site should provide a minimum of 9,000 sq ft (840 sq metres) of accommodation. On 15 December the Chamber of Commerce noted that 'a very suitable site' in Broad Street had been offered. The Postmaster General confirmed this rumour on 16 February

1904, stating that the new building would be on the site of 15 and 17 Broad Street.

From the late 1880's onward, the GPO had provided a temporary telegraph office at various locations around the Weighbridge to supply services for exporters during the potato season. This was a popular facility and there were often calls for this to be made a permanent office. The Chamber of Commerce frequently petitioned for this but received little response from the PMG. In 1903 a new sub-post office was opened at Hulbert & Co, Conway Street. From then on the temporary telegraph facility was operated from there, but no permanent telegraph office was established. The telegraph traffic during the short potato season, which usually extended from early May to the end of June, was considerable. A report in the *Jersey Press* from 6 June 1906 illustrates this well. The GPO installed two duplex machines at the office which could handle 500 messages per hour at a transmission rate of 27 words per minute. Up to 4,000 messages were handled each day and the local telegraph office was supplemented with 16 extra staff brought in from Portsmouth and Southampton, making a total of 27 in all. This office closed permanently for telegrams when the new office at Broad Street opened on 22 June 1909. That year a temporary office was operated from the Commercial Street entrance of the new building prior to the official opening.

The new office at Broad Street was built by contractors Corbett and Co of Gray's Inn Road, London. The building is still the site of the main post office today and the building is substantially the same, save for the unfortunate alterations to the entrance made in the early 1970's. The original building had fine Edwardian features which are still retained on the upper floors. Entrance was by two double oak doors to the left and right of the edifice which is constructed of Portland stone. The building is over three stories at the front and to the rear, backing onto Commercial Street, there is a yard for vehicle storage. The main activities of the postal authorities have long since been transferred away from the building to more suitable sites, firstly at Mont Millais, St Helier and now at Rue des Pres, St Saviour. The telegraph was transferred to a second floor office, which was 40 feet square (150 sq metres) and contained all the telegraph equipment and printers. The battery room and connection frame were located on the ground floor and the cables trunked up through the building. A pneumatic tube system connected the counter area to the telegraph clerks' office.

The telegraph equipment was also updated at the time of the move to the new premises. The Wheatstone 'stick' perforators were replaced by the more modern hand operated 'Gell' type, which appeared more like the familiar typewriter keyboard, and were faster to operate. At the same time a 'gummed paper' system was introduced using a Wheatstone Morse printer, which directly transcribed the incoming Morse code signals onto gummed paper tape in text form[48]. This allowed for the operator to cut up the tape and stick the message onto the telegraph form without the need to transcribe the code first. This was seen by the traditional telegraph operators as an affront to their skills. Nevertheless, it increased production and accuracy and was a boon, especially during the busy potato season.

Wireless telegraphy was already in the Channel Islands. Alderney had a station operated by the Admiralty and there had been a station at Fort Regent since 1902. The Post Office opened its first ship-to-shore wireless coast station at Bolt Head, Devon on 14 December 1908. Bolt head is a promontory some 450 ft above sea level and a high mast was erected with an equipment room for the transmitter and receiver for the princely sum of £2,000. Tests made to the Fort Regent station at that time demonstrated that the circuit was reliable. The Post Master General in announcing the new service stated that although its primary function was for ship to shore traffic, it could also be used in the case of cable failure to the Channel Islands. This feature was tested only 3 months later when one of the regular cable failures interrupted traffic between Jersey and Guernsey on 10 March 1909. The normal procedure in these cases was to divert urgent traffic via

---

48 The Telegraph and Telephone Journal, October 1914

Image permission of the Société Jersiaise

The picture above shows the telegraph office at Grove Place (Now Halkett Place) which is now the home of the Mechanic's Institute. Morse encoded paper tape is prepared off-line ready for transmission. The picture appears to have been taken around 1909, just prior to the Post Office relocation to purpose built offices in Broad Street, St Helier.

France or use the mail boat to transfer written scripts to Guernsey. However, on this occasion, the Fort Regent station was used to provide the service. This was noted to be a considerable saving on the cost of telegrams, which via France were charged at 2d per word while from the wireless station the charge was just ½d per word.

These frequent cable interruptions had long been a bone of contention with local businesses. There were regular demands for the cable to be replaced from the local Chamber of Commerce and others. There was even a petition presented to the PMG in April 1905 by the Lt Governor. This was rejected by the PMG, and in a statement he said that the maintenance services offered to the Channel Island was sufficient and that alternative routes (via France) ensured continuation of vital services. The cable continued to exhibit trouble for some time after, but the alternate routes did alleviate the problem.

The next major development of the telegraph, the new Creed reperforator system[49], was introduced in the spring of 1914. This was an advance on the existing system enabling speeds of up to 200 words per minute. These machines still used Morse code but the keyboard was similar to the modern equivalent and the operator produced an output tape encoded immediately into telegraph code. This output tape could then be sent on the line using a Creed sender. The existing Gell and Wheatstone systems were retained as backup. The introduction of the new system enabled a 10% increase in traffic during the busy potato season to be realized.

When the First World War began in August 1914, a two additional telegraph cables were laid to the UK and to France (see below). These were primarily for the use of military communications but were also available for domestic traffic.

At the time of the introduction of the telegraph into Jersey, the island was in an era of change. The continental unrest had brought about a larger and more permanent garrison at Fort Regent and the island was gradually becoming more English-centric. In 1834 the States had adopted English

---
[49] ibid

currency and trade with the mainland increased. English was being used more and more and local papers were now being printed in that language as opposed to the traditional French printed word. It was in this light that the telegraph was introduced.

The coming of the telegraph changed many things almost overnight. It was now possible to communicate almost instantly with trading partners in the UK and this enabled fortunes to be won and lost on the London stock market. The first telegraph cable to the mainland was highly celebrated, a special parade being organized to announce its opening. The second cable to France a year later attracted rather less attention.

The new communication medium also enabled news to be published in the local press more or less at the same time as in London. An interesting aspect of this was the prominence given to the English racing results in the local press. A considerable amount of the telegraph traffic was dedicated to news transmissions. Indeed, over time, these news circuits became so important that dedicated circuits were provided with permanent connections directly installed at the newspaper offices.

Despite the high cost of early telegraphic communications, it was well used by trade. The effect on local business was also considerable. The tourist industry benefited as did local farmers, who were able to decide when or whether to ship crops in order to gain the best market price. This became so important that during the main potato season (May to August) a temporary telegraph office was set up at the harbour weighbridge each year, although the Post Office resisted petitions to make it a permanent office. Later, a post office was opened in Conway Street, close to the Weighbridge, but the telegraph service only operated seasonally. After the main post office moved to Broad Street in 1909, the Conway Street office was closed.

The telegraph became an indispensable tool for business and an important means of communication for ordinary citizens. Sub-telegraph offices were installed in most parish post offices, which were connected back to the main office. The local offices used simple ABC telegraph type transmission devices that required the minimum of expertise. After the takeover of the Nation Telephone Company's network in 1913, the Post Office connected the sub-post offices to the main telegraph transmission centre at Broad Street by telephone, thus simplifying even more the task for sub-postmasters. However, it was not until the States of Jersey took over the telephone system in 1923 that it was possible for the general public to use a telephone to send a telegram. The telegraph went into decline with the introduction of submarine telephone cables to connect to the UK in 1933 and into sharper decline after WWII when the telephone trunk network was expanded and the cost of trunk calls began to be reduced. The telegraph service was finally closed in the autumn of 1982.

Chapter 6 - The Arrival of the Telephone

The Victorians embraced technology and enterprise. They followed new inventions with enthusiasm and vigour. Their romance with the electric telegraph was now to be re-invigorated by the invention of the telephone.

In April 1877, shortly after the filing of Graeme Alexander Bell's patent application in the United States, the local press reported at length on the invention. There was much speculation as to when such technologies would be deployed in Jersey.

The first recorded occasion of telephone usage was during the repair of the STC cable to Pirou in December of 1877 when it was reported that a telephone had been used aboard the *CS Lady Carmichael* so as the electricians could maintain contact with the Fliquet shore station. This was just over a year after the original patent, filed on 7 March 1876. It seems that the usage of such means of communication became commonplace thereafter.

The first land telephone was installed on the Jersey Eastern Railway. In September 1880 the company contracted the India Rubber, Gutta Percha and Telegraph Company Ltd. of London, to install wires and instruments along the line between Gorey and St Helier. This would probably have been a single wire circuit utilising the earth to complete the circuit, the type commonly used by railway companies at that time, in which distinction between stations was achieved by coded ringing sequences. The Jersey Railways and Tramways Co used a telegraph system on the western line but did not use telephones until 1899.

The following year it was reported in the press that a Gower-Bell loudspeaking telephone had been used in Jersey by the Government Survey. A similar instrument had also been used by the Post Office during tests on the telegraph cable to Alderney.

The age of the public telephone system finally arrived on 18 June 1888. The Western Counties and South Wales Telephone Company ('WC & SWT Co'), which had been floated on 17 December 1884 with a capital of £400,000 and whose head offices were at 16 High Street, Bristol, took out a lease on 22 Bath Street, on the corner of Minden Place (now 66 Bath Street, the houses were renumbered three times between 1841 and 1927). The WC & SWT Co had obtained licenses from the Post Office to install telephone exchanges in a number of places including Bristol, Cardiff, Cornwall, Devon, Dorset, Wiltshire, Gloucestershire and the Channel Islands[50]. They installed a small 50 line magneto switchboard, possibly of the then much favoured Edison design, and opened the offices to the public. Mr John Durham was appointed as the local superintendent. As far as is known, the company made no attempt to install a system in Guernsey. The Chariman of the WC & SWT Co board was Mr C Naish, the board also contained two NTCo nominees, Mr J Staats and Mr JW Batten.

A large gantry or derrick was erected on the roof of the building and a number of subscribers secured. Shortly after the opening of the exchange, the first recorded advertisement including a telephone number was entered in the *Jersey Times and British Post's* 18 August 1888 edition and for some time thereafter:

---

50 The Electrical engineer, Volume 5 27 June 1890

> **H Elliot's
> Springfield
> Nursery**
> **& 34 King Street**
> Telephone No. 18

Figure 1

The only other documented subscriber was Mr F De Gruchy of David Place, this is known because lightning struck the telephone wires during a storm on 21 October 1891, some months after the suspension of service. Although his telephone number was not revealed at the time of the incident, Mr De Gruchy was later allocated the number 231 by the NTCo.

The WC & SWT Co, in common with many of the UK telephone companies at that time, suffered from the restrictions imposed by the Post Office's strict implementation of the Telegraph Act 1869 and had financial problems. It constantly struggled to raise capital for expansion and as a result, on 25 May 1891, announced the closure of the Jersey exchange. This was presumably due to financial consolidation. The local superintendent was relocated to Weston-Super-Mare which had proved a more profitable investment.

The following January the WC & SWT Co was absorbed into the National Telephone Company.

The Great Western Railway Co installed a private telephone circuit at St Helier harbour. This is noted in a Letter from John Wimble, District Superintendant of the Great Western Railway addressed to the President & Members of the Committee of Harbours. The letter asks permission to erect a telephone line over the pier and was dated 17 April 1891.

In June 1896 the St Helier Town Hall invited tenders for a private

Image by the author

The last remaining Police Alarm Box now mounted on the Picquet House in the Royal Square, St Helier. There were a number of these boxes installed around St Helier and keys were held in nearby shops for the use of the general public.

telephone exchange and the following month another tender was invited for the provision of police telephone boxes to be installed around St Helier. This was to enable the police to report back to the station during their patrols or if specific help was needed. This tender was awarded to Mr F A Jubel, described as an electrician. It is not known whether the telephone circuits were provided privately by the contractor or by the NTCo. The first cast-iron wall box enclosing the telephone was installed at the Harbour Master's office, Albert Pier on 1 October 1896. The boxes were manufactured by the local foundry company Grandons of Bath Street and Commercial Buildings, St Helier. A number of other boxes were installed around the town. These boxes were available to the public as the keys needed to access them were often available from nearby shops. This system of communication was abandoned in 1923. The last surviving box, No 2, originally installed in David Place, is mounted on display in the Royal Square, St Helier.

In August 1898 a private telephone circuit was provided between the Viscount's chambers, Hill Street and Rozel Manor. This is likely to have utilized private wires provided by the National Telephone Company.

In August 1899 a demonstration was given at the Town Hall, by a certain Mr King of London, of an automatic telephone system. There was the usual press speculation about it being adopted in Jersey including the promotion of the 'Jersey Automatic Telephone Company'. There was no clear description given of the equipment but as the first telephone system to incorporate a dialling mechanism had been invented in 1892 by Almon Brown Strowger, a Kansas undertaker, it is likely to have been this or a derivative of this system. Indeed, a number of such private exchange installations had been made in England by the Automatic Telephone Exchange Company of Chicago in the late 1890's. However, the automatic systems in use at that time bore only superficial resemblance to those eventually adopted by the GPO for the UK and eventually Jersey. The article included a report from the New York *Freeman's Journal and Catholic Register* that there were currently over 4000 automatic telephone lines installed in the United States. However, Jersey would have to wait a further 60 years before automatic telephony finally arrived in the shape of this

Strowger system, by which time the technology was almost 70 years old.

Chapter 7 - The National Telephone Company 1895 - 1911

The telephone companies in the United Kingdom operated under licences issued by the Postmaster General. Although the earlier Telegraph Acts (1869 and 1870) contained no reference to telephones, by virtue of their non-existence at the time, a court judgement issued on 20 December 1880 determined that the Post Office had exclusive rights to provide 'electrical communications' in a landmark legal action[51]. The judgement laid down that a telephone was a telegraph, and that a telephone conversation was a telegram, within the meaning of Section 4 of the Telegraph Act, 1869. The following year the Post Office began converting some of its telegraph offices to include telephone exchanges.

After this judgement independent telephone companies were obliged to obtain a licence to operate from the Postmaster-General, for the arbitrary period of 31 years. The Post Office demanded a royalty of 10 per cent of gross income and the licences included the option to purchase a telephone undertaking at the end of ten, 17 or 24 years. It was Post Office policy to issue licences for the few existing telephone systems, restricting these systems to areas in which they were operating, and to undertake the general development of the telephone itself. The net result of this policy was to undermine confidence in investment in telephone operators, leading to financial difficulties for many small companies.

---

51 Attorney General vs. Edison Telephone Company of London Ltd. - Law Report 6 Q B D244

The National Telephone Company ('NTCo') was formed in March 1881 from the amalgamation of a number of independent operators, thereby enabling economies of scale. On 17 July 1882 the then Postmaster-General, Henry Fawcett, decided to issue licences to operate telephone systems to all responsible persons who applied for them, even where a Post Office system was established - reversing the previous policy 'on the ground that it would not be in the interest of the public to create a monopoly in relation to the supply of telephonic communication'.

Further relaxation of the restrictions in 1884 allowed telephone companies to operate outside their original licensed areas. Thus the Western Counties and South Wales Telephone Company ('WC & SWT Co')was able to enter the Channel Islands.

The NTCo Purchased the WC & SWT Co in January 1892 and assumed control of all its assets and subscribers. The company announced on 18 December 1894 its intended restoration of the public exchange in St Helier. Although the WC & SWTCo had suspended operation in 1891, it retained the lease on the offices at 22 Bath Street and the switchboard and exchange derrick remained in situ. The NTCo was therefore able to reopen the St Helier exchange from the same address.

On 11 March 1895 NTCo engineers arrived in Jersey with 60 telegraph poles measuring between 6.5M and 21M (20 feet and 70 feet), the largest consignment of poles ever received in the island to date. Mr T A Bates was appointed as the local superintendent and he set about installing the necessary equipment and cable for opening the system.

The NTCo refurbished and extended the St Helier exchange and installed a new switchboard at offices in St Aubin, located at Bel Vue on the High Street and leased from Mr John Peter Le Fol. They obtained a way leave from the Jersey Railways and Tramways Company (see also below) and erected telegraph poles to provide junction circuits between the two exchanges and for subscriber service provision.

In St Helier the NTCo provided a number of tall telegraph poles and metal roof-mounted standards in strategic sites to enable the rapid deployment of wires over rooftops from their exchange derrick in Bath Street. Some of the wire spans must have been considerable but this was common practice at the time in the UK. Obtaining way leaves on private property was less onerous and far quicker than negotiation with local authorities for road digging. Telephone line connections were often of many spans of open wire looping between poles, derricks and chimney brackets. Peppercorn rentals were offered to subscribers for the use of these way leaves, although other larger settlements were sometimes negotiated.

Bates gave an interview to the *Jersey Times and British Press* on 5 August that year and described the intentions of the NTCo. The article also contained a description of the 'mahogany switchboard' (ex-WC & SWT Co) which was equipped for just 50 lines but was 'easily extendible' and two types of telephone: 'a Blake[52] transmitter, induction coil, double pole Bell receiver and cord, magnetic electric generator, automatic switch and backboard' and 'a Hunnings[53] transmitter, watch receiver and cord, automatic switch, push button and trembling bell'. The article also mentioned the roof derrick and the main distribution frame (MDF) which was equipped with lightning arrestors.

Later that year the NTCo announced new exchanges at St Peter, St Ouen and Millbrook. La Rocque and Gorey were added later and a way leave was obtained from the Jersey Eastern Railway (see also below) to erect poles along its line from Snow Hill to Gorey harbour. At first sight these exchange locations may not seem particularly well selected, being, as they were, largely near the coast. However, with the exception of St Ouen's and St Peter's exchanges, they were all placed close to the railway lines. This proximity to good transport links enabled easy logistical movement of plant by rail in preference to horse and cart, the only alternative at the time. The remaining exchanges were built along the road to Plémont, which provided good transport links for the military which had garrisons at St Peter and Greve De Lecq. An existing route of poles along this road, now belonging to

---

[52] Francis Blake Jr, inventor see: Lewis Coe, *The Telephone and its Several Inventors*, McFarland Publishers, 1995
[53] Reverend Henry Hunnings, inventor see: Baldwin, F.G.C. 1938. *The history of the telephone in the United Kingdom*. p.72

the Post Office telegraph, had been erected in 1870. The NTCo erected its own poles along the same line, although, in general, these were somewhat taller. The NTCo obtained storage space for their growing network supplies at 34 The Parade, St Helier. This was rented from a Mr A Gulliver. The telephone stores remained at this location, commonly known as Gulliver's Yard, for almost 70 years.

The St Peter's exchange was located at *Elmfield,* Rue de Fliquet (neither the road nor house now exist as they were cleared to make way for the airport runway extension) which was leased from Mr Philip Le Gresley. The St Ouen's exchange was in a house called *Sans Ennui* opposite Rue de la Forge and leased from Mr John Matthew Perie, this being located immediately in front of the current exchange buildings. La Rocque exchange was located in the Post Office, Rue de la Lourderie near La Rocque harbour. The Gorey exchange was leased from Mr Philp Le Vesconte at 2 Hillgrove Terrace, Gorey Village and Millbrook exchange was in Millbrook House on the corner of Route de St Aubin and Chemin du Moulin.

The first newspaper advert displaying a telephone number under the new system was placed in the *Jersey Times and British Press* on the 28 May 1895, just 5 days after the opening of the exchange:

> J T Baker
> Waterloo
> Pharmacy
> Halkett Place
> Telephone No 15

Figure 2

During 1896 the NTCo ventured into Guernsey to assess the prospects of developing a telephone network. The General Post Office appears to have authorized this project as a switchboard and a number of overhead wires were installed, including lines to the St Peter Port post office. The NTCo even went as far as publishing a list of telephone subscribers. However, the States of Guernsey sent workmen to cut the wires down and there was a stand-off between the two sides. This venture was finally extinguished by the decision of the Guernsey States to apply to the Post Office for a licence to establish its own telephone system on 18 June of that year.

In the meantime the NTCo continued to develop the local network in Jersey. At first the uptake of the telephone was slow and it was only towards the end of the year that major businesses started to see the advantages of the new technology. Voisin & Co, a large local retailer, finally placed an advert in the Jersey Times and British Press on 6 November 1896 displaying their telephone number as 60.

The Victorians had by now become much more used to advances in technology and much less was written in the press about developments of the telephone than had been written about the telegraph some 40 years earlier. However, newsworthy incidents were reported such as the gas fire explosion at the exchange building in St Helier at 3:00AM on 30 March 1897, which caused little damage but caused a temporary interruption to operator duties.

The telephone system was subject to the similar interruptions and faults as the telegraph. Lightning was a regular problem which could cause extensive damage to lines. Such incidents were occasionally reported in the press, such as a lightning strike at Millbrook exchange on 17 October 1898.

The NTCo replaced the local District Manager, T A Bates, with John Lemon in 1898. At the time of the handover the local network had expanded to some 90 lines. Shortly after this appointment, in May 1899, the NTCo decided to reduce telephone installation and call charges. This had an immediate stimulating effect on the growth of the network, as by the 15 August the NTCo announced that the system had expanded by 100 lines since the introduction of the new tariffs. This rapid increase in lines led the company to begin printing a local directory which was first published in the the *Jersey Times and British Press* on the 18 September 1899. This advert also carried instructions that subscribers should request the exchange and number when placing a call and not simply ask for the subscriber by name as had been the practice.

Such was the growth in business, that the company opened a further exchange at La Moye Post Office in June 1900. In August the local District Manager was again replaced. Lemon was moved to Barrow-in-Furness, where he replaced the local manager who had, unfortunately, been killed in a motor accident – surely one of the earliest victims of motor traffic. The new incumbent was Howard Eady from the NTCo Bournemouth office. During the period of John Lemon's duty, the number of subscribers had increased to over 400.

Eady resided at 4 Windsor Terrace, Val Plaisant, St Helier which was also the NTCo published office address until around 9 February 1901 when the NTCo moved to offices at 2 New Street, above the retail shop rented from Wilfred Le Chemiant (later occupied by the 50/- Tailor; the original building was demolished in the 1970s). It is known from contemporary photographs that a derrick was erected on the roof of this building but it is not clear when this construction was made, the property having been leased since 30 December 1899.

On 16 April 1901 the NTCo opened an exchange at Victoria House near Le Geyt Farm, Five Oaks (this building also was redeveloped in the 1970s). By the end of that year there were 25 subscribers connected. In December the NTCo shut down the exchange at La Rocque and transferred the 9 subscribers to the Gorey exchange, prefixing their numbers with a '0' to distinguish them from the identical Gorey subscribers' numbers. Although the exchange was closed, the lease was retained.

On 17 May 1902, following a prolonged campaign by the Chamber of Commerce, a telephone box was opened at the Weighbridge, next to the police baroque. Its existence was short lived as it was removed by Christmas that year, but no reason for its removal was given, nor did it appear to cause the Chamber of Commerce to complain. The only public telephones available were on private premises, at the NTCo exchanges and offices, as well as most stations on both railway lines.

By now the NTCo provided a reasonable service over the most populous areas of the island. Prompted by increases in the cost of line installation announced in March 1903, the Chamber of Commerce noted that the NTCo

Image by permission of Société Jersiaise

Although not proven, it is possible that this picture shows the ill fated telephone box installed at the Weighbridge. It displays the characteristics of contemporary telephone kiosks installed elsewhere in the UK in that period.

Image by permission of Charles Watts

The photo above shows States Telephone Department technicians during the dismantling of the derick above the former exchange at 1 New Street, St Helier around the early 1960s. This derick had been erected by the National Telephone Company when it relocated to this building the St Helier exchange from Bath Street around 1902. As can be seen in this picture the derick was constructed of substantial metal components and was a precarious working environment for the technicians.

pricing structure was more expensive than in Guernsey, where a uniform rate applied and questioned whether the States should intervene.

On the 18 March 1903, Howard Eady announced the opening of a new 800 line magneto exchange in the NTCo offices at 2 New Street. This would probably have been of the 'standard' National magneto design which incorporated a subscriber multiple allowing several operators to work the board simultaneously. This new board would also have been extendible beyond its installed capacity and thus would have utility for several years.

On 29 April the exchange at the corner of Bath Street and Minden Place was closed and all subscribers were transferred to the new switchboard at 2 New Street. This process had begun earlier in the year as customers were gradually migrated from the old MDF to the new. Tie circuits between the exchanges permitted the gradual change without interrupting service. The Bath Street offices were too small for expansion and the switchboard had reached its practical limit. An underground cable laid in ducts had been installed the previous year between the exchange buildings.

During that summer, new lines were being erected across St John and St Mary in preparation for the opening of further exchanges. In the meantime, some customers in these parishes were temporarily connected to St Ouen or Millbrook exchanges.

In July 1905 a new exchange was opened at a house called *Fairview*, Rue es Picots, Trinity. A further exchange was opened at *Newcastle Cottage*, Sion, to serve the St John's area. The same year, on the 12 May, St Mary's exchange opened in a property next to St Mary's Hotel. There was a further exchange open during April at Samares Post Office, Grande Route de St Clement and plans for another at St Lawrence.

In July 1910 an article from the well known electrical engineer Alfred R Bennett MIEE, was published in the *Jersey Evening Post* (JEP). In his article Bennett discussed the future possibilities of the telephone service following on from the ending of the current GPO licensing regime for the NTCo and other telephone operators in the UK.

In this article he selected four areas for comparison. These were, Anglesey (operated by the GPO), Jersey and the Isle of Wight (operated by the NTCo) and Guernsey (operated by the Guernsey States). These areas were chosen because of their similarities in size and geography. It was noted that the GPO operated a system whereby the exchange was only open during business hours (8:00AM – 8:00PM) Monday to Saturday and for only two hours on the Sunday. The NTCo and Guernsey States both operated a 24 hour service. Bennett considered it would be a retrograde step if all exchanges were to adopt the GPO model, proclaiming that 'the utility of the telephone is lost'. The statistics used by Bennett[54] (current for 1909) are as below:

	Guernsey States	**Jersey NTCo**	**IoW NTCo**	**Anglesey GPO**
No of Telephones	1760	1150	1000	110
Population	40200	56000	90000	50000
Penetration	4.4%	2.1%	1%	0.2%
Tariff (flat rate)	£5 per annum for up to 4000 calls then 5 calls per 1d. No addition charge for distance.	£6-10-0d per annum within 1 mile of exchange, 30/- per each ½ mile thereafter	Standard NTCo country tariff	Standard GPO country tariff
Tariff (calls metered)	30/- per annum 1000 calls at 1d per call then 5 calls per 1d.	30/- per annum plus 1d per call. 30/- for each ½ mile beyond 1 mile from exchange		
Tariff (calls metered)	45/- per annum 1500 calls at 1d per call then 5 calls per 1d.			
Party Lines	No	No	Yes	Yes

Table 5

---

54 History of the Guernsey Telephone Department 1896 -1925, Bennett A R

Bennett noted that telephone density increased with lower rates and that traffic growth increased with lowered charges. This was demonstrated in all but the GPO exchanges. He recommended that Local Authorities should take over the operation of the telephone system as they provided the best service as demonstrated by the Guernsey States system and that of Kingston upon Hull Corporation (both of which Bennett had been instrumental in establishing).

He also observed the low growth in Anglesey and this being besides the fact that the GPO had trunk lines connecting to the main national routes, a facility denied to the Channel Islands. It is also worth noting that the Channel Islands did not employ party line working, a practice which was common elsewhere.

The debate over the hours of service continued for some time in the press. Local businesses were alarmed at the possibility of their telephone service being downgraded following the GPO takeover. However, in a letter published in the JEP on 23 December 1910, the local Post Master, Mr C Fenton, quoted from Hansard the answer to a question on the subject of NTCo exchange operating hours after takeover asked by Mr John Wood MP in which the response was 'mostly in the affirmative'.

During 1911 the Rector of St Saviour, the Reverend Canon Luce and Jurat de Carteret were charged with investigating on behalf of the States, the offer by the Postmaster General to purchase the assets of the National Telephone Co in Jersey. On 11 August an initial report was presented to the States for lodging for debate. The Rector had met with the PMG and reported that the matter would require careful consideration. Jurat de Carteret reported that the likely cost would likely be in the order of £30,000. On the initial presentation some discussion ensued on the right of the NTCo to dig roads, however, it was pointed out by the President of the sitting that the telephone company had a special Act[55] enabling this. Deputy Nicolle considered that the takeover of the telephones would be unsound as the British Government, he claimed, had lost money ever since taking over the

---

55 Jersey Archives D/AP/AD/8/13 Act-Report regarding the conditions imposed on the company known as 'The National Telephone Company Ltd' on the subject of the placement of underground telephone lines. Presented by Jurat Briard, lodged au Greffe.

telegraphs and it was unlikely that Guernsey, on close examination, would be making any profit. Furthermore, he contended, telephones were a luxury. The Rector pointed to the popularity of the instrument with local tradesmen, but the Deputy rebuffed this saying that likely they should closely study their accounts. Besides, he asserted, if the States purchased the system at costs of thousands of pounds it would likely soon be out of date. The matter was lodged au Greffe.

On 10 October there was another meeting at the Chamber of Commerce where a presentation on the success of the Guernsey telephone system was given. Mr J H Wimbole, the Chamber president announced a public meeting to be held at the town hall on 25 October to discuss the public takeover of the telephone system. The ensuing meeting was well attended and a petition letter to the States was proposed by Jurat Payn and seconded by Jurat Godfray the following day. The letter was lodged au Greffe for later debate. At the same time both Deputy Crill of St Clement and Deputy Bois of St Saviour asked the president of the States when a debate was to be held, but none was fixed at this time.

The matter next came up for debate on 2 February 1912. The Rector of St Saviour reported the latest progress to the States, which in essence was very little. Jurat Payn considered that it may be too expensive to purchase and would require a substantial loan which could jeopardise island economy. The Rector of Trinity supposed that the States could operate the system cheaper than the GPO but gave no justification for his view. The States voted to continue investigations but Jurat Payn proposed no takeover and was supported by the Deputy of St Clement, Jurat Falle. The matter was again logged au Greffe.

There the matter rested for a whole year before the next debate. This time the States were driven to debate the matter as a result of the Railways and Canals Commission final arbitration report in the UK, which provided the settlement of terms for the Post Office purchase of the NTCo assets. The Rector of St Saviour presented to the States the investigation committee's report on the telephone question. It would appear that very little real research had been completed and the pricing seems to be guesswork, offering a supposed cost of 'around £30,000'. Deputy Giffard said that the States would need a loan to buy the system and that it needed completely renewing anyway, thus it would be impossible. Deputy Le Cornu thought all communications should be better placed under one body. After a short debate the States decided unanimously that it would be too expensive and directed the Greffier to write to PMG confirming that decision.

That was not quite the end of the matter, however. A final attempt at bringing the telephone system into Jersey States control was made on 11 March. Deputy Nicolle of St Saviour presented a petition[56] to the States proposing private funding. This was lodged au Greffe, although the house President said States would not be party to any such proposition. The Greffier wrote to the PMG and received a reply which was presented in the States on 18 April. The PMG's response was that such a proposal was not in line with the Telegraph Act, which permitted purchase only by a local government authority. Therefore, such an offer could not be entertained. That ended the debate on the ownership of the local telephone system for some time.

---

56 D/AP/U/97 Petition to Major General A N Rochfort, Lieutenant Governor; William H V Vernon, Bailiff and the States of Jersey from the Jersey Incorporated Chamber of Commerce regarding the acquisition of a locally owned telephone service for the Island, with a short history of the proceedings to date, reference to the National Telephone Co and the Railway Commissioners. Presented by Edmund J Nicolle, Deputy of St Helier, lodged au Greffe. 1913

Chapter 8 - Telephones and Telegraphs under the Post Office 1912 - 1923

The National Telephone Company's licence expired at midnight on 31 December 1911. The PMG exercised his right to purchase the system but before that could be executed, the purchase price needed to be determined. For some time the two parties had been in discussions to determine the final purchase price, but on the expiry of the license, this still had not been settled. The Railways and Canals Commission was appointed as arbitrators and in order to ensure the continuance of service, a Memorandum of Understanding was signed between the NTCo and the PMG. This enabled, in effect, the system to be operated by the Post Office. As a result, nothing changed overnight; the NTCo local superintendent, Howard Eady, being appointed as the local GPO telephone superintendent, continued to manage the day-to-day operations.

When the PO took over the telephone system in Jersey, the system consisted of several buildings, way leaves, public call offices and considerable line plant. The building list is in the table below:

Designation	Address	Lease Date / Term	Lessees	Annual Rental
Five Oaks Exchange	Victoria House			
Bel Royal wayleave		1 year	Mary and Anne Charlotte Gibault	£2 - 0 - 0
Gorey Exchange	2 Hillgrove Terrace	1907 7 years	Philip Le Vesconte	£12 - 0 - 0
La Moye Exchange	La Moye PO			
La Rocque Exchange	La Rocque PO			
Millbrook Exchange	Rue du Galet			
St Aubin Exchange	Bel Vue, High Street	29/9/1906 5½ years	John Peter Le Fol	£24 - 0 - 0
St Helier Stores	34 The Parade	5 years	A Gulliver	£18 - 0 - 0
St Helier Exchange	2 New Street	14 or 21 year	Wilfred Le Chemiant	£40 - 0 - 0
St Helier	field at Belozanne*			£3 - 0 - 0
St Lawrence Exchange	St Lawrence PO			
St Mary Exchange	House next to St Mary's Hotel	23/3/05 7 year	John Percival Edward Le Gresley	£11 - 0 - 0
St Ouen Exchange	Sans Ennui, Route du Marrais		Charles John Matthew Perrie	£16 - 0 - 0
St Peter Exchange	Elmfield, Rue des Minquiers		Philip Le Gresley	£10 - 0 - 0
Samares Exchange	Samares Post Office			
Sion Exchange	Newcastle Cottage	Yearly	John Edwin Pinel	£12 - 0 - 0
Trinity Exchange	Fairview			

Table 6

* Pole store, at field known as Biles' Field, First Tower.

At the time of transfer, there were 1313 subscribers and 30 private circuits together with 24 public call offices, mostly in NTCo exchanges and at railway stations, although there were some located in shops and one at the head Post Office, Broad Street. The liquidator of the NTCo assets for Jersey was Mr George Franklin.

Although La Rocque exchange had closed at the end of 1901 when the subscribers were transferred to Gorey exchange, the Post Office found it expedient to re-open the switchboard in December 1913 when Mr A Page was appointed as the supervisor and night operator.

All the telephone exchanges at the time of transfer were of the NTCo 'standard' small magneto type which would have had a capacity of 50 lines, expandable in some instances, to 100 lines. The exchange at St Helier, however, was a larger magneto board which, although initially designed for a maximum of 800 lines, it was catering for some 900 lines at takeover.

The system remained much as it had under the NTCo during the first twelve months after the transfer while the negotiations were finalized on the financial settlement. The outcome of the arbitrations was announced in the States on 13 February 1913. The original claim by the NTCo for their whole system was £20,000,000 but the finally agreed sum was somewhat less at £12,500,000. The value attributed to the local system was just £23,566.

In May 1914, the telegraph office in Jersey was upgraded for more efficient handling of traffic during the potato season. At this stage, Morse code was still used over telegraph links but this was improved by the inclusion of off-line message preparation and automatic senders. The off-line message tapes, which used a hole punch method, were up until then prepared on Gell perforators, which were a three key device each of which were struck with a rubber hammer to create a hole in paper tape, rather like a drummer these devices required considerable skill by the operator. These were replaced with Creed re-perforators which were more akin to a typewriter keyboard. Incoming Morse code was automatically converted into tape which could be fed into a printer which would output a continuous tape with the message printed in capital letters. These are cut to size by the telegraphist and gummed to a telegraph slip. The off-line tapes were fed into the Wheatstone transmitters and sent to line. The standard speed for off-island working was set at 150 words per minute.

Shortly after the takeover of the NTCo, the First World War broke out, in August 1914, and the GPO was placed on war alert. The Channel Islands were viewed as strategic locations and as a consequence very little of the GPO's covert operations on behalf of the military were reported in the press. The islands were used as a staging post for communications with France, following the disruption of communications in Northern France and Belgium during the early part of the war (this is discussed more fully in Chapter 10 below).

This was the first time that the GPO had been directly involved in a major conflict and the expectations of the government and the military put a considerable strain on the GPO resources. Many of its staff were either seconded into military service or, in the case of 'less essential' staff, conscripted onto the front line. Consequently, the development of the domestic telegraph and telephone networks took a back seat for some time. This was evident in Jersey, especially so as it was a remote and rather small part of the GPO system.

Many of the local telegraph operators and other GPO staff were conscripted into the army during the early part of hostilities, since working for the GPO required allegiance to His Majesty's laws, which appeared to have little regard of local legislation. Later, however, after much debate, the States implemented its own method of conscription and other Jerseymen[57] faced the prospect of the front line. As a consequence, the GPO began to employ women into the jobs normally the preserve of men and by 1917 their permanence was recognized as specially designed uniforms were issued for 'Telegraph Girls'.

Although new cables were installed during September 1914 to both Dartmouth and St Malo (see Chapter 10 below), these events went unreported in the press. The only consequence of the outbreak of war was the notification on 20 August 1914 that all telegrams would be subject to censorship.

During March of 1915, German prisoners of war were brought into Jersey to be billeted at a special camp built at Les Blanche Banques, St Ouen. This had required considerable work by the GPO to provide the necessary

---

[57] *'Ours'. The Jersey Pals in the First World War.* by. *Ian Ronayne* The History Press ISBN: 9780752451459. Publication Date: 03/08/2009

military telephone and telegraph circuits for the management of the camp.

On the 10 May 1915 Howard Eady announced that he was being moved to a new appointment as district manager for Exeter after 15 years in Jersey. His new area 'has 60 exchanges and over 6000 lines compared with 13 exchanges and 1300 lines in Jersey' according to a JEP article. Under his management eight rural exchanges had been opened. He had also been chief officer of the St Helier fire brigade since 1901, a service that he had helped to establish. He was replaced by Mr A G Mackie, who had formerly been stationed in Portsmouth.

Cable faults during that year's potato season caused more than the normal disruption because of the priority given to military traffic. The Jersey telegraph office was used for traffic from the UK to France following the disruption of the cables in the north of the English Channel caused by the war[58]. The radio system was also unavailable being less secure and in any case was in use for sending encrypted Admiralty signals to the fleet.

In October 1915, the price of a call from a Public Call Office was doubled to 2d (1p). This was in line with GPO policy elsewhere in the UK. At the same time the flat rate charge for a telegram was increased from 6d to 9d for 20 words.

By 1917 the war was beginning to affect everyday life more and more. There were food and fuel shortages and newsprint was in short supply. The GPO announced that the normal regular quarterly telephone directories would be postponed for 3 months until the following January. There was also a spate of 'spam telegrams' which were being sent to random addresses requesting money transfers of 10/- (50p) to a London address. The police advised precaution and to report any incidences. The war ended in November 1918 and a semblance of normality reappeared. De-mobilized troops began to return to their posts and the everyday matters of civilian and business life.

In June 1919, Makie was again relocated, this time as Telephone Manager

---

58 The Invisible Weapon: Telecommunications and International Politics, 1851-1945. Daniel R. Headrick. Oxford University Press US, 1991

at Hanley, Stoke-on-Trent. His replacement was Mr A H Darker. At about the same time the local Postmaster, Mr Ford was transferred to Middlesborough and replaced by Mr A O Forrest, formerly of Truro. Later, in 1921, Ford joined the Marconi Company in Lima, Peru. In September, Mr Angel, the superintendent of telegraphs retired. He had started his career with the Post Office at the age of 18 in 1874 and had been due for retirement at the age of 60, the normal age for British civil servants at that time, but because of the skilled labour shortage during the war, had remained at his post. He was presented with a long service award by Forrest on behalf of the GPO.

In December, the cable hut at Plémont was broken into and some test equipment stolen. The detail of what was stolen was not reported in the local press, but it is hard to imagine what a thief would do with any specialist telecommunications equipment. The attraction may have been the quality oak chest in which Post Office test equipment was normally mounted or the scrap value of the brass instruments.

In January of 1920, the Chamber of Commerce petitioned the Postmaster to extend the opening hours of the telegraph office as it was claimed to be affecting export trade. The GPO finally agreed in June to open one hour earlier at 7am until October. A bombshell for business landed in November when the GPO announced that after a review of costs, the price of telephone usage would be increased. Jersey, they asserted, attracted the same operating costs as any other rural area. The proposed changes would increase telephone rental and usage costs by an alarming 600%. The current cost was £6-10-0d which included the making of 5000 calls per year. The new proposed rate would, claimed the Chamber of Commerce, increase the cost of a line to £38-15-0d per year. In addition, a new charge of 25/- per 1/8 mile would be charged for lines further than 1 mile from the exchange. The JEP ran a scathing editorial on the new proposals at the beginning of December.

In January, Mr Guiton of the Chamber of Commerce reported that he had investigated and discovered that the Post Office would be prepared to sell the telephone system to the States or to a private company. This latter proposal was contrary to their previous decision in 1912 when this option was excluded. On the 24 January there was a general meeting called by Guiton at the Chamber of Commerce at which the Farmers Union and the Farmers Co-operative Association also attended. The details of the new charges were given to the meeting. The current line rental of £6-10-0d would be replaced by a new charge of £7-10-0d. In addition, each call would be charged at 1½d. This would increase the current charge six-fold and it was compared to Guernsey where charges were only £8-7-6d per annum including 5000 calls. On 2 February 1921, the Chamber of Commerce published figures from its business survey which indicated that most respondents would give up the telephone service if the new charges were imposed.

On 16 February, the President of the Chamber of Commerce, Mr Le Masurier went to London to meet with the PMG's secretary (telephone department), Mr Kinder, at the GPO offices at St Martin-le-Grand. At this meeting they successfully argued that Jersey was a special case and that costs on the island could not match that in the UK rural areas. Furthermore, they contended, Jersey had no trunk line connections with the rest of the Post Office network. The secretary agreed to suspend the proposed increase pending further review. It is clear from this meeting that the Post Office had determined that the Jersey – and indeed the Manx - telephone systems were something of an embarrassment to them, being situated off-shore and in jurisdictions outside the UK. However, the Manx government did not pursue the offer to purchase the system at that time. The policy adopted by the GPO may also suggest a carrot and stick approach to encourage the States to buy the system. The temporary reprieve of the increase, coupled with the Chamber of Commerce's survey indicating a mass desertion of the system may have encouraged the States to believe that they could, after all, operate the system profitably.

There is no indication that any private enterprise initiative for the purchase of the system was forthcoming but the States revived the telephone committee in March, this time under the presidency of Jurat Lemprier. The committee appointed Alfred Rosling Bennett to conduct a survey of the local telephone system. Bennett was an engineer with much experience in telephone systems, including having advised the Guernsey States prior to them setting up its own system in 1895. Bennett reported back to the committee at the beginning of August. The press reported that the survey was thorough and satisfactory. The actual value put on the system by Bennett was £32,000. This sum was agreed as a fair price in a document published by the Greffier, John Edward Le Huquet signed on 15 August. The TelephoneCommittee subsequently set up a sub-committee to negotiate with the PMG on the matter of salaries and other running costs. They also visited the Guernsey telephone department to learn how the Guernsey States had managed its system.

The Alfred Bennet Report

The report from Bennett was a complete examination of the Post Office telephone network on the island. As part of his inspection, Bennett visited all the GPO exchanges and made particular note of a random sample of facilities. These were named as:

Distribution poles were inspected in Museum Street, Commercial Street, Elizabeth Lane and at West Park railway station.

The standard on the exchange standard at New Street was declared to be in good condition.

Manholes were inspected in Minden Lane, Outside the British Hotel, Broad Street, the New Street exchange, Snow Hill, the Town Hall and James Street.

The linesman's room at Millbrook exchange. (Bennett noted that as early as November 1912 the GPO had contemplated closing Millbrook exchange but had never progressed the matter)

Eight sections of telephone line on telegraph poles that will remain with the GPO.

He also reported on the staffing costs:

Office staff	30/6 per week
Part time telephonist (nights)	32/- per week
Exchange Attendant (E R LeGros, Millbrook)	£1 per line per annum
Inspector (Mr L O Hales)	£180 per annum
Skilled workman (P Perrin, established 5 years)	42/- per week
Skilled workmen (unestablished)	30/- to 39/- per week
Youth in Training (Charles Gottrel)	20/- per week

Table 7

At this time Bennett noted that the inspector spent 22% of his time on telegraphs.

The total wage bill was £7,320-10-8d per year.

Operators worked a 45 hour week, engineering staff 51 hours. Female employees' employment was terminated on marriage.

The cost for internal maintenance of the system was estimated at £1068 per year.

Exchange maintenance costs estimated at £252 per year.

Exchange repairs for the previous year were £177.

(Although the report did not specify the number of staff employed at this time, this can be deduced from the wage bill and the average wages to be around 85 employees).

A survey of calls taken at St Ouen's exchange showed the following distribution:

| Local Calls || Junction Calls Out || Junction Calls In ||
Weekdays	Sunday	Weekdays	Sunday	Weekdays	Sunday
30	9	97	23	193	50

Table 8

Bennett concluded that a fair price for the complete system would be £32,000. This sum compares with the sum of £23,566 determined by the Railways and Canals Commission for the NTCo network in 1913. This increase in value could probably be attributed almost exclusively to monetary inflation brought about by the Great War.

In the meantime, it was business as usual at the local Post Office. Following a study, it was decided in November 1921 to install electric lighting at the main post office in Broad Street and the St Helier exchange in New Street. This would be installed by the local company of electricians Messrs A L Lowke and Sons of Vine Street, St Helier and installation supervised by local postmaster Darker. At that time there was no public electricity supply, although the question had long been extensively debated, various proposals for parish or other municipal supplies having been projected. Much of the town lighting, such as it was at the time, was provided by the Jersey Gas Light Company. There must have been a considerable lobby operated by them, since almost every public debate on the subject ended in disarray.

There was, however, a thriving business for local electrical engineers who supplied bespoke systems, usually provided as petrol or paraffin stationary engines driving dynamos providing a DC supply. Some installations incorporated lead-acid batteries for more resilience, for example that supplied by Messrs Lowke to the Jersey Railways and Tramways Company at the Weighbridge terminus the same year. The magneto switchboards systems would have operated on Leclanché cells, since they would not have required the large currents necessary for the more advanced Central Battery (CB) switchboards then being employed extensively in main towns everywhere in the world. Each subscriber's telephone had a separate battery and each exchange switchboard had its own battery box to supply the current feed for the operator's telephone set. Ringing was of course supplied from the switchboard hand operated magneto. Similarly, the telegraph systems would have used a battery power supply.

The telephone sub-committee started discussions with the Post Office, particularly with regard to the conditions of transfer of staff and pensions to the States. The GPO had had similar issues when incorporating the NTCo personnel into its system in 1912, hence the delay of complete transfer till 1913. The NTCo staff rights had been honoured by the GPO but staff employed subsequently had been employed on the standard GPO terms. This would consequently complicate staff contractual terms on transfer.

Discussions were slow and in response to a question raised in the States on 6 April 1922 by Jurat Le Marquand, Jurat Le Boutillier (who was acting president of the telephone committee due to the indisposition of Jurat Lemprier) confirmed that they were waiting for the GPO to respond. There were also continuing discussions with the GPO staff that would be transferred to the States department. On the 3 May J W Bowen of the Post Office Union came to Jersey to discuss the staff issues with the committee. The transfer conditions of the staff seemed to be more of a problem than the handover of the system.

By 18 August Jurat Lemprier was able to confirm to the States that the final report on the purchase could be lodged *au Greffe*. The report was then debated on the 18 September as the first item of business. The report was taken as read and Jurat Lemprier went on to say that the expert reports were 'full and ample', and he recommended that the States adopt the report and instruct the Finance Committee to provide the necessary funds. He went on to say that he had received correspondence from the Postmaster General to the effect that he was anxious to dispose of the telephone system. Charges had increased in England but these were disproportionate for Jersey as there were no trunk lines. The price asked by the GPO was £32,000 and an expert report from Mr Bennett, in whom they had great faith, indicated that there was about 17 years useful life in the current system and that the price asked indicated a good bargain. He admitted that the system should be run profitably but was confident that the proposals in the report would enable this. The staff had been asked to reply yes or no as to whether they wished to transfer. The question of wages had been gone into with Bennett in consultation and an agreement had been arrived at. The only outstanding question was the issue of a royalty, or licence, which the PMG insisted upon, which currently amounted to a yearly sum of £800. Correspondence with the Corporation of Kingston upon Hull had indicated a similar issue and it seemed that it was a matter that could not be dispensed with under the Telegraph Act 1872 as extended by Order in Council to the Channel Islands.

The States would need to find a suitable manager, and they would seek a person who would be able to develop and improve the system. He noted that Guernsey ran its system ably and profitably and there was no reason why Jersey should not do the same. It was important for business that there was an efficient telephone service and he recommended the report to the States. The proposal was seconded by the Deputy for St Saviour. He said that the members may have thought that this matter had taken some time, but it was important that it was thoroughly researched. Guernsey, he noted, also had to pay a licence to the GPO. The proposed pricing structure developed by Bennett would be simpler and cheaper than that proposed by the GPO, and the report also indicated that there was a prospect of strong growth at this price. It would be necessary, therefore, to raise a loan more than the purchase price in order to service this demand.

Jurat Payn submitted that the full facts of the cost were not placed before the States and that the proposal should be referred back to the committee. He further questioned whether the States should be involved in commercial ventures. The Deputy of St Clements seconded this proposal as he noted that the royalty was based on nett takings and therefore could be far more than £800 per year. Deputy Middleton noted that Mr Bennett had suggested a further £20,000 would be required for system development and he thought the committee should give full details. This view was supported by Deputy Gray (St Helier) who asked for assurance that the current subscribers would continue to use the service.

The Deputy of St Saviour said if the report were referred back, they would only be able to report what they already knew; the GPO had been very patient and it was not up to the States to abuse that patience. The Deputy of St Peter supported the motion and said the very fact that there could be a requirement for up to £50,000 meant that the committee had not put all their cards on the table. Jurat Le Cornu noted that over the proposed licence period of 20 years there would be at least a further £16,000 due to the GPO, he did not think that it looked as cheap as the committee had proposed. However, he thought it was worth it and the States should look upon it along business lines.

Concluding the debate, Jurat Le Boutillier submitted that it was absolutely essential that the States purchase the system, the GPO was pressing and no more information was available that was contained in the printed reports. The vote on the proposal to refer back was defeated by 34 votes to 9 and the proposal to buy the system was therefore adopted.

On 17 October the Finance committee lodged a proposition in the States to raise a £52,000 loan to complete the purchase of the telephone system and to provide funds for system development. It was proposed to issue bonds at 4% and the telephone committee would repay the loan at £1,500 per annum out of its profits. Jurat Payn, as ever, was concerned that there may be no profit but the proposition was nevertheless Lodged au Greffe.

At the States sitting on 1 November the proposition was authorized for the Finance committee to raise a loan of £52,000 to fund the purchase and development of the telephone system. The loan bonds would be issued at 4% but Jurat Payn proposed that the telephone committee should be charged at 4.5%, a suggestion which somewhat surprised Jurat Lemprier as his committee had not been made aware of this beforehand. However, the proposition was passed as Jurat de Cartaret of the Finance committee supported the proposal. A letter to the Lt Governor from the Hull Corporation congratulating Jersey on taking over the telephone system was read out to the States. The letter noted that Jersey would be the second municipality to do so.

The following year the States were asked at the sitting of the 16 January

1923 to vote £100 to cover the expenses of the telephone committee to travel to London to finalize the trades union matters with the PMG. The telephone committee noted that there were several unions involved and it seemed that the many agreements involved produced a complex problem. This was agreed and on 9 February Jurat J E Le Boutillier, Deputy J T Fergusson and the Constable of St Saviour, J A Perrée, travelled to London to meet with the GPO. The negotiations progressed well and on the 17 February Jurat Le Boutillier returned and reported to the States on 20 February that the other two committee members had remained in London to finalize the details. Jurat Lempriere reminded the States of the progress made since the setting up of the committee a year ago and announced that the telephone system would be transferred on 31 March. He recommended to the States that telephone committee be reappointed to manage the new States Telephone Department. This would enable the smooth transition from the investigating committee to a permanent management committee for the telephones. This was seconded by Jurat Renouf who also asked that house thank the committee for their work in this difficult matter.

Jurat Le Boutillier explained that it was necessary to have a management committee, since at present the seat of management was in Bournemouth and it would also be necessary to establish at once a local head office. Jurat Le Marquand opined that in his view many a business had failed because of bad accountancy and he urged the committee to appoint someone with both engineering and accountancy experience. The President pointed out that that question was not before the house. Further discussion continued on the cost of salaries and whether or not such officials should be approved by the house. Finally, the proposal to reappoint the committee was adopted.

On 6 March 1923 the following notice appeared in the Jersey Evening Post:

> **States of Jersey Telephone Department**
>
> Notice to Telephone Subscribers
>
> From the 31st March 1923 the Telephone System in the Island of Jersey will be transferred from the Post to the States.
>
> Any Subscriber who desires to continue the use of the Telephone after that date must sign an Agreement to that effect with the States Telephone Department.
>
> Forms will be sent to all Subscribers and must be completed and returned to this Department on or before the 25th March, 1923.
>
> SCALE OF CHARGES
>
> **TARIFF A** – For a first Business Connection without limitation of distance: £3 15s. per annum and 1d per call for up to 1,000 calls ½ d for the second 1,000 calls and ¼ d for the third 1,000 and after.
>
> **TARIFF B** – For a second Business Connection without limitation of distance and not necessarily to the same premises: £3 10s. per annum, with the calls as in (A).
>
> **TARIFF C** – For a Private Residence Connection without limitation of distance: £3 per annum, with the calls as in (A) and (B).
>
> Subscriptions, as regards the Annual charges, to be payable in advance. Accounts for the calls to be rendered quarterly and to be due for settlement immediately.
>
> All charges to be payable at the States Telephone Office.
>
> R. R. LEMPRIERE,
>
> President,
>
> States Telephone Committee.
> 6/3/1923

Figure 3

The public accounts committee report in the States sitting on 13 March showed an estimated expenditure for the coming year of £8,000. No details were given on how this would be allocated.

On 27 March the JEP reported that public call office calls would remain at

2d following the takeover.

The GPO had offered the States a 30 year licence with an annual commission of 10% on revenue to be paid to the PMG. This was accepted since it was substantially the same as the terms offered to both Guernsey and Hull. At midnight on 31 March 1923, the telephone system passed into the ownership of the States of Jersey.

Chapter 9 - The General Post Office Telegraphs and Submarine Cables 1923 – 1940

After the sale of the telephone system to the States of Jersey, the Post Office retained control of the telegraph system and all submarine cables. Although all the telephone overhead and underground plant was handed over to the States, the Post Office retained much of the main telegraph distribution network. Some of the minor post offices were, however, connected to the central telegraph office using line plant now owned and operated by the States Telephone Department (STD). Therefore a working relationship with the STD was necessary albeit somewhat strained at times.

The GPO thereafter relied to a greater or lesser extent on the STD for the provision of new overhead plant, since it did not retain sufficient staff to carry out such works locally. Where special circuit requirements were necessary, however, engineering staff from Bournemouth were drafted in.

The telegraph system continued to be the only means of off-island communication and, as a consequence, came under pressure from business to be more accessible. Petitions were occasionally raised to extend telegraph services to minor post offices but were usually resisted by the GPO. This situation became easier after the agreement in the summer of 1923 between the local postmaster and the STD to permit telegrams to be dictated over the telephone directly to the telegraph clerks.

At this stage the development of the telegraph system had almost peaked. Although printing telegraphs had been introduced in the 19th century, and they had been introduced into Jersey before the Great War[59], the system was not yet fully automated. The only major change to the system had been the introduction of the through circuit to France for military purposes which had been operated by the GPO and used for overflow traffic when not required by the War Department.

In 1925 the Post Office resisted an organized petition demanding a new telegraph sub-office in Beresford Street near. The Postmaster claimed it was too near to the Broad Street office and that 50% of the petitioners were businesses with access to the telephone and so could use the service provided in conjunction with the States of Jersey Telephone Department.

In 1927 the GPO adopted the new Creed [60] teleprinters for the telegraph system first introduced in 1922 by the Creed Company of Croydon. This was what is called a start-stop machine and it used a different transmission code from the previous Morse used in the Wheatstone and earlier Creed machines. This new machine which the GPO designated the Model 3 operated at 65.3 words per minute and printed the messages directly onto the gummed paper tape. The new code was based on the Baudot code which had been modified by Murray and finally incorporated into the GPO's own version for the Inland Telegraph Service. Unlike Morse, which used between 1 and 5 elements for each character, the new code used a standard of 5 elements per character. This enabled the machines to work asynchronously and without supervision, thus greatly improved the efficiency of the service and naturally reduced the number of operators required at each station.

The new printers were introduced into Jersey during October 1928 and installed only at the head STD Broad Street. This required minor changes to the transmission equipment on the submarine cables to ensure that higher speeds could be successfully transmitted. At this time the sub-offices now used the telephone for the passing of messages rather than the slower old

---

59 The Telegraph and Telephone Journal, October 1914
60 The Worldwide History of Telecommunications by Anton A. Huurdeman. John Wiley & Sons ISBN: 0471205052

ABC telegraphs. The same year the unreliable Alderney cable originally installed by W T Henley in 1870 was replaced with a wireless telegraphy circuit to Guernsey. This work was completed on the 9 February 1929.

The Post Office's attention was now turned to the problem of connecting the local telephone systems to the mainland. Questions on the matter had been raised in Parliament during 1928. Developments in telephone technology enabled the Isle of Man to be connected to the UK telephone trunk network in July 1929, albeit as adjunct of the new cable laid by the GPO *CS Faraday*[61] between Port Erin and Ballyhornan in Northern Ireland and then between Port Grenaugh and Blackpool, primarily to provide a link to Northern Ireland. This raised hopes in both islands that a solution could soon be found. However, this was not a matter for the local officials, since there were no suitably qualified personnel to work at what was at that time leading edge technology. Instead, the issue was passed to the Post Office research department, then based at Dollis Hill, north London.

More fundamental issues diverted the attention of the local postal authorities. A great storm during early December severed both northward telegraph cables including the inter-island link. This meant that all traffic had to be diverted via the Fliquet-Pirou cable to France. The French network was also badly affected and thus extreme delays of up to 14 hours were experienced. In response the Post Office opened a temporary wireless service from Fort Regent (see below), this enabled the service to be restored to near normal. The service operated until the cables were restored on 16 January 1930.

Meanwhile, the Guernsey Press reported that Deputy Kitts was in negotiations with the Post Office and that the 'secret' former German Borkum to the Azores telegraph cable was being tested for its suitability as a telephone circuit. This cable would be diverted via Saints Bay to provide a link to Guernsey as well as Jersey.

At this time the research department engineers were testing the cables in order to ascertain whether they would be suitable for telephony. The earlier

---

61 Cableships and Submarine Cables by K R Haigh. Adlard Coles Ltd, London 1968

cable laid by the Post Office in 1884 was considered unsuitable because of the cross-talk problems. The cable construction was three-core which meant that it was difficult to create an electrical balance either between individual cores or between any core and earth. Therefore, it was not possible to use it for telephony because of the interference from the telegraph circuits which would continue to share it. However, the German cable, formerly used for the Borkum-Azores telegraph route until captured by the British at the outbreak of the Great War proved to be suitable. This cable was a single core cable over which an unbalanced circuit between the core and earth could be constructed. Tests on this circuit had begun as early as 1928 but at this stage many of the necessary sophisticated amplifiers and building-out circuits were still in development (see below). Tests continued throughout 1929 and 1930 in order to make suitable arrangements for the cable's usage for telephony.

A component part of the delay was the continued use of the German cable for military purposes. This cable was linked with the cable from Greve D'Azette to St Malo and carried traffic to Rennes. However, during the early part of 1930, the GPO Telegraphs Department indicated that they would not require the cable for telegraphs after the middle of the year. This enabled plans to be made for the conversion of the cable for telephony. Part of these plans included the diversion of the cable via Saints Bay, Guernsey in order that the cable could serve both islands.

The *CS Monarch* assisted by the Guernsey States' tug laid the necessary cable lengths at Guernsey during the autumn of 1930. *CS Monarch* also laid suitable 'sea-earth' sections at all cable ends to ensure a good low-noise transmission path. The final cabling changes were carried out before the end of the year. Testing of the new cable sections was completed by mid February 1931 and the necessary final balancing of the amplified section was continued during March, ready for the opening later that month. The telephone circuit was finally opened in the early spring of 1932.

Meanwhile, the state of the northern cables continued to cause concern, there were frequent faults and interruptions, often noted in the press when the news links failed. The telegraph continued to have priority over telephone and whenever it was necessary, the trunk circuit was suspended in favour of the telegraph. To improve the availability of the telegraph under fault conditions it was decided to provide a sub-audio telegraph circuit on the German cable. This would ensure that both the trunk circuit and the telegraph could remain in service under fault conditions on the old three-core. This work was carried out in late 1932.

This effective 'spare' circuit gave the telegraph division comfort in the event of a cable failure and enabled the news telegraph circuit to be permanently connected through to both islands. The receiving telegraph machines were located in the offices of the local newspapers. This service was effectively unidirectional, although an uplink to the London agency was possible using a Morse circuit. The system used the latest Creed 7B page printer teleprinter machines that had been introduced into the Post Office telegraph service in 1931.

The number of telegrams began to decline after the opening of the telephone trunk service. This was probably because of the relative cost compared to the available information channel width. A phone call of three minutes could deliver more data density than an equivalent cost telegram. In reality, this led to the long slow decline of the telegram as a means of business communication in common with other areas of the developed world. This process was exacerbated as the GPO followed a policy of continuous reduction in the cost of trunk telephone calls as traffic increased.

On 30 May 1935 saw the first reduction in telegram charges since the introduction of the flat rate of 6d for 20 words had been introduced in 1885. Previous rate changes had been upwards, in 1915 it was increased to 9d while in 1920 it was further increased to 1/-. The new rate limited the telegram to nine words, with additional words charged at 1d. This was no doubt in response to the decline in the use of telegrams which had begun with the introduction of lower trunk telephone charges and compounded by the economic depression of the 1930's. The GPO telegraph department regarded itself as the senior network and was now struggling against the junior telephone system.

The first telegram under the new rate was sent from the wife of the Lt Governor, Mrs Mary Martelli, to HRH Prince of Wales. The telegram (which exceeded the nine word limit!) was written with a silver pen produced for the King's Jubilee and read:

'I have the honour to address to your Royal Highness the first telegram from Jersey at the new rate of nine words for sixpence. Mary Martelli'

This received a suitable response from the Prince and was followed by an exchange of nine word telegrams between the PMG, Sir Howard Kingsley Wood, and the JEP.

However, on the 31 July the Post Office celebrated its 300th anniversary and the JEP reported receiving a commemorative telegram from the new PMG George Clement Tryon. The telegram read:

'Greetings from the Post Office on its 300th birthday. Tryon PMG'

Note that this succinct message conformed exactly to the new nine word limit.

On the telephony side, questions were being raised in Parliament over the long delays in trunk calls to the Channel Islands. The Rt Hon CWH Glossop (Conservative) asked the PMG what was being done about the two hour delays currently being experienced.

This question was somewhat fortuitously coincidental with the introduction of a new wireless telephony service to Guernsey and an upgrade to the inter-island circuit that provided an additional channel (see below).

The existing three core cable continued to give trouble; it was by now over 50 years old and had been subject to continuous and frequent repair. This cable was, however, maintained in service and with the improvements in technology its useful life was extended. The inclusion of a telephony circuit in addition to the telegraph (see below) improved its utility between the islands, although it was still not possible to extend telephony over the Guernsey - UK section.

Eventually it was decided that a new cable would be necessary. In April 1938 the provision of a new coaxial cable was approved. The cable laying began in August and the sea sections were completed by early September. Further civil works were required to connect the cable from the shore cable huts to the main termination equipment. The Guernsey cable transited the island from L'Ancress Bay to Saints Bay via the States' exchange at St Peter Port. This work was not completed until 1939.

In February 1939 a prolonged outage of the telegraph cable resulted in disruption of the Creed Direct teleprinter service to the local newspaper offices. The GPO engineers set up a temporary service using one of the wireless telephone circuits at Fort George, Guernsey. This circuit was further extended to Jersey using the newly completed land sections in both islands. This circuit used only the voice frequency capacity of the coaxial cable.

The next month the long anticipated abolition of the Cable Charge identified in Bridgeman Committee report[62] on the efficiency of the General Post Office as being an impediment to commercial development of the telephone system was announced. The Assistant PMG Sir Walter Wormersley declared its abolition in a speech while opening a new Head PO in Scunthorpe. During the speech he announced the investment of £4M in new underground cables and that the charge would be removed for calls to the Scottish Islands from 1 April and to the Channel Islands from 1 May. This would have a significant impact on the cost of trunk calls to and from the Channel Islands. It was in anticipation of this abolition that the PO Engineering Department was given approval for the installation of the new cable. Table 9 below illustrates the effect of this change.

---

62 Report of Committee of Enquiry on the Post Office (1932) HMSO Cmnd 4149. (The main recommendations in this report were largely ignored until the 1960s when the Wilson government revisited the matter of incorporation of the Post Office and splitting it into separate post and telecommunications divisions.)

Radial Distance	9am – 2pm		5am – 9am & 2pm – 7pm		7pm – 5am	
	Old Rate	New Rate	Old Rate	New Rate	Old Rate	New Rate
Guernsey	1/9	1/-	1/3	9d	1/-	6d
75 – 125ml	3/3	2/-	3/-	2/-	1/9	1/-
Over 125 ml	3/9	2/6	3/6	2/6	1/9	1/-

Table 9

The above rates include the removal of the 1d per minute terminal charge as per the new GPO Inland Call Tariff.

The GPO presence on the island with regard to both telegraph and telephone engineering was terminated on 1 July 1940 with the German occupation.

Chapter 10 - The 1914 Telegraph Cable and Military Telegraph Circuit

Immediately after the declaration of war against Germany on 4 August 1914, the British government decided to cut the German telegraph cables in the Baltic, the North Sea and the English Channel as a strategic move against the enemy. The operation was completed between the evening of 4 August and the morning of the 5 August 1914. The Channel section was probably cut by the cable ship *CS Alert*, although this is not completely clear from the available records[63]. This action effectively isolated Germany from the rest of the world outside Europe, especially with North America, where it had maintained a close relationship with the United States government. This also forced the German government to either use telegraph cables controlled by British interests (such as the Great Northern Telegraph Company which ran through Russian) or to signal to the USA using recently developed wireless telegraphy, which was wide open to interception by the British intelligence services.

It is interesting to note that all this activity went unreported in the national and local press, although the cutting of other cables, such as the Germany-Sweden cable, was covered at length.

The cable, which ran from Borkum to New York via the Azores through the English Channel, was picked up in early September 1914 by the Post Office Cable Ship, *CS Monarch*, and the ends were diverted to provide a new link between Dartmouth and Plémont. This was in addition to the existing 3-core cable, which had proved to be somewhat unreliable over time, and would be used to provide a back-up in case of failure. The German cable was of substantial construction, because of its length, and consisted of a 500lb/mile copper core insulated with 300lb/mile gutta percha insulation, the final cable being armoured with 10 No 2 gauge galvanized wires.

Jersey was extensively garrisoned at the outbreak of war and, later, prisoner-of-war camps were provided on the island. This second cable, therefore, was of immediate importance to the War Office in order to maintain regular and consistent communication with their forces based on the island.

The cable was terminated in the cable hut at Plémont alongside the existing cable and had been hauled up the beach using the same path such as the two cables lay together along most of the shore end. The circuit was connected to St Helier Head Post Office (HPO) at Broad Street with an additional circuit provided along the existing overhead open wire route to the Western Railway terminus at the Weighbridge. There it was connected via a short underground cable to the post office building in Broad Street. A duplex telegraph circuit was supplied, connected via Creed relays.

In the following weeks, it was decided to install a further link into France via St Malo. This was as a result of the interruptions to cables across the Channel in northern France. The considerable correspondence between the War Office and the Post Office suggests that this decision was made on both military and commercial bases[64]. The proposed cable was to be provided in conjunction with the French authorities who would fund half its expense. The French decided that St Malo would be a better point of interconnect with their network than Pirou and so it was agreed that the cable would be laid across that stretch of sea. Initially, it seems that a telephone circuit was proposed but as the work was urgent and the only available cable ship was the Eastern Telegraph Company's *Electra*, it was decided to use a single core cable which was also immediately to hand. Correspondence between

---

64 Collected papers from the British Telecom archives.

the South West Area and Post Office HQ[65] indicates that two-core cable could have been sourced within a fortnight, however, the *CS Electra* was an old ship and could not carry the heavier load. The cable laying was started on 9 September 1914 and completed within a few days. The cable was laid from Cancale, near St Malo to Greve D'Azette in Jersey and connected to the Head PO via the existing telegraph poles along the Jersey Eastern Railway.

A letter from the South West Area superintendent to GPO headquarters confirmed that the circuit was working on 27 September, having been extended to Rennes, using a Wheatstone omnibus circuit[66] and a Hughes telegraph relay in Jersey. The French indicated that they would extend the circuit to Bordeaux. Further correspondence between the South West Area superintendent and headquarters showed the breakdown of the cost of the circuit provision as £12,826 - 12 - 4d, half of which was to be recovered from the French authorities, however, this part of the exercise appears to have been postponed 'pending cessation of hostilities'. It is not known if costs were eventually recovered. The total length of cable used was 39.71 knots for deep sea and 2.89 knots at the shore ends.

The result of the circuit being provided over a single core submarine cable meant that the circuit operated only as a telegraph circuit. Although it was originally intended to route the old German cable via Guernsey, this was not pursued because of the urgent need to expedite the work, and the circuit operated from London via Bristol and Dartmouth through Jersey (although a breakout was made at the St Helier telegraph office) and onto Rennes.

---

65 ibid
66 Ibid. An omnibus circuit is one that serves several telegraph offices on the same line. The messages are distinguished by a prefix code so that the telegraphist knows whether to record the message, although on automatic recorder lines each office would receive all the messages.

Chapter 11 - The first submarine telephone circuit

In early 1928, following enquiries from the States of Jersey and States of Guernsey Telephone Departments, the GPO directed its research department located at Dollis Hill in north London, to investigate the feasibility of providing a telephone circuit on one of the submarine telegraph cables. Initial tests were carried out and the results of the survey were reported in a research document published on 26 March of that year[67].

Tests showed that of the two cables, only the German cable could be utilized as a telephone circuit since the 3-core telegraph cable exhibited too much cross-talk owing to its method of construction. The German cable, however, responded well in tests, and although it had been extensively repaired since 1914 with 10 new sections of various cable types being spliced in along it's route (see below), it seemed to be satisfactory for the provision of a telephone circuit. Transmission tests indicated an overall characteristic impedance of approximately 70 ohms at 800Hz, the standard frequency adopted by the Post Office for measuring voice frequency circuits.

The only significant problem encountered during the tests was a degree of cross-talk from the 3-core cable which lay alongside the German cable at the shore ends. This problem was, however, somewhat mitigated, by removing the earth bonding at each terminal. The engineering report

---
67 Collected papers from the British Telecom archives.

indicated that an end-to-end speech test had been made from London to Jersey and that apart from the fact that the circuit was 'drummy', an intelligible conversation was possible. The reference to the term drummy would be on account that the submarine cable section, which was some 91.62 nautical miles following the latest repairs in 1926, was completely unloaded and thus its high frequency transmission quality would not be good. The tested section from Jersey to Dartmouth was unamplified but the UK shore end circuit was connected over existing amplified trunk circuits from Taunton to London which would have helped the overall quality. The cable itself was, of course, a single wire with an earth return path, effectively an unbalanced co-axial pair.

Although these tests were carried out in 1928, it was some 3 years before the telephone circuit was finally provided to the States' operated island networks. The reason for this delay was probably due to the continued use of the telegraph circuit for military purposes. However, in November 1928 PMG Sir William Michell-Thompson in answer to a question in the House of Commons stated that there were still 'some financial and technical obstacles' which were being assessed by the GPO.

Then in March 1929 the question of telephone circuits to the Channel Islands was raised again in the Commons by Sir Nicolas Gratton Doyle. The PMG, now Vicount Wolman, replied that the 'the GPO was still testing cables and that more information may be available at Easter'. The local press and both the States of Jersey and States of Guernsey were keen for the GPO to provide a telephone circuit to the UK and Guernsey wished also to have a circuit to Sark.

Coincidentally, at the beginning of 1930 the GPO telegraph department announced their intention to abandon the telegraph circuit to France, thus relieving the single core cable. The tests on the cable were revived and an investigation was established to determine the stability of the overall circuit. A test circuit was first set up on a London-Newport circuit using an artificial submarine cable to test out the practicality of the project using 2 to 4 wire amplifiers and voice-operated stabilizers. These latter devices are necessary to prevent 'singing' or feedback on the amplified sections of such circuits where transformation from 2 to 4 wires is required. Two wire amplified circuits are notoriously difficult to balance and the stabilizers effectively switch off the receive path of the circuit when speech in the send direction is detected. This prevents 'spillover' occurring at the 2/4 wire termination and thus, when correctly aligned, prevents the circuit from singing. These devices are similar to modern 'echo suppressors'. Later, the tests were carried out on the cable itself between Compass Cove and Jersey.

Before the circuit could be fully utilized the cable would be diverted via Guernsey. It was first necessary to pick up the submarine section west of Guernsey and splice in suitable lengths in order to loop the circuit via Saints Bay in Guernsey. This increased the overall cable length from Dartmouth to Jersey to 98.63 nautical miles, 79.8nm to Guernsey and 18.83nm from Saints Bay to Plémont. Tests showed that the core resistance of the total cable was 78.6 ohms. The experiments continued throughout much of 1930. On 5 September the *CS Monarch* laid a 'sea earth' at Plémont to ensure adequate earthing arrangements for the audio circuit to prevent crosstalk between the telephone and telegraph circuits. This was in the form of a 2-core submarine cable bonded to the deep-sea section of the telegraph cable sheath about a mile offshore. Similar arrangements were carried out at Guernsey and Dartmouth. The Jersey Evening Post reported that the shore ends of the cable necessary for the diversion were dropped by the Guernsey States' tug *Sarnia* on 15 September. After the changes to the cable had been made, the equipment was installed on the cable ends to finalize the circuit tests. The circuit was a simple 'ring-down' junction employing 16Hz (ringing frequency) inter-operator signals.

The submarine section was terminated on line matching transformers at either end of the circuit to match the unbalanced 70 ohm cable to the land sections of 600 ohms in order to reduce the line noise and cross-talk with the adjacent telegraph circuits on the 3-core cable. Ringing was extended via the centre-tap 'phantom' circuit. In order to make a trunk call for many Central exchange subscribers and those outside the Central exchange area, it was necessary for users to visit the telephone exchange to use one of the recently introduced 'advanced' telephone sets manufactured by Siemens

Brothers and designated Telephone No 162 by the GPO. These telephones had sufficient transmission quality to enable a reasonable conversation to be maintained with a distant subscriber, albeit by shouting!

The German cable make-up at the time of the feasibility test is detailed below:

	Status	Original	New	Re-inserted	Original	Original	New	Re-inserted
	Core/insulation	500/300	107/150	500/300	500/300	500/300	107/150	500/300
Dartmouth	Armouring	10/2	10/2	10/2	10/2	10/2	10/2	10/2
	Length (nml)	0.425	0.188	0.743	0.186	5.358	0.383	0.094

Original	New	New	Reinserted	Original	New	Original	New	Original
500/300	107/150	500/500	500/300	500/300	107/150	500/300	107/150	500/300
10/2	10/2	10/2	10/3	10/2	10/2	10/2	10/2	10/2
28.143	0.576	5.596	4.076	0.100	0.527	7.304	0.067	7.259

Orig-inal	Orig-inal	Orig-inal	Re-placed	New	New	Reinserted	Original	Original	
500/300	500/300	500/300	500/340	500/320	500/250	500/300	500/300	500/300	
10/2	10/2	10/2	10/2	10/2	10/1	10/2	10/2	10/2	Jersey
5.518	5.045	15.804	0.519	0.695	0.044	1.377	0.330	0.670	

Table 10

As table 10 above shows, the cable had been extensively repaired with both recovered (reinserted) lengths of the original cable and other sections using any available cable type at the time. This must have had a detrimental effect on the overall impedance matching of the cable, but as it was only intended to use this section at voice frequencies, that probably did not matter, especially with the overriding importance of having any voice connection from the islands.

The circuit was terminated at the HPO on each island and then extended to the States' exchanges over local circuits. In Jersey the overhead route from Plémont to St Helier had originally been requested from the States

Telephone Department. However, the GPO engineers decided that the chosen route was unsuitable because of the potential for crosstalk with local telephone circuits and a new construction of wires was taken from the Plémont cable hut then joining existing GPO telegraph poles along the route of the western railway line to St Helier. Amplifiers and 2/4 wire terminations were installed at the Guernsey Head PO and the circuits extended to the Guernsey States' exchange. It was reported that the tests were complete at the end of December 1930 but the circuits were being delayed by the installation of a new exchange in Guernsey.

Meanwhile, correspondence between both islands' telephone departments and the GPO indicated that an opening ceremony was extensively planned, with dignitaries including the Bailiffs of both islands and the Lt Governor of Jersey Major-General the Lord Ruthven CB, CMG, DSO, taking part in an initial call with the Lord Chancellor. Special telephones were provided including amplifiers so that all present could hear the inaugural call. A letter from J Stanhope, the Jersey Telephone Manager who at that time was still at offices at 2 New Street, written to the Post Office authorities requested that this call should be 'free of charge'.

The circuit to London was eventually opened at a special ceremony in both islands at 14.25 on the 26 March 1931. After the inauguration, during which it was reported that the Lord Chancellor's speech was practically inaudible as he was reading from a script and not talking into the microphone, the circuit was opened for public traffic at 14.50. The public were quick to use the service as there were already pre-booked calls waiting: 14 from Guernsey subscribers and a further eight from Jersey.

The trunk was shared between Jersey and Guernsey Telephone Departments by using an elaborate system of delay working. This was termed 'modified special attention working' in the GPO switchboard operator instruction manual. This was necessary as there being only a single circuit there was no possibility of using an 'order wire'[68] which was common practice elsewhere on the GPO network. The Jersey calls were

---

68 An order wire enabled distant operators to communicate separately from the normal circuits used for telephone traffic. This enables each junction to be optimised as a planned route and timetable can be determined without interrupting the call circuits.

transited via the Guernsey switchboard operator. Calls between the islands could be managed while Guernsey-London calls were in progress but not *vice versa*. Although only a limited service, it was much in demand, being the only means of communication other than via the telegraph. However, the telegraph remained the main means of business messaging for some considerable time.

The following table shows the charges levied by the GPO at the time of opening for each call of a minimum period of 3 minutes:

Rate Code	Distance miles	Call Charge Periods		
		0700 - 1400	1400 - 1900	1900 - 0700
I	75 - 100	3/9	3/-	2/-
K	100 - 125	4/5	3/5	2/3
L	125 - 160	4/9	3/9	2/6
M	160 - 200	5/9	4/5	3/-
N	200 - 250	5/9	4/5	3/-
O	250 - 300	6/9	5/-	3/6
P	300 - 350	6/9	5/-	3/6
Q	350 - 400	7/9	5/9	4/-
R	400 - 450	7/9	5/9	4/-
S	450 - 500	8/9	6/6	4/6
T	500 - 550	8/9	6/6	4/6
Irish Free State	-	T + 1/-	T + 9d	T + 6d
Person to Person	-	+ 1/-	+ 6d	+ 6d

Table 11

Note that these tariffs were as charged by the GPO to the local Telephone Departments. Subscribers in Jersey paid an additional 3d per call which covered the Telephone Department administration costs. The GPO repaid Jersey and Guernsey 2d for every call originated or terminated on their systems to cover the cost of the operator and equipment used for the trunk connection in each island.

Inter-island calls were charged by the GPO to the local authorities thus:

Call Charge Periods		
0700 – 1400	1400 – 1900	1900 – 0700
1/9	1/3	1/-

Table 12

In addition each island administration charged 3d for call handling, thus a full rate call charged to a subscriber was 1/9 + 3d + 3d = 2/3. An addition of 1d per minute or part thereof for calls over 3 minutes was also charged.

In 1931, following a request from the Telegraph Department, a further investigation was undertaken by the GPO Research Department, Dollis Hill into the feasibility of adding a sub-audio telegraph circuit onto the telephone circuit from Dartmouth to the islands. Following the closure of the military circuit to Rennes, it was decided to make an additional telegraph circuit available as a reserve in the event of a cable failure on the 3-core. The submarine cable to St Malo, however, was not included in this plan and remained abandoned. The tests started in December 1931 but a cable fault delayed their completion for a month. Tests, however, were carried out on board the *CS Monarch* while the cable was being repaired.

The reconfiguration of the cable meant that it would be necessary to introduce frequency filters at either end of the submarine sections in order to separate the telegraph and telephone circuits. The telegraph operated at below 80Hz while a telephone circuit's frequencies are in the band 300Hz to 3400Hz, although in practice, the maximum realistic frequency on the unloaded submarine cable was in the order of 2,200Hz. Thus it was feasible to operate both a telegraph and telephone on the same cable.

A consequence of introducing filtering into a circuit is signal attenuation. The signalling circuit from the islands to Dartmouth had formerly been purely metallic and ringing currents were the same as used on any other junction call. The introduction of the filter meant that the standard $16\frac{2}{3}$ Hz ringing signal could no longer co-exist with the telegraph circuit, since this would cause interference with the telegraph code. The ringing was, therefore, converted into a 500/20Hz signal[69]. Since the submarine sections were effectively a 2-wire circuit, the Compass Bay terminal used a 2 to 4 wire termination which was then backhauled into the Torquay repeater station over a 4 wire circuit. All work was completed by March 1933.

The tests indicated that the circuit would work with little noise between the telegraph and telephone circuits. The filters at the different stations had to be adjusted in variance with the standard settings to meet these requirements, however. The Guernsey Saints Bay telegraph filters requiring non-standard settings. This low-pass filter bridged the two submarine cables together for the telegraph circuit, it being shared between the islands by teeing the Guernsey telegraph circuit into the Dartmouth-Jersey circuit. Originally a duplex 24V teleprinter circuit was provided as a reserve for the 3-core cable. Later, however, this circuit was later reported to be carrying the Central News agency connection to the offices of the Jersey Morning News and the Guernsey Press. This was a duplex circuit with the receive half at both islands being connected to an automatic printer while the return circuit for supervisory control was by means of a Morse key. This circuit was worked using a +40V and -40V signalling system. At night, this circuit was also used to send weather messages, in Morse, between the islands.

The circuit terminated at the Head Post Office in Broad Street, St Helier where the signalling and repeating equipment was located. This was provided using an open-wire pair from Plémont to the Weighbridge railway terminus, then on underground cable to the HPO. From the HPO the telephone circuit was extended to the States' exchange over a cable connected to the distribution pole at Pier Road, St Helier where the GPO and States wires met. The telephone circuit signalling was a standard ring-down circuit thus a converter was necessary to change the $16\frac{2}{3}$ Hz ringing signal

---

69 This is a standard method of ringing signalling over amplified circuits. The normal ring signal is detected at the transmission bridge and converted into an amplitude modulated signal and inserted into the transmission line via the amplifier. A carrier of 500Hz is modulated with 20Hz in step with the incoming ringing signal. This signal is detected at the distant amplifier using a selective filter circuit and converted back into a standard $16\frac{2}{3}$ Hz ringing signal.

into a higher frequency for signalling over the audio circuit. Unlike the Guernsey to Dartmouth circuit, the signalling system between Jersey and Guernsey used a single 500Hz tone, generated by a single triode valve oscillator. The Guernsey Head PO housed the receiver and also two GEC manufactured 500/20 ringers. Some problems, however, were encountered on the island link, manifested largely as ring signals failing to be detected in the Guernsey to Jersey direction, and so the ringers were later replaced with standard 500/20 units. The valve circuits in both cases utilized the 120V and 24V battery supplies associated with the existing telegraph signalling equipment.

At the Guernsey States exchange one switchboard circuit was modified for the use of through switching to the UK from Jersey. This was necessary to remove the transmission bridge components which introduced too much attenuation on Jersey calls. There were also reports of cross-talk on calls when a UK trunk call from Guernsey was in progress at the same time as a call from Guernsey to Jersey. This was solved by instructing the operators to used non-adjacent cord circuits. Additionally, the overhead circuits were rebuilt at both islands using a transposition plan[70]. This also reduced the incidences of complaints.

Reports of 'howling' were common and the amplifiers, which set high because of the long submarine section, needed frequent adjustment. The improved circuit did, however, permit some Central subscribers to be connected directly to the trunk, rather than visit the exchange to make calls. However, the Jersey Telephone Department at this time was also upgrading customers, especially business subscribers, with newer telephones when they were frequent users of the trunk system. Reports of howling were also associated with the use of the telephones No. 162 at the exchange since these were more efficient and connected over a much shorter line than most of the subscriber telephones. At the UK shore end the circuit was connected through to the Taunton repeater station and then onward to London.

---

[70] Transposition on overhead transmission lines is used to reduce crosstalk. The wires of each circuit are effectively twisted around each other and other adjacent circuits by shifting their relative position on the pole crossarms and insulators.

Later that year, in November 1933, further investigations were carried out into the possibility of providing a further inter-island circuit over the 3-core cable. Earlier tests had ruled this cable out over the mainland section but tests by the Dollis Hill research team concluded that it would be possible to provide a single circuit over any two of the 3 cores but that cross-talk problems precluded the use of the sea-earth return circuit which had worked satisfactorily on the German cable. It was noted that at the time of the tests the following circuits were operated over the cables:

    TS(Taunton) – JE1   Duplex teleprinter voice frequency (VF) circuit connected straight through at Guernsey terminal and terminated at Jersey HPO

    TS – JE2 as above

    TS – GY Duplex teleprinter VF circuit terminated at Guernsey HPO

    GY – JE Inter-island circuit, at night the TS-GY circuit extended to Jersey HPO over this circuit

The TS1 – JE circuit, over the German cable, was connected at all times to the Central News service and terminated on automatic printer receivers in both islands.

This circuit was substantially the same as that employed on the German cable, except that it used wire 1 and wire 2 of the 3-core cable. Line filters separated the telegraph sub-audio circuits from the voice circuit. The new junction was used for inter-island traffic and in the evenings and night, the Guernsey HPO telegraph was extended to Jersey, telegrams were passed to the duty telegram boy via telephone. Although unamplified the same type ringers were used as on the other circuit.

Following the opening of the telephone circuit to the UK, there was a noticeable decline in telegraph traffic from the islands. This is illustrated in the following table:

	Press		Phonograms		Other	
	Sent	Received	Sent	Received	Sent	Received
1930/31	2080	58460	56511	34550	223178	193849
1931/32	2359	64853	53485	32534	205161	179989
	+13.5%	+10.9%	-5.4%	-5.8%	-8.1%	-7.1%

Table 13

It was noted that the overall decline in traffic was 7.8%. However, the GPO was also running the telephone service at a loss. This loss was being subsidized from other services as was traffic to other offshore destinations, such as the IOM, and the Scottish islands that required a submarine cable. As a consequence there was a 'Cable Charge' applied by the Government to cover the cost of provision of such cables. This further reduced the profitability of these circuits for the GPO, since this had to be paid back to the Treasury.

Traffic on the telephone circuit was nevertheless increasing following the restructuring of the tariffs in 1933, especially during the low charge period. This was because of the demand of holidaymakers making calls home during the evenings at the lower rate. The traffic increase was particularly noted in Jersey.

In 1935 an agreement was signed between the States and the GPO for the provision of telephone traffic between Jersey and London. This agreement was to run from 1 January 1936 and to last for 25 years. The terms of the agreement allowed for the PMG to unilaterally decide whether to repair any cable failure on the submarine cable, thus exposing the islands to the possibility of cessation of telephone traffic via that route. The States Telephone Department provided a trunk position on their St Helier Central exchange switchboard for which they were paid £45 for the initial provision and a further £10 per annum thereafter for the ongoing operational costs. Calls were to be charged a premium of 3d for a minimum of three minutes with an additional charge of 1d per minute or part thereof thereafter. The agreement also allowed the States to charge up to 2d more per call for the connection of call office telephones. The States were to receive 20% of the total revenue for their operational costs in handling the calls. This agreement, which also allowed for suitable renegotiation of rates and the revenue sharing at appropriate intervals, was signed by the States Greffier H Le Riche Edwards and on behalf of the GPO by Thomas McDonnald Banks.

In July 1935, following the successful opening of a radio channel to the UK from Guernsey, the GPO opened a 1 + 1 carrier system (one physical audio circuit plus a single carrier circuit) in place of the single audio circuit on the 3-core cable between Jersey and Guernsey. This increased the number of available circuits between the islands to 3. By early 1936 there were 5 telephone circuits available in Guernsey following the upgrade of the radio service. Consequently, on 15 July 1936 the carrier circuit between the islands was upgraded to a 1 + 2 system resulting in the availability of 4 circuits to Guernsey.

After the upgrade to 1 + 2 carrier working on the 3-core cable, the cross-talk and noise on the German cable increased to a level which virtually turned it into an emergency only circuit. It was effectively used only as an order wire between the Jersey and Guernsey operators. However, as the rate of traffic increased in Jersey following the introduction of the new simplified tariff structure in 1937 (table below), three circuits were permanently 'patched through' from Jersey to London. These were all on the 1 + 2 carrier system. This reduced the inter-island service to a single low-quality trunk. Charges were set as in Table 14.

Charge Band	0700 - 1400	1400 - 1900	1900 - 0700	Island supplement
Rate Jersey - Guernsey	1/6	1/-	9d	3d
I	3/3	3/-	1/9	3d
K	3/9	3/6	1/9	3d

Table 14

This new arrangement was introduced from May 1937, following complaints from the Guernsey Telephone Council that they were not being fairly compensated for the increased demand from the Jersey operators.

This change of circumstance again prompted the GPO to investigate the provision of additional circuits to the islands. A radio link to Jersey was contemplated and tests carried out during June 1937, but the additional distance required to reach Jersey ruled this out on grounds of reliability. The Guernsey service was also occasionally very poor because of fading in certain weather conditions.

So once again the issue of a new cable was contemplated. However, in December 1937 the newly opened airports in Jersey and Guernsey made a request for a private teleprinter circuit which would require the use of one of the inter-island circuits. This request had to be urgently met because of air safety concerns and thus the usual GPO procrastinations were sidestepped. In order to meet this request one of the carrier circuits was required since the audio circuit on the German cable was not of sufficient quality.

It became apparent that the only feasible solution was a new cable. Fortunately, circumstances changed as the government announced that the Cable Charge was to be abolished on all domestic cables by 1 May 1939, following on from its abolition on continental cables in the previous year. This tipped the balance on the economics of provision. On 27 January 1938 authority was given for the planning of a new cable between Dartmouth and Jersey. This would carry traffic from Jersey and thus make more circuits available between the islands and from Guernsey to the UK. A non-competitive tender was received from Submarine Cables Limited for a Paragutta co-axial cable, a new technology which had recently been installed on the UK – Netherlands service. The 12 channel multiplex terminal equipment, designated Carrier System No 6 by the GPO, was supplied by Standard Telephones and Cables Ltd.

A continental call service was opened via London in July 1938. Calls were charged according to British Zone II. A note from the GPO showed that according to their records calls could only be delivered to Central exchange subscribers. However, after clarification from John Stanhope the exchange records were modified to show all Jersey exchanges. Correspondence between the States Telephone Department and the GPO[71] shows that telephones No 162 were being installed on all exchanges lines where subscribers required regular trunk calls. The correspondence also indicates that the telephone No 162 was locally modified for local battery working on the Jersey country exchanges. These telephones must have been connected in addition to the existing instrument since the caller would still have needed the hand generator to originate and terminate calls and such devices were not available for the telephone No 162.

The cable was not laid until 9 August 1938 and the submarine sections were completed by the 14 August by the *CS Faraday* assisted by the seagoing barge *Moultanian* of Littlehampton. Two 34nm lengths of co-axial submarine cable, each weighing more than 300 tons, were laid between Dartmouth[72] and Fort Doyle, Guernsey and spliced mid-channel. A channel was dug up the beach at l'Ancress in which the shore end was laid to the cable hut. The Guernsey – Jersey section was laid on the 13 August between Saints Bay and Plémont.

The landings were connected with an underground cable between Fort Doyle and Saints Bay and the Plémont Cable Hut and St Helier using Paper Core Quad Trunk (PCQT) 14pr/40lb/ml cable. The Jersey land cable was completed around May 1939. The Guernsey cable passed via the States Telephone's St Peter Port telephone exchange in Guernsey. The reasoning behind this route was to make the cable accessible in Guernsey in the event of failure of the Guernsey wireless circuits, although under normal working conditions, the cable would be connected straight through to Jersey. The line amplifier equipment in both islands was provided by GEC Limited of Coventry. The cost of the complete system was as in Table 15 below:

---

71 BT Archives
72 POEEJ Vol 31 p 237 – Landing of the Jersey – Guernsey – Compass Cove Submarine Cable at Dartmouth

Paragutta submarine cable	£55,710
Contractor laying charge	£6,400
Land cables Jersey and Guernsey	£11,200
Terminal equipment STC & GEC	£14,500
Total	£87,810

Table 15

While the submarine sections were laid in the summer of 1938, the equipment necessary to interconnect the cables was not ready until the late spring of 1939. Thereafter, testing continued into the summer.

In Jersey, it had become apparent that the HPO at Broad Street no longer had sufficient space for the necessary multiplex equipment and thus it was decided to build a separate Repeater Station. A suitable location was found at Trinity Gardens, St Helier and pending the construction of a permanent building, temporary huts were erected on the site to contain the multiplex and amplifier equipment.

Before the new cable could be implemented in Jersey, it was first necessary to expand the available operator positions on the Central switchboard. The CB10 exchange only had 4 trunk and junction operator positions and with increased local traffic it was becoming difficult to manage. It was therefore decided to install a separate trunk 'island suite' in the exchange switchroom. A request was made to the GPO by John Stanhope for the provision of switchboard equipment. The GPO made available seven CB10 switchboard frames which were installed by States personnel. This new suite had by necessity to be incorporated into the existing multiple. The trunk operator positions each had access to 6 London circuits, 4 Guernsey circuits and 10 transfer circuits to the main switchboard. Note that two of the Guernsey circuits were normally patched to the inter-island airport teleprinter circuits.

The cable was opened to traffic on 1 August 1939 with six circuits provided over the 12 channels, each channel was unidirectional and therefore two were required for each trunk. Signalling conditions over the circuits were the same as before, being simple ring-down.

The new tariffs in force at that time were as in Table 16:

Charge Band	0700 - 1400	1400 - 1900	1900 - 0700	Island supplement
Rate Jersey - Guernsey	1/6	1/-	9d	3d
CI -UK	2/6	2/6	1/-	3d

Table 16

Consequent to the changeover of the inter-island circuits to the new cable and the backhaul to the repeater station over the new land cable, the GPO sold part of its overhead wire routes from Plemont to St Helier to the States Telephone Department. This is recorded as a sale taking place around 8 June 1940.

Chapter 12 - Wireless Telegraphy

Wireless was first taken up by the General Post Office following successful demonstrations and trials by the Marconi Company. Indeed, William Preece, the Post Office chief engineer, had given much encouragement to the technology through research support of the young Guglielmo Marconi. Wireless first attracted the attention of the War Office and it began to be installed in warships very shortly after the first successful demonstrations, the British Army first used wireless telegraphy during the Boer War.

Although the Admiralty had established a Radio Station in Jersey as early as 1902, the first continuous use of wireless in the Channel Islands was from a Royal Navy station based in Alderney which was set up in the autumn of 1904, shortly before the introduction of the Wireless Telegraphy Act (1904) which came into effect on 1 January 1905. This Act vested the control of wireless radio frequencies in the Post Master General.

The Post Office installed a commercial ship-to-shore station at Bolt Head, Devon in 1908 and the following year took over the Marconi Company coastal stations. The Post Office also installed a wireless station, which it operated on behalf of the Admiralty, at Fort Regent in December 1908 from which it was able to communicate with Bolt Head as well as ships equipped with radio equipment. The station proved to be an asset when the telegraph cable again failed in March 1909. The Admiralty officer in charge, Chief Officer Ainsworth, was moved to a long distance radio station on Cromarty later that year. A similar station was installed at Fort George in Guernsey in

1909 which provided comparable services including the sending and receiving of telegraph traffic during a cable fault in February of 1910. In 1911, wireless telegraphy equipment was installed on both the Great Western Railway and London and South Western Companies' mail boats. The Fort Regent station received a new mast during renovation work in June 1913, the station now being in regular use with local shipping.

During the First World War the station played a vital part in the War Office communications network, particularly with reference to relaying military messages between the UK and France as well as the Royal Navy. It was also a vital component in the management of the merchant shipping under the control of the Admiralty.

The Admiralty ceased to have any interest in the station at Fort Regent from the end of the war, largely because technology had changed and the Channel Islands were now of less strategic importance. After the war, improvements in technology reduced the requirement to operate stations in the Channel Islands for both the War Office and for commercial shipping. The Post Office instead relied on more powerful mainland based stations for the majority of its maritime wireless services. The station appears to have been closed, sometime after 1919, indeed the Guernsey States purchased the Admiralty's old station at Fort George in 1922 for local shipping purposes.

The local shipping companies, The Great Western Railway and the Southern Railway companies requested wireless facilities similar to those available at Southampton. The GPO considered the requests and at the beginning of 1925 funding was approved for a new station at Fort Regent and a second station at Corbiere. However, it wasn't until 1927 that these stations came into service. They were operated not by the GPO, but by personnel seconded from the railway companies that were trained by the GPO. Stanley Harris of the GWR being the first non-GPO operator in December 1929. He was assisted by Arthur Davis of the Southern Railway Company, which had been formed in part by the LSWR in 1923 under the terms of the railway grouping as a result of the Railways Act of 1921.

The station was refurbished in 1928 at a cost of £1,000 when it was also reported that the annual operational costs, including spares and staffing, were in the order of £1,400, part of which was funded by the States of Jersey and part by the railway companies. The States Telephone Department was involved in the refurbishment as interconnection with the telephone system was included. The States voted £700 additional funding to the Telephone Department for the project in January 1928.

Paradoxically, shortly after this handover, the GPO found it necessary to install a temporary wireless station at the former War Office site at Fort Regent in order to provide a telegraph circuit as a result of yet another cable failure.

In February 1930 wireless telegraphy equipment was installed for use by the States Harbours by the States Telephone Department. In 1936 a number of wireless systems were installed at Jersey Airport for the management of air traffic. Post WWII the States Telephone Department ceded control of wireless to the newly formed Jersey Airport Telecommunications Department that took over management of the ever increasing complexity of wireless communications and RADAR.

By late 1933 the GPO was coming under increased pressure from the island governments to provide more circuits to the UK. An investigation into the provision of a new telephone cable was undertaken. However, after some investigation, the GPO decided that the cost at £85,000 was too expensive at the time to consider.

Concurrent with the investigations on the cable in 1934, another investigation was initiated in January by the GPO research establishment at Dollis Hill into the provision of a wireless telephony circuit to the Channel Islands. Such an installation would be considerably cheaper than the provision of a submarine cable and quicker to implement.

The shortest distance from the UK to the Channel Islands is between Alderney and Portland, but as the main traffic sources were from the larger two islands of Guernsey and Jersey, thus circuit provision to either of them would be more logical. The nearest point, therefore, was from Dartmouth to Guernsey, a distance of some 128km (80 miles). The highest point on

Guernsey is about 75 metres above mean sea level and at Dartmouth about 180 metres. Over this distance and height there is no optical path.

Feasibility tests were carried out during the spring and summer of 1934. A test site was established on the cliffs above Compass Cove south of Dartmouth to test field strength from a transmitter sited at Fort George west of St Peter Port. Fort George was an abandoned garrison establishment vacated by the British some few years earlier and was sited on one of the highest points in Guernsey. Field measurements were carried out in the 3, 5 and 8 metre bands (37MHz – 100MHz) from a 10 Watt transmitter. These frequencies were at the time referred to as "Ultra Short Wave", (not to be confused with the modern definition of UHF). Tests were carried out over a period of some weeks to determine the reliability of the circuit. On account of the tendency of signal fading, the tests were continued for over two months. Finally, it was concluded that if Koomans[73] array antennae were used at either end of the link, it should be possible to operate a reasonably reliable service.

In the event, it was not possible to find a permanent site at Dartmouth and it was found that similar results could be obtained from a test site 40km (25 miles) inland and 240 metres above sea level. A new site near Shaftesbury was found and tests during the summer confirmed the earlier findings at Compass Cove. It was decided to implement a two way circuit using 5 and 5.5 metre bands (54.5MHz and 60MHz). The two-way service was established in December 1934 but initial results were disappointing. However, testing continued into the spring of 1935 and by March the field strength had improved such that the circuit could be used for traffic. The circuit was temporarily brought into service in April 1935 but considerable fading was experienced during the warm weather. Secrecy on the circuit was by the use of simple speech inversion which was introduced in July, not wholly satisfactory, but sufficient to block casual radio hams.

Further tests during the summer at a point 40km (35 miles) nearer the coast showed that the fading was less and the average field strength better.

---

73 Patented by Prof N Koomans in 1934, a configuration of directional transmission or receiving aerials

Monitoring of the signal continued while the circuit was carrying traffic in order to establish its reliability. In January, discussions took place between the GPO and the island authorities with a view to increasing the number of circuits both between the islands and between the islands and the UK. Later, in February 1936, the station was moved to Chaldon Herring on the downs near Lulworth Cove, about 136km (85 miles) from Guernsey, the site being some 190m above sea level. The original Koomans arrays were replaced by rhombic arrays, 65' stout telegraph poles were erected at Fort George in Guernsey to support them. The receiving equipment was changed to superheterodyne and the transmitter output power increased. The first channel was officially put into service on 11 May 1936 the first call being between A H Brice (Guernsey Telephone Committee) who spoke with Sir Donald Banks, Director General of the Post Office. It was noted that this was at the time of opening the highest power short wave telephony link in world, operating on the 5 metre band. A third inter-island cable circuit was opened on the same day. A further three wireless channels were brought into service between Chaldon Herring and Guernsey later that summer, having been authorized by the GPO Engineering Department only on the proviso that circuits could be taken for telegraphy in the event of cable failure.

A wireless telephony system continued after the war, when the islands were reconnected to the UK mainland. Later, the system was extended to 6 circuits, although they were always subject to interference, especially in early summer, they continued in service until the 1939 cables were upgraded in 1952/3. The circuits used frequencies later allocated to the BBC for television broadcasts on the 405 line system and so it was acknowledged that they had a limited life.

Chapter 13 - The War Office Cables

In January 1939, some time before the outbreak of war, the War Office commissioned the GPO Dollis Hill research department to investigate the possibility of installing a cable from the UK to France via the Channel Islands in order to provide communications to west France for intelligence and, subsequently, for communications with the British Expeditionary Force (BEF). The investigation concluded that a cable should be laid from Compass Cove, Dartmouth to Fort Doyle[74], Guernsey, then from Saints Bay, Guernsey to Plémont, Jersey and on to Pirou, Normandy from Fliquet. The French had wanted the cable to be landed in St Malo, but the sea route from Jersey was not clear and it was known that the existing Pirou route was especially suitable for cables.

The investigation was completed by early 1939 and it was initially decided to extend the existing cable route from Fliquet to Pirou. The cable used was of the recently developed concentric paragutta type. This was a cable made up of a central copper core with paragutta insulation all overbound with copper tape, nowadays referred to as a co-axial cable. This section was laid by September 1939, shortly after the opening of the new CI – UK cable (see Chapter 11) and replaced the old telegraph cable. A 40km overhead route was constructed by Standard Telephone and Cable Company Limited contractors from the Pirou cable hut to the French telephone repeater station at St Lo. Between Fliquet and Pirou a Siemens Brothers 1 +

---

74 Fort Doyle was a radio station, it is likely that the GPO report meant l'Ancress as the landing point

4 carrier system was supplied. The through circuit was tested successfully by the end of September.

However, with the outbreak of war on 3 September 1939 it was decided to duplicate the existing cables to France. The necessary preparations were put in place and a cable was ordered from Submarine Cables Ltd of the same type as used on the other cables. GPO staff were despatched to France to manage the necessary preparatory work in January 1940. The cable ship used for the sea sections was the Siemens Brothers vessel, *CS Faraday*, and 168km (105 miles) of cable was laid in just a few days during early May 1940. The GPO cable ship *CS Ariel* was used to lay the shore landings which were completed between 31 May and 11 June. Meanwhile, on the French side, armoured cables were ordered to replace the overhead route from Pirou to St Lo. In the meantime, the German advance was causing chaos with supply delivery and the GPO engineering group commissioned for the work was continually being mistaken for foreign spies by the French authorities. The armoured cable was eventually landed at St Malo and had to be transferred to St Lo, some 150km distant. However, this was completed in time to connect with the submarine cable on 10 June.

In the meantime, as the terminal equipment had been greatly delayed, it was decided to reinstate the abandoned Greve D'Azette – St Malo cable as a voice frequency circuit. The cable ends were picked up and tested by the GPO engineers, and surprisingly, the cable was found to be in good order, but the military events were by now changing on a daily basis and although the cable was available it was not possible to utilize it. All GPO personnel were evacuated from France by a small paddle steamer by 14 June. It was reported that a London – Cherbourg circuit was still operational on 12 June 1940, this being serviced over the earlier cable.

The Dartmouth shore end was connected back to Dartmouth using a 4-wire balanced pair 40lb/mile trunk cable and then on to Torquay repeater station. The submarine cable was a single coaxial pair which was terminated in a 2 to 4 wire matching termination, the characteristic impedance of the submarine cable being 50 ohms. It had originally been intended to provide a new type of carrier equipment of 24 channel capacity across the entire link. The equipment was delivered to Guernsey on 28 May and GPO Engineer-in-Charge of the installation staff from the GPO research department, Dollis Hill, arrived on 6 June. In the meantime, temporary transmission terminal equipment on the Dartmouth - Guernsey section was supplied by Siemens Brothers of the 1 + 9 system type, a base audio circuit and 9 carrier channels.

All haste was made in setting up the circuits, and engineers from Dollis Hill were despatched to Jersey to supervise the installation of the 24 channel system. Considerable co-operation with the French authorities was required and much bullying was done to overcome the normal administrative delays. Testing was started at Fliquet on 6 June but was hampered by the late provision of the line matching transformer on the French side. Tests were completed without the transformer and fortuitously, when it did arrive on 10 June, the circuit performed as expected. Unfortunately, the absence of the carrier equipment meant that it was not possible to fully commission the system.

In Jersey, the amplifier equipment was installed the St Helier repeater station and carried out to Fliquet over wires provided by the States Telephone Department. This circuit used a mixture of cable and overhead lines between St Helier and Fliquet as shown in Table 17 below. Bel Val was a telephone pole Distribution Point (DP) in St Catherine's Bay.

	Head Post Office		States Central Exch		Gorey Exch		Bel Val DP		Fliquet Cable Hut
Cable Type		10lb/ml unloaded cable		20lb/ml unloaded cable		10lb/ml aerial cable		70lb/ml cadmium-copper OH	
Distance		0.65ml		4.75ml		4.6ml		0.34ml	

Table 17

This mix of cable types caused some difficulties with the line transmission characteristics and a special equalizer was built by the installation engineers.

At Pirou cable hut the line was terminated on the matching transformer and taken back to Maison Pirou on the recently laid 1.36 mile long armoured 40lb/ml 6 pair Paper Core Quad Trunk (PCQT) unloaded cable, thence on to Coutances on a 12.22 mile 38 pair 20lb/ml cable loaded with 88mH coils at 1.136 mile intervals. The circuit was amplified at Coutances and then connected to the Rennes repeater station.

The layout of the racks at either end of the submarine circuit is shown below:

Fliquet

Balancing Network
Terminating Set
Insulation Test
Cable Termination

Pirou

Balancing Network
Terminating Set
Insulation Test
(24 Circuit test tablet mounted at rear)
Cable Termination

Figure 4

Image by the author

The photo above shows the cable hut at Fliquet as it currently stands. It was originally constructed by the Submarine Telegraph Company as a convenient test point between the sea cable and the land line. This cable hut contained cable terminating amplifiers from 1939 until the closure of the Fliquet – Pirou cable in the 1980s. It was latterly used as a summer icecream and drinks kiosk serving the nearby beach.

The Cable Termination panel held the line matching transformer 2 wire to 4 wire termination and the 10uF line isolation capacitor.

On 30 June 1940, the German Army occupied the Channel Islands and all telecommunication traffic northwards ceased. The cables to Pirou were cut by departing GPO engineers, but quickly repaired by the German army engineers and both them and the cables to Guernsey remained in service throughout the war carrying German military and some civilian traffic.

Prior to the occupation and because of the deteriorating situation, the GPO decided to withdraw all staff from the islands, clearly such skilled workers would be of use in the war effort. However, the local authorities on both islands insisted that at least a skeleton staff remained. All installation staff and some of the local engineers were withdrawn with the exception of one engineer on each island. The staff and the salvaged 24 channel system sailed for Southampton aboard the *SS Biarritz* at 9:00am on 20 June. On 2 July an attempt to withdraw the remaining Guernsey linesman, L Le Hurray, was made by an RAF speedboat supported four *Blenheim* aircraft, but because of reports of German strafing the officer making representations to the Bailiff for the staff release was forced to withdraw without success. Two of the escorting aircraft were lost in the attempt. The linesman remaining in Jersey was Mr P G Warder who was ordered to sever the northbound cable by the occupying forces on 1 July minutes after the last telegram was sent to the UK by telegraphist Vera Le Dain. Warder continued to maintain the telegraph and transmission systems on the island throughout the occupation, under the watchful eye of the German military.

Chapter 14 - Telephones under the States of Jersey 1923 – 1940

The new era for the telephone system came into being without any noticeable effect to subscribers. The systems continued to operate as before on the stroke of midnight and telephone calls were handled as they had always been. Almost all existing subscribers had signed up to the new service and there was a growing list of applicants attracted by the new tariff pricing structure.

The renowned telephone engineer Alfred Rosling Bennett was appointed the first Managing Engineer, although at that time he commuted between two homes, one in Cabot, Guernsey, and the other at Lake on the Isle of Wight. He was still acting as a consultant to the Guernsey States Telephone Council and also to Kingston-upon-Hull Corporation. In this capacity Bennett offered the services of the Guernsey Telephone Department[75] and this was accepted. A number of personnel were then loaned to the new Jersey Telephone Committee to assist in the preparation of accounting and stores management procedures. The system at that time consisted of 15 exchanges as listed in Table 18 below as was recorded by the Greffier, Ernest Le Sueur.

---

75 History of the Guernsey Telephone Department 1896 -1925, Bennett A R

Exchange	Address	Lease Term	Date of Lease	Annual Rental
St Helier (Plus 4 rooms for engineers)	2 New Street	21 years	29/9/1913	£76 & £366
Five Oaks	Le Geyt Farm Mr J le Masurier Exchange Attendant	21 years	24/6/1918	£30
Gorey	'Ringarooma' 2 Hilgrove Terrace, Gorey Village. A E Andrews, Postmistress	14 years	29/9/1917	£25
St Aubin	Bel Vue, High Street, St Aubin	5 years	25/12/1920	£25 (less £5 for sub-let)
St Mary	Nr St Mary's Hotel, M F Brideaux, Operator/Caretaker	Yearly		£17
St Ouen	Nr Haut du Marais, K Acort, Operator/Caretaker	Yearly		£16
St Peter (including 1 room for	Elmsfield, Rue des Minquiers A Syvret, Operator/Caretaker	14 years	29/9/1915	£22
Sion	Newcastle Cottage, Sion. M J Reynard, Operator/Caretaker	Yearly		£12
Trinity	L. Goldsmith, Operator/Caretaker	21 years	24/6/1917	£30
Millbrook	Millbrook House Mr E R Le Gros Exchange Attendant			*
La Moye	La Moye. Mrs A D'Aubert, Exchange Attendant			*
St Lawrence	St Lawrence Mrs W Norman Exchange Attendant			*
St John	St John PO, Temple Villa. J P Le Moignan, Postmaster			*
La Rocque	La Rocque PO, Rue de la Lourderie. E A Page, Postmaster			* †
Samares	Samares PO. E Le Coutier, Postmistress			*

Table 18

* On takeover, these properties had arrangements with the local postmaster and/or exchange attendant for use of the premises as a telephone exchange, the costs of which were met from the wages paid.

† Note that La Rocque exchange had been taken out of service in 1901 but re-instated in 1913.

Other properties included the telephone stores at 34 The Parade and the pole yard at Belozanne (Biles' field).

The capacities of the exchanges were also recorded. The St Helier exchange switchboard capacity was 900 with 793 lines in use. the remaining exchanges are detailed in table 19 below:

Name of Switchroom	Capacity of Switch-board	Number of Subscribers Connected	Number of Junctions Connected	Spare Capacity for Subscribers	Spare Capicity for Junctions	New Junctions Comissioned
Millbrook	100	78	4	18	6	-
St Aubin's	100	86	6	14	15	-
La Moye	26	19	1	7	3	1
St Peter's	60	46	5	13	5	-
St Ouen's	50	34	3	16	7	-
St Mary's	50	26	1	24	9	1
St Lawrence	50	32	2	18	8	-
St John's	40	24	1	16	4	-
Sion	50	29	1	21	9	-
Trinity	50	28	1	22	9	1
Five Oaks	60	43	2	12	3	1
Gorey	100	83	4	16	5	1
La Rocque	100	47	3	53	16	1
Samares	70	56	2	14	-	1
Totals	906	631	36	264	99	7

Table 19

Bennett's report also included the average traffic of each exchange in Table 20 below:

Exchange	Originating Local Calls		Originating Junction Calls		Originating Calls Total		Incoming Junction Calls		Total Traffic	
	Week-day	Sunday	Week-day	Sunday	Week-day	Sunday	Week-day	Sunday	Week-day	Sunday
Five Oaks	18	8	89	16	107	19	65	8	172	51
Gorey	160	58	245	57	405	115	185	11	590	126
La Moye	6	9	71	11	77	20	64	10	141	30
La Rocque	17	7	114	22	131	29	100	14	231	43
Millbrook	40	5	140	20	180	25	128	18	303	43
Samares	23	41	100	31	123	72	72	38	205	110
Sion	3	1	31	9	34	10	26	8	60	18
St Aubin	132	48	264	90	396	138	213	81	609	219
St John	12	6	37	11	49	17	35	7	74	24
St Lawrence	5	1	45	20	50	21	29	11	79	32
St Mary	25	5	43	7	68	12	30	8	98	20
St Ouen	30	9	97	23	127	19	66	18	123	50
St Peter	56	17	169	19	225	36	147	25	372	61
Trinity	12	10	37	8	49	18	20	4	69	22
St Helier	2706	49	875	204	3581	703	1304	29	4885	1002

Table 20

It is interesting to note from this table that la Moye and Samares exchanges generated more local traffic on a Sunday than during the week. There is no obvious explanation for this anomaly.

Revenues recorded at the time of the takeover were:

Source	Revenue
Subscriber rentals	£4866-1-11d
Local calls	£2319-12-0d
Removal charges	£245-10-0d
Surcharges	£720-0-0d
Total	£8159-3-11d

Table 21

The data recorded here was deduced from the GPO revenue account and under its regime rental and calls were bundled together up to a maximum of 5000 calls per line per year. As there were 1598 subscribers on takeover, the above figures would appear reasonable if calls were split from an assumed rental of an average of approximately £3-10-0d per line.

Referring back to the Bennett report, it can be seen that the revenue was barely able to cover the £8076-0-0d running costs on wages and other overheads under the Post Office. This no doubt contributed to the GPO's decision to make increases in rental and call charges since considerable necessary investment in upgrading line plant was foreseen. Bennett also noted that many of the GPO staff had received salary increases due to the war bonus system. This resulted in Jersey telephone staff wages being higher than their counterparts in Guernsey. It is not clear what changes were made by the new department to reduce the overheads, given that there appears to be little change in the overall revenue streams under the Bennett proposals except that all calls would in future be paid for individually rather than prepaid bundling. However, there is an indication from the foregoing that the issue of salaries and pensions had been an important factor. It is also not known what allowances the GPO made for 'overheads'. In general this would appear to be high, since GPO telephone charges were above those of the NTCo even before the takeover[76].

As the Bennett report had identified a lack of investment in the telephone system, firstly by the NTCo as its licence neared its end and subsequently

---
76 History of the Guernsey Telephone Department 1896 -1925, Bennett A R

since the takeover by the GPO, the new States' department wasted no time in making amends.

The first quarterly report on the new department outlined the progress made in the first three months of States' control. The report highlighted that in the past three months there had been a further 104 subscribers added to the system. In addition there was a waiting list. Despite difficulties in obtaining materials and the fact that the department's workforce was now largely manned by 'learners', this had been achieved quickly and efficiently. The reference to learners must imply that some of the more experienced staff chose to remain with the GPO and had probably been transferred to England. Two new junction circuits had been installed. The first was between St Helier and St Aubin and the second between St Aubin and La Moye. These additions relieved the congestion on these routes which had often resulted in calls being delayed. The report also noted that the managerial reorganization of the States Telephone Department was complete.

Shortly after the report a new telephone directory was issued with an additional 120 subscribers entered, including orders not completed. The new directory was produced with 'clear type and additional useful information' which included a service for subscribers to pass telegram messages over the telephone. This was in arrangement negotiated with the GPO, a facility for which the PMG requested a supplementary payment from the subscriber. Such telegrams would be passed to the post office telegraph clerk at the end of the day. It was also noted that two new telephone kiosks had been provided. The first near the cab rank in Broad Street which was allocated the number 1003 and the second in the central market at the corner of Beresford Street and Halkett Place which was allotted the number 1004. It is interesting to observe that these boxes could be called to arrange cab journeys as each one would be answered by the nearby rank cab drivers from Jehou's cab company. Both these kiosks were equipped with electric lighting, in the absence of a public mains supply at that time the Broad Street box possibly being fed from the nearby telephone exchange at 2 New Street.

Later that month it was announced that extensions to the switchboards at

La Moye and St Aubin had been completed. At La Moye the capacity was doubled while at St Aubin an additional 50% was added. Further public call offices were added soon after. These were at Snow Hill, near the bus terminus and at West Park at the corner of Pierson Road. These two kiosks were provided with electric light although the source of supply is unknown. Two further boxes were opened at the harbours; at the GWR landing stage on the New North Quay and a further kiosk was constructed near the Southern Railway landing on the Albert Pier. This latter box was, unusually, served by an automatic gas lamp installed by the Jersey Gas Company that lit at dusk and extinguished at dawn. Another public telephone was provided at Beaumont railway station.

A new junction circuit was also installed between St Peter's exchange and St Aubin. This was a marked improvement in service between the two adjacent exchanges as formerly all calls had to be completed via St Helier, thus occupying two junction circuits per call with the resulting loss in transmission performance.

At the States Sitting of 2 October Jurat Lempriere presented a progress report on the first 6 months of States ownership. After the first 3 months of operation £5,124-6-11d had been collected from subscriptions and the new department had been able to pay £200 to the Treasury out of profits. The number of calls handled in the previous week had been 24,252. Underground cables had been installed from St Helier (now renamed Central) exchange to First Tower and to St Clement. A notice in the JEP gave the route of the St Clement cable as: Snow Hill, Colomberie, St Clement's Road, Havre des Pas, Greve d'Azette, St Lukes, Samares, Le Hocq, Pontac, La Rocque. The notice announced that prompt connection was available for subscribers in those areas and also in Rouge Bouillion. This new cable provided 300 pairs of wire and thus sufficient capacity to enable the closure of Samares exchange. All 60 subscribers were transferred to Central exchange and reallocated numbers from 1031 to 1091.

The Telephone Committee also announced a deal had been struck with the Markets Committee to take over from December the Toy Market in Minden Place at an annual rental of £50. This building, they claimed, would be ideal for refurbishment into a new single story exchange building and provide adequate space for the department's offices. At the same time Jurat Lempriere announced that a tender had been requested for the provision of an automatic exchange to replace one of the rural switchboards and this was due shortly. In his opinion the department should move towards an all automatic system.

By the year end great progress had been made in the upgrade of the system. Around 300 new lines had been installed. There were also 10 new public telephone boxes, several of which were lit by electricity. As no public mains supply was available at this time, this must have been in arrangement with nearby private installations or with batteries. Six new junction circuits had been completed, relieving call delays between Central and St Aubin, St Peter and Sion as well as between St Aubin and St Peter and La Moye exchanges. In October Salem chapel in Gorey had been purchased to provide a new location for the local switchboard. A further debate in the States on 4 December requested a transfer of £800 from the Telephone Committee's revenue to capital accounts for system development, this was approved. At the same sitting the problem of pensions was raised. The States had a commitment to the GPO to provide pensions for the transferred staff and the committee felt that this should be extended to the other staff in order that there was equity. However, this was hotly contended in the chamber and the States voted to refer back for further reports.

In the New Year the States were requested to approve a further £30,000 loan for the telephone system enhancement. Needless to say this caused considerable debate in the chamber, especially as the first time the proposition was introduced the Finance Committee had only been made aware of the request at 10am that morning. The main question was how had the original £20,000 asked for in addition to the purchase cost been disposed of? The Telephone Committee argued that it was necessary to upgrade the system as it had suffered years of neglect under the GPO and the move to Minden Place was needed to provide space for expansion of the switchboard and to provide offices and stores. The system had been expanded by over 300 subscriber lines in 9 months, more than in 12 years of GPO ownership, and income was increasing steadily. It was noted that in Guernsey, the system had already expanded to 4000 lines. Jurat Lempriere said that the new exchange could be automatic and a quotation was expected soon. Jurat Renouf though that ill advised but the President pointed to the demonstration given at the town hall by the late constable Baudins and said he personally endorsed an automatic system. The proposal was lodged *Au Greffe*.

The expenditure of the original loan was detailed in a report from Jurat R Lempriere published in the JEP and detailed in Table 22 below:

Item	£ s d
342 new subscriber stations average £23 each	7,800 - 0 - 0
Work not contemplated in original estimate detailed below	4,926 - 0 - 0
Stores, materials and poles on hand	2,500 - 0 - 0
Sub total	15,294 - 0 - 0
Balance unexpended on hand	4,706 - 0 - 0
Total	£20,000 - 0 - 0

Table 22

The detail of the work carried out so far, including that not planned in the original estimate was as follows:

Item	£ s d
Bringing 60 Samares subscribers into Central Exchange	1,500 - 0 - 0
Removal expenses for GPO employees	250 - 0 - 0
Preliminary expenses up to date of transfer	350 - 0 - 0
Purchase of Salem chapel and cottage, Gorey	618 - 0 - 0
Repairs and alterations to ditto	150 - 0 - 0
Spare wires in new cables laid to reduce future connection costs	2,000 - 0 - 0
Total	£4,928 - 0 - 0

Table 23

It was estimated that the spare cable capacity would reduce connection costs of new subscribers by as much as £6 per station, that is, £17 instead of the then £23.

The proposed new loan would be allocated as in Table 24 below:

Item	£ s d
Works paid out of old loan but not planned for as above	4,928 - 0 - 0
New building in Minden Place	3,570 - 0 - 0
Erection of offices and stores	2,500 - 0 - 0
Electric light and power plant for new building	650 - 0 - 0
Hot water and central heating for new building	250 - 0 - 0
Diverting wires and cables from 2 New St to new building, ducts and manholes	2,000 - 0 - 0
Adding 500 new subscribers at £17 each average	8,500 - 0 - 0
New switchboard and test room	8,500 - 0 - 0
Total	**£29,998 - 0 - 0**

Table 24

The need for the urgent move of the exchange to Minden Place was precipitated by the landlord of 2 New Street serving notice that the lease would not be renewed at the next break point of 1927. The saving on rent at 2 New Street would, however, provide £442 annually towards servicing the new loan. The cost of the new switchboard was more than had been anticipated, but this was attributed to the GPO monopoly and the restrictions on manufacture[77] which meant that only three companies in Great Britain could make the equipment to the size required and the time available. The United States was identified as an alternate source but in a

---

77 The GPO introduced the Bulk Contracts Committee in 1923 which was effectively a government sponsored cartel that fixed prices of telephone equipment in return for a guaranteed supply to the GPO. This agreement lasted until 1969 when it was terminated by the Monopolies and Mergers Commission.

recent tender, the Guernsey Telephone Council had found them to be dearer, largely because of the great demand from the American home market, and the Automatic Telephone Manufacturing Company of Liverpool was at present preparing them an alternative quotation. There possibly were considerable savings to be made through the reduction of operators but regard needed to be had for the condition of the former NTCo and GPO line plant which would probably require considerable upgrade in order to achieve required line insulation standards for automatic working.

The state of the line plant was the breaking point in the debate. The cost of an automatic exchange was at that time twice that of an equivalent sized manual switchboard. Bennett had a quotation from Messrs Siemens Brothers, who at that time were supplying their No 16 automatic system[78] to the GPO. The price of a 1,500 line installation was £16,500, which included facilities for 60 junction circuits but an additional £1,550 was required for subscriber meters. The installation would be extendible to 3,000 lines.

In addition to the switching equipment, all existing subscriber telephone instruments would need replacing with dial telephones at a cost of £4 each, plus about £500 for their installation, making a further £6,500. It was noted that the resulting surplus of some 1,500 magneto telephone sets would not easily be able to be absorbed locally, and thus would need to be disposed of. The resulting loss of capital investment would be in the order of 38/- per instrument or £1,840. There was no certainty that such a number of instruments could readily be disposed of in the current market given the trend to automation elsewhere.

A further problem was the condition of the line plant. Under the NTCo and the GPO, lines had been installed in a haphazard manner and the quality of the insulation was indifferent. This was not particularly noticeable on magneto exchanges but became important for automatic working. The quality of the lines affected the dial signals and thus the insulation standard had to be high. This was particularly true of the Siemens No 16 system

---

78 The Siemens No 16 System utilized a 60V DC battery which applied extra constraints on line plant insulation standards, compared with the 22V or 40V manual switchboard battery voltages. See Telephony: A Detailed Exposition of the Telephone Systems of the British Post Office : Volume II – Atkinson J

which operated at 60 volts. The cost of upgrading the line plant was difficult to estimate as the extent of the work was not easy to define, however, Bennett guessed at not less than £1,500. This made the overall cost of a single exchange £27,000 and allowing for unforeseen problems this was perilously close to the requested loan. The staff savings as regards operators was also not easy to determine without the exact number and cost of maintenance engineers required together with their training costs. It was these factors that finally decided the question. The overall cost of the proposed new system was too high at that time and thus the proposal for an automatic system was abandoned.

With the way now clear to move forward with a less expensive manual system, the States approved the £30,000 loan. There was still considerable debate in the chamber as to the future prospects of the system and the fear that the States may have to provide subsidies in the long run. But the Telephone Committee stuck to their guns and assured their critics that the system could be run profitably. The loan would be funded serviced from revenue and that nobody who did not use the system would be expected to fund it. A loan act was passed to the Finance Committee who would prepare the Order in Council for approval by the Privy Council. The loan was raised at 4% and the Telephone Committee would repay at 4½ %. Because of the delays brought about by the Privy Council process, at the following sitting the States were asked to approve a temporary advance of £10,000 from public funds. This was hotly debated in the chamber but the opposition was mainly among a small number of telephone sceptics, including Jurat Payn and the Rector of St Martin. The proposal was nevertheless adopted by a large majority.

In August of 1924 Alfred Bennett tendered his resignation as Managing Engineer of the States Telephone Department because of failing health. In his resignation letter he heaped praise on his colleagues in the department, E F Guiton and W Symonds and the others in lower management who had assisted in the establishment of the system under States control. He also thanked the GPO for their generous assistance in supplying stores for repairs and maintenance. He further recognized the input of the Guernsey

States Telephone Council and its staff which had made available both supplies and expertise to Jersey and had opened their books to enable the establishment of a suitable management structure and accounting system for the new department. He heavily criticized the NTCo for their piecemeal development of the local system, noting that establishing 15 exchanges was ridiculous in such a small island, and their system of junctions was totally inadequate. He also remarked on the lack of investment from the GPO over the period of their stewardship. He commented on the state of the line plant, where wires had been erected 'on the twist'[79] everywhere and were in frequent contact with trees. Under the States ownership, much of this had been corrected and improvements were being made in every area. (Bennett had introduced his own patent insulator[80] which, to a large extent, had come via the Guernsey stores. His 'corrugated' insulator was cheaper and, in his opinion, better than the 'double-shed' type favoured by the GPO). He noted that over 600 new lines had been established thus far, about twice the number connected under GPO control. He also was pleased to see that even after GPO licence fees and money paid to pension funds, there was a trading surplus of £1,000 annually. Finally, he wished Guiton well in his position as acting Managing Engineer. Later that year John H Stanhope was appointed as Managing Engineer having formerly been a manager with the GPO with which he had already completed 44 years service.

The decision not to install an automatic exchange was perhaps an opportunity missed. However, the newly formed States Telephone Department had to be pragmatic. It was funded by public loans and there was a limit to the extent that the States would provide resources for every investment necessary. The legacy of the previous operators' of the system had to be considered and in practical terms, there was little choice, given

---

79 This refers to the principle of shifting the relative position of one telephone pair with the others on an overhead route, which in theory reduces the amount of cross-talk. The method favoured by the NTCo and GPO rotated a pair of wires along the route, the 'A' wire going from top left to bottom right top left and the 'B' wire from bottom right to top left then bottom right along the route. Bennett favoured the cross-over which reversed the 'A' and 'B' wires at regular intervals along the route while maintaining their relative position on the pole. This method was later adopted by the GPO.

80 The Electrician, London, England, Saturday, February 17, 1883. p. 319-320, col. 2,1

the line plant improvements necessary, other than to continue with the manual system. Further, the manual system was simple and easily manageable within the available local expertise. Bennett was a realist and recognized that it was more important to instil confidence in the system than to venture into the unknown. He had devised a sound footing for the continuation of the service. His bequest to the newly formed department was a sound structure of pricing and practical guidelines for progress. His insight into the pricing structure, first deployed in Guernsey, separating rental from call costs, was a positive move towards the 'user pays' structure of all progressive businesses. The pricing structure was carefully balanced so as to attract both business and residential subscribers alike. The Telephone Department policy was also more socially conscious in the provision of services. This was demonstrated by the development of a comprehensive public telephone kiosk network. He left a sound platform for the States Telephone Department to move forward and develop its network.

Having made the decision to remain with a manual telephone system, the States Telephone Department went out to tender in late 1924 for the provision of a new Central Battery Exchange with sufficient capacity for 3000 subscriber lines and an opening capacity of 2000 lines.

The contract was won by Messrs Siemens Brothers and Company Limited of Woolwich. They proposed a Common Battery Number 10 system switchboard. This system was originally developed in the United States by the Western Electric Company, a subsidiary of American Telegraph and Telephone (AT&T), and the technology transferred to the UK. It was adopted as a standard by the GPO and designated CB10[81]. The CB system provided all the necessary electricity for the operation of the telephone from the exchange, as in modern systems. This was a divergence from the then practice of local battery magneto exchanges that had been installed by the NTCo. The CB10 exchange was designed for installations where 1000 or more lines were required as initial capacity and could be extended to up to about 2,800 lines and exceptionally by removing the switchboard cornice, up to 4,200 lines. The basic suite consisted of a 3 switchboard set over

---

81 Telephony : A Detailed Exposition of the Telephone Systems of the British Post Office : Volume I : General Principles and Manual Exchange Systems – Atkinson J

which the subscribers were multipled, that is repeated, every 6 panels and outgoing junction circuits repeated over every 6 panels. These panels contained jack plug fields that used a standard ¼" jack plug. There were two types of switchboard, an 'A' position, which connected local traffic and a 'B' position, which dealt with incoming junction calls from other exchanges. Each switchboard could manage a maximum of 17 simultaneous calls. The total number of A and B positions would reflect the initial capacity. In addition there were monitor positions mounted in the centre of the exchange floor and a supervisor's desk. The exchange opened with 4 'B' positions and 20 'A' positions to the left of them, all in a straight suite.

Initial equipment started being delivered on 22 June 1925 and installation work commenced two days later in the new purpose-built exchange in Minden Place. Construction was completed at the end of October. Thereafter, it was necessary to make the necessary connections between the old exchange and the new to enable a seamless changeover for the subscribers. A new Main Distribution Frame (MDF) was installed in the separate equipment room that was equipped, on the line side, with 3,100 fuse terminal blocks and connected via beeswaxed silk and cotton insulated copper wire, lead covered cables to the cable chamber beneath the MDF. The beeswax was used to prevent the silk and cotton from unravelling once the outer cable covering was removed. The underground cables were paper insulated copper, bound with paper overlay and sheathed with lead and were joined to the internal cables in the chamber. Four external cables containing a total of 2,100 pairs were taken out of the building through a 7-way duct into Cattle Street and three cables of total capacity 1,200 pairs through a 6-way duct into Minden Place. The exchange side of the MDF was equipped with lightning protector mounting blocks with heat coils and carbon lightning protectors[82] of total capacity 2,480 lines, which, in addition to subscribers, would have included capacity for junctions and private circuits.

---

82 ibid

Image by permission of Société Jersiaise

The above picture shows the extent of the original installation of Central CB10 exchange at Minden Place. In the picture the junction switchboard is nearest the camera and the test position voltmeter can be seen next to it. The supervisor's desk is seen behind the pillar. The building was of a single storey construction. The switchboard was later greatly extended and an additional 'island suite' was installed in the space seen on the left of the picture.

The new exchange building was able to benefit from a public mains electricity supply provided by the Jersey Electricity Company which had begun operations in July of 1925. Nevertheless, the exchange required a central battery of 40 volts for its operation. Two secondary cell batteries were supplied by the Hart Accumulator Company[83] of Marshgate Lane, Stratford, London and each had an initial capacity of 297 Ampere Hours (AH). In those days, such batteries could be extended in capacity by the addition of extra plates into the cell containers and this battery could be increased to an ultimate capacity of 500 AH as the exchange grew. These secondary cell batteries need charging and for this purpose a 420 volt 50 Hz three phase AC 7.5hp squirrel cage motor was provided to drive a 50V 100A dynamo set. Charging was done off-line, thus the batteries would be used alternately to supply the exchange. There were also two ringing dynamo machines providing both ringing current and tones to the switchboard. All the electrical power equipment was controlled via a two-panelled enamelled-slate switchboards on which were mounted the control switches for charge and discharge functions, voltmeter, ammeter, rheostat, automatic circuit breaker and main fuse. The automatic circuit breaker was used to prevent the battery from discharging into the dynamo if a public mains failure occurred during the charging cycle. This would have caused irreparable damage to the dynamo and it was thought that a fuse would not provide sufficient protection.

In the equipment room there were racks providing 2000 subscriber line equipments. These consisted of the line (L) and Cut-off (CO) relays, one per subscriber. These relays were mounted on 2' 6" wide racks. In addition, there were the repeater coil and condenser (capacitor) racks that provided the junction circuits to other exchanges. All other exchanges in the island were of the magneto type. In addition to the MDF there was an Intermediate Distribution Frame (IDF) for cross-connecting the line and cut-off (L & CO) relays to the switchboard answering positions. Theoretically this would allow groups of consecutively numbered busy lines to be distributed across several switchboard positions although in practice this was rarely done. It is

---

[83] The Hart Accumulator Co became part of the Chloride Electrical Storage Company in 1929.

interesting to note that no subscriber meters were installed as it was decided to continue with the current call ticketing system. Alongside the exchange equipment there was a one-position test desk. This would not be permanently manned but used by the exchange engineer to liaise with linemen during fault repair.

The exchange was brought into service on Thursday 21 January 1926. It was reported in the JEP that work on the transfer would commence at 1:30pm and be completed by 3:00pm. It is curious to note that this was carried out in the middle of the day, unlike modern practice, however Thursday was early closing day and it was probably supposed that telephone traffic would be low. After the transfer, subscribers were warned not to use their generator handle to originate and terminate calls, as this would not be necessary as the system would work 'automatically'. The existing magneto telephones would work on the new CB system but would be inefficient. The magneto telephone had a separate local battery for the transmitter and the receiver would have been placed across the line in series with the telephone's induction coil when the handset was lifted. This would have provided the necessary calling loop for the CB exchange but would have meant that the receiver would have received a small DC current which would have provided a magnetic bias on the receiver thus reducing its efficiency. Departmental technicians were despatched to all subscribers after the cut-over to remove the generator handles from telephones to prevent callers from operating their telephone incorrectly. However, a much longer term plan must have involved gradually replacing magneto instruments with more modern CB telephones. The new exchange was opened for public viewing between 1 and 13 February.

The first Private Branch Exchanges supplied by the Telephone Department were installed during the same year. Private manually operated switchboards were provided for Messrs Le Riche at Colomberie and for Messrs Orviss in Beresford Street. Of course, such switchboards had also been provided by the earlier system operators and private suppliers.

Following the changeover of Central exchange from magneto to CB10 it would seem that the budget was capped, since there were no further upgrades to the switches throughout the island for some time. Most of the funds seem to have been spent on developing the network. The number of subscribers increased steadily and reached 3000 in August 1927, doubling the size of the network in just four years.

Also, in late 1927, it was decided to close St Mary's exchange since it was a small switchboard and only had a few subscribers. Existing lines were transferred to a new magneto switchboard installed at St Ouens exchange. At this time it was also decided to prefix all numbers with the digit 6 that is, for example, St Ouen's 23 became St Ouen's 6023. This decision was taken in order to alleviate the confusion in sounds over telephone circuits between 'St Ouen's' and 'St John's' when subscribers were requesting numbers. By making St Ouen's numbers longer, it enabled local operators to distinguish more easily between the two exchange destinations. Reports around that time suggest that the old St Mary's switchboard may also have been of a poorer quality than that of the other exchanges; possibly it was a second-hand exchange from another NTCo region.

The following May the former Telephone Manager AR Bennett MIEE MILocoE died at the age of 78. Bennett had been a great figure in Channel Islands telephony, having worked in both Guernsey and Jersey. Much of the success of both the islands systems could be attributed to his skills and foresight.

The same year the question of off-island telephone circuits to connect to the UK was raised. Answering a question in Parliament the PMG Sir William Michell-Thompson stated that there were financial and technical obstacles which need to be investigated before any prospect of a cable could be envisaged.

The States of both islands, however, continued to lobby the GPO and Parliament for a resolution to the problem. In March 1929, a further question in the Commons by Sir Nicolas Gratton Doyle, MP for Newcastle upon Tyne North, evoked a response from the PMG who stated that tests were continuing on the existing telegraph cables to evaluate their suitability for telephony. He further stated that more information should be available at Easter. Further questions later in the month produced no further progress.

The pressure on the GPO was increased with the introduction of a telephone link to the Isle of Man at the end of June. However, the GPO insisted that it was still assessing the problem and working with the Chief Engineers Department for an economic solution.

In December, Stanhope announced that the Telephone Department would 'broadcast' the Jersey Green Room Club's pantomime. This would be available to a number of subscribers for a fixed fee. It is not known what the fee was or how many subscribers availed themselves of the service but it does demonstrate a certain air of commercial realization by the Department.

A few days later on 7 December 1929, one of the severest winter storms for many years hit the English Channel. As a result, the telegraph cable again failed and communication with the UK and Guernsey was interrupted. UK traffic was limited to the French cable which naturally resulted in delays, up to 14 hours in some cases. Several other continental cables had been damaged in the storm and consequently the GPO cable ships were fully employed repairing the most important routes first. Therefore, the GPO decided to install a temporary wireless telegraphy system between Jersey and the UK. This was installed by Mr Coates, Assistant Engineer from the Bournemouth area and Mr Britton from the Chief Engineers Department. The Jersey station was installed in the former military canteen at Fort Regent after obtaining War Office permission. Messages were telephoned from the Head Post Office to the temporary station and incoming messages passed by teleprinters directly to the Broad Street telegraph office. The outgoing messages were transmitted via the radio station at Stonehaven, near Bournemouth, while incoming messages were received from the Dartmouth transmitter. The GPO made it clear that this equipment would only remain operational while the cables were out of action. The States Telephone Department assisted in erecting the aerial on an 80' telegraph pole mounted on the ramparts. In the Commons there was a question from Sir Brandon Falle (Conservative) regarding the Channel Islands telegraph via France and the resulting delays and costs. The PMG in a written reply stated that: 'the French post and telegraph authorities were doing their utmost under the circumstances'. In January, it was reported that the PMG and his wife were aboard the *CS Monarch* during cable repairs in the Channel. The Guernsey to Guernsey cable was restored on 10 January and the UK to Guernsey section returned to service on the 16 January, The PO engineers returned to Jersey on the 20 January to recover the temporary wireless equipment.

The storms continued throughout the month and on 30 December 150 telephone lines were reported out of action. The Gorey and La Rocque exchange junction circuits were cut when a falling chimney at Le Hocq railway station brought down the lines which ran along a pole route beside the railway track.

The New Year also saw progress on a submarine telephone connection. At the end of February the PMG announced in the Commons that the 'negotiations were almost complete' with the Channel Island authorities and that the technical difficulties were nearly resolved. However, further signs of the connection did appear until 5 September 1930 when *CS Monarch* laid a 'sea earth' off Plémont and on 15 September the Guernsey States' Tug *Sarnia* dropped the shore ends for the cable diversion at Saints Bay. No further progress was evident but the PMG announced that the circuit would be ready 'next year' in a written answer in the Commons on 26 November. Later the next month, however, the GPO announced that the circuit was nearly complete but that its implementation was being delayed by the installation of the new switchboard in St Peter Port.

On 12 February 1931, the testing of the cable was complete and final circuit balancing was being carried out. The JEP published the proposed scale of charges for trunk calls (see below) on 15 February and the inaugural arrangements for the official opening were published on 5 March, together with a list of possible trunk call destinations which curiously omitted Weymouth, the important GWR mail boat port. The JEP article went on to describe the process for making a trunk call.

Because of the limitations of the single trunk circuit it was necessary to restrict availability from the outset. The standard telephone in use by subscribers was still largely the old magneto type that had existed before the cut-over of the new exchange in 1926 and thus the transmission characteristics of such telephones were unsuitable for the making of trunk calls. Thus such calls were initially limited to certain call offices and to those subscribers who were willing to make an additional deposit payment for a new telephone, a GPO Telephone Type 162[84], and sign a new rental agreement.

The instructions told subscribers to ask for 'Trunks' and then to wait for the trunk operator. They were then asked for their number and the required number. When this information had been passed, the operator would instruct the caller to replace their receiver and await a ring when the trunk line became available. The subscriber would then be asked to verify their number and the call would be connected by the operator saying, for example: 'your London call'. It was also possible to book a person-to-person call for an extra fee. This would ensure that the caller did not waste time and money waiting for the desired party to reach the telephone.

The same day, 5 March, it was also reported that the new Guernsey Central exchange switchboard was opened for service and that the Guernsey system now had over 4,000 subscribers and made over 2,000,000 calls annually. The official opening date would be the 26 March. The opening of the trunk circuit was an important event in the islands. For the first time it would be possible to make a telephone call not only to Guernsey but to many UK, European and even world destinations.

The trunk telephone service opening ceremony was on 26 March 1931 at Central telephone exchange in St Helier and simultaneously at Central exchange St Peter Port. Among the esteemed guests were the Lieutenant Governor, General Willis and his wife, the Judge Delegate Reginald Malet-De Carteret and his wife, Jurats J H and J E Leboutillier, and P E Bree, The Rectors of St Saviour, Grouville and Trinity, Deputies Orange, Le Quesne and Belford, the Constables of St Helier and St Saviour, Major Hindson, Viscount Le Gros, the Evening Post reporter A Harrison and Telephone Department officers Stanhope, Guiton and Rumfitt. The special telephone rang at 2.31

---

84 Telephony : A Deatailed Exposition of the Telephone Systems of the British Post Office : Volume I : General Principles and Manual Exchange Systems – Atkinson J

PM exactly and the British Secretary of the Home Department, the Right Honourable Mr John Clynes MP, was announced. He read out the following inaugural speech, which was reported to be almost inaudible since he insisted on reading it with his head away from the special telephone on his desk.

Mr Clynes message:

'On the occasion of the inauguration of the telephone service to the Channel Islands, I wish, on behalf of His Majesty's Government and the people of Great Britain to send a message to the people of the Channel Islands expressing our hopes that the new means of communicati0on, by bringing those islands for the first time within speaking range of us, will strengthen the many ties of friendship and pleasant intercourse which already bind us together, and will further our common interests and activities.

Apart from the personal relationships which we shall now be able to make closer and more intimate, I hope that the telephone will help the trade of the Channel Islands by making it easier for them to keep in touch with the British markets for their agricultural produce. Farming and growing in the Channel Islands has, I am sorry to learn, been passing through lean times, partly on account of falling prices due to the agricultural depression which is unfortunately at present a worldwide phenomenon, and partly on account of increased competition form the continent; but I sincerely hope that, with the aid of scientific methods and enlightened marketing, neither of which can now be neglected with impunity, the Channel Islands will succeed in regaining the prosperity which their favoured climate should make possible.

It is also not without importance that the Channel Islands can now offer to British businessmen the possibility of a holiday in remote and beautiful surroundings without the disadvantages of being inaccessible to telephone communications on urgent business affairs.'

General Willis replied as follows:

'On behalf of the Bailiff, the States, the Royal Court and the islanders, I wish to express the inhabitants' deep appreciation of the message which the inauguration of the Channel Islands telephone service has enabled you to send to the people of Jersey, and to assure you that strenuous efforts are being made to deal with the agricultural depression which I sincerely hope will reap a great reward.'

At this point there was an interchange with the Guernsey officials in which the Judge Delegate spoke with the Guernsey Bailiff A W Bell. The general tone of the exchange was that of increased cooperation and prosperity as both islands pursue their own independence. There followed a short conversation between the Lieutenant Governors of both islands, in which General Willis expressed his hope that the Muratti Cup would return to Jersey this year.

The JEP reporter was then able to conduct the first public trunk call to the Jersey Society in London. It had been hoped to make the call to the Jersey Society of Canada, but that proved to be too technically difficult at the time. The call was conducted between A Harrison and Mr De V Payen-Payne of the Society in Ilford, Essex, and both extolled the advantages of telephonic communications over the telegraph.

Finally, John Stanhope spoke with the Administrator British Telephones GPO, London thanking him and his staff on behalf of the staff of the States Telephone Department for their assistance in realising Jersey's dream, and that trunk calls were now available to the general public.

With the service now in operation, businesses and private individuals quickly took up the service. In the first month of operation some 500 calls were made from Jersey and over 650 inbound calls were received. This was a remarkable number considering there was only a single circuit which was also shared with Guernsey. Each call had to be booked and passed via the

Guernsey operator before progressing to its final destination. This naturally extended the time required to set up each call and, in addition, more calls were generated by the larger number of Guernsey telephone subscribers which added to the difficulties experienced in Jersey.

Local business was quick to use the service for making long distance business calls. The JEP reported that Sir John Buck-Lloyd of the National Provincial Bank, St Helier and Financial Director of the Anglo-Persian Oil Co (now BP), made a trunk call to South America. This involved the use of the GPO wireless telephony service based at Rugby. It was reported that the conversation was not clear but nevertheless audible.

A month after the opening of the circuit, on 28 April 1931, the JEP offered an unusual service using the new cable. The Muratti Cup match between Guernsey and Jersey was 'broadcast' from its offices in Charles Street, St Helier. On this occasion, a JEP reporter, L M Bourke ran the commentary from *the Track*, in Guernsey. In Jersey, the Constable of St Helier allocated extra police to the area and diverted traffic so that crowds could listen to the commentary which was audible over amplifiers and speakers provided by Mr T Cane of the Radio-Electric Service Company, Charring Cross. The broadcast lasted over two hours and the JEP praised the unprecedented level of cooperation between the two islands' telephone authorities. Perfect weather ensured that an audience estimated at 'thousands' crowding Bath Street, Charles Street, Peter Street and Beresford Street heard Jersey win a thrilling 6-goal match.

Shortly afterwards, however, one of a number of frequent interruptions to the telephone service occurred on 18 May, when a cable fault resulted in the telephone circuit being seconded for use by the GPO telegraphs department which exercised its priority on communications. The new trunk call service was nevertheless a boon to local business and quickly established itself as a means of commerce ensuring that the Channel Islands had a telephonic route to the UK, albeit that it was not guaranteed in times of need by the telegraph system. These intermittent problems did not, however, repress the growth in trunk call traffic, since by July, the number of trunk calls originating in Jersey (951), for traffic both to the mainland and inter-island, had grown to exceed that of Guernsey (902). A second trunk circuit was opened between Guernsey and Jersey at the end of November 1931, which took pressure off the UK route and enabled separate traffic to be delivered between the islands.

Development elsewhere in the Jersey telephone system was less exciting. Following the installation of Central exchange, it would seem that the Telephone Committee had been somewhat reluctant to request further funding for network development. Certainly, this investment programme cannot have been helped by the prevailing economic depression which was general throughout the world at the time. As noted above, Jersey agriculture, then the main driver of the island economy, was suffering from the recession. The other island exchanges remained as before and overall network development seems to have been confined to making more connections and extending the line plant.

Line plant was largely overhead open wire at that time. Some underground cables had been put in place to feed distribution poles since the days of the NTCo but structured distribution cable ducts were still not widely used, apart from those feeding Central exchange and those installed when Samares exchange was closed. Underground cables were mostly simply laid directly into the ground, usually without any protection. In urban areas, the location of the cables was often marked by the engraving of the initials 'STD' on granite kerb stones, a practice also used by the NTCo. In rural cables were sometimes marked by small rounded granite 'tombstones' or square concrete locators, similar to boundary stones. Contemporary cable was of lead sheath and paper insulation construction and cable joints were made using lead sleeves which were plumbed to the cable using blow-lamps and lead solder to form a water-tight seal. These joints were also buried directly into the ground after completion. A new cable between Central exchange and La Mare slipway was completed on 16 February 1932 utilizing the ducts installed in 1923. This was to cater for growth due to residential and commercial development along the St Clement's coast road.

The JEP again arranged a broadcast for the inter-island Muratti cup semi-final between Guernsey and Jersey at the Track, Guernsey on 16 April 1932.

The cooperation of the local telephone departments, the Electric Services Company, which supplied the Marconi amplifiers and speakers, and the JEP again resulted in a successful day out for the hundreds of people who thronged the streets around the JEP offices. This time, the commentator, L M Bourke, used a grid reference of the pitch as a means of conveying the positional play. This was printed in the previous evening's edition of the JEP which also distributed leaflets on the day. The success of these broadcasts was short-lived, however, since at the inter-island Muratti Committee meeting held on16 June, it was decided not to allow further broadcasts as the committee feared that the attendance to future matches would be detrimentally affected.

At the end of October 1933, building began on the remainder of the former Toy Market site for the construction of new administrative offices for the Telephone Department. This was necessary as the offices in New Street were becoming too small and the lease was unlikely to be renewed. Meanwhile, the network continued to develop and an increasing number of trunk calls were placed between the islands and the UK. The number of calls to and from Jersey gradually exceeded those from Guernsey, especially during the tourist season. This inevitably led to longer delays, especially during the busy periods. For example, the total trunk traffic over the period 1932 to 1934 is shown below:

Route	**July 1932**	**July 1933**	**July 1934**
GY – UK	372	456	497
UK – GY	429	583	532
GY – JE	641	745	557
JE – UK	619	637	1374
UK – JE	876	852	2002
JE – GY	592	486	526
Totals	3529	3759	5488

Table 25

While the traffic from Guernsey remained fairly static, that from Jersey grew rapidly.

This would have been consistent with the growth in the number of subscribers on the Jersey system which had now grown to 4,625 by mid-summer 1934, almost three times the number that existed on the takeover from the GPO and rapidly catching up with telephone density in Guernsey. Indeed, in September 1935 it was noted that Guernsey now had 5000 telephone lines with around 11,000 calls being handled daily. The JEP reported that on inquiry from the Telephone Department there were now 4,820 lines in Jersey with a daily call rate of some 12,000, with new lines being added daily. Central exchange now had well over 2,000 lines and was approaching its original design capacity. Around this time, an extension of the Central exchange switchboard must have been commissioned, since by the middle of 1935 new numbers were being issued in the 25XX range.

During 1934 and continuing into 1935, the GPO started tests for the introduction of wireless telephony in order to increase the number of circuits available between the islands and the mainland. This was in lieu of providing a new submarine cable which had been discounted because of the cost. Wireless had been successfully tested elsewhere in the UK, particularly across the Bristol Channel and between Scotland and Northern Ireland. In the end it proved not to be technically feasible from Jersey and the service was connected via a wireless station in Guernsey and by providing additional multiplexed circuits on the inter-island submarine cables.

In July 1937 the States bought and then transferred to the Telephone Department a new exchange building at St Aubin. This was necessary as the Department was beginning an overall upgrading of telephone exchanges from magneto to Central Battery. In addition, the exchange building would appear to have been too small for the new switchboard to be installed next to the existing, thus a move was necessary. In some cases the upgrades would have been driven by the difficulty in expanding the older magneto boards to cater for more subscribers. It seems that this work was done using local skills and labour and carried out over a period of time. The other exchanges immediately affected were Gorey, Millbrook, Trinity and

Five Oaks. Further exchanges would be scheduled into the plan as mains power became available in their areas. These changes would certainly have improved the service to the subscribers, allowing the use of more modern telephone instruments on the outlying exchanges. The ability to install new switchboards and associated equipment next to the existing switchboards would have been an advantage, since little change would be necessary to underground cable systems. However, is likely that some external cabling would have been necessary in order to ensure that the line plant was of sufficient quality for the new switchboards to operate satisfactorily.

At the same time the world was changing. There were rumours of war and suitable preparations were being made in the event of belligerencies. The States set up an air raid patrol (ARP) with the view to ensuring the safety of citizens. The Telephone Department reacted by setting out to install more public telephone boxes around the island, particularly in built-up areas. New telephone kiosks designed locally and made by two local companies Farleys and Le Selleur were introducedThis would assist both the ARP and the man in the street to report fires and other damage in the event of air raids. During the year 15 new public telephone kiosks were installed at various locations in time for the blackout and heightened security that was declared following the Czechoslovakian crisis in September.

In the spring of 1938 the popular broadcasts of the Muratti matches were revived around the JEP offices in Bath Street. Commentary was provided by Messrs L M Bourke and H G le Cocq. The public address system was provided by Grant's Radio Ltd. and the circuit from Guernsey by joint cooperation of the local telephone departments. On this occasion, the semi-final match with Jersey was won by Guernsey 4 – 3. The following year the exercise was repeated when Jersey regained the cup 1 – 0.

A new location was found for Millbrook exchange at 2 Boulevard Avenue necessitated by the conversion to CB10[85]. The post of caretaker/night operator was advertised but one applicant by the name of J P Bartholomew unfortunately collapsed with a suspected heart attack while queuing in

---

[85] Telephony : A Deatailed Exposition of the Telephone Systems of the British Post Office : Volume I : General Principles and Manual Exchange Systems – Atkinson J

Image by the author

This photo at La Rocque harbour shows the 1938 design in its original form with Crittal steel window frames intact. These were later modified to 5 larger panes of Perspex to combat vandalism. The boxes were constructed of hardwood and had a zinc covered lantern roof.

anticipation of an interview outside the Minden Place offices. Sadly, he died on 3 October. The post was instead given to Mr and Mrs Alfred Le Marquand, existing employees of the Department.

In December, during a particularly cold spell with snow and freezing temperatures, Mr Rumfitt of the Telephone Department reported that calls during the 24 hours of the worst conditions were almost double that normally expected at 26,220, some 12,760 more than average. All available staff was called upon to handle the traffic. At the end of 1938 the number of lines had grown to 5,746, which meant that 1 person in 9 in Jersey now had a telephone, and for the first time the number of lines was greater than in Guernsey, which had 5,657.

In March 1939 the new CB10 switchboard at Five Oaks exchange opened. At the same time about 70 subscribers in the Bagatelle, Wellington Road and St Saviour's Hill areas were transferred from Central exchange to the new switchboard. This relieved pressure on Central, which had already been extended to beyond its original design capacity.

By now the island was on a war footing and preparations in case of air raids were being put in place. The JEP reported that 234 ARP Wardens had been trained of which 119 were appointed in St Helier. At the same time air raid sirens had been installed at various strategic points around the island to which the Telephone Department supplied control circuits. Sirens were located at Fort Regent, the St Saviour's Mental Hospital, the Jersey Miniature Rifle Club, St Peter and Sion. It was also necessary to connect a public mains electricity supply to these locations to operate the siren. A control panel was installed at the main paid police station in St Helier town hall and a test was carried out on 10 May 1939. A switchboard for use in emergency was installed in the Telephone Department workshops in Minden Place.

During the summer of 1939 the large overhead telephone routes along the now defunct western railway track were partly removed and substantially replaced with underground cable, leaving only a few distribution poles in place. This route had formerly carried both junction circuits and subscriber lines from Central, Millbrook and St Aubin exchanges, the latter two having recently been moved to new premises. The GPO also removed its telegraph poles which extended to Beaumont as a new underground cable had been installed from Plémont to St Helier as part of the preparations for the the new submarine cable to the UK which was opened at the beginning of August. The new cable passed via Guernsey and increased both the quality and quantity of the trunk circuits available.

With the declaration of war in September, the status of the ARP was enhanced. The States determined that certain telephone lines should be treated as priority during air raids (and during tests) and that these should be separately marked on all switchboards. Instructions were given that other telephone lines were to be ignored during these times and the exchange operators were trained accordingly. A notice was published in the JEP to make the public aware of the arrangements. At this time the war was still a distant reality and life in Jersey continued more or less as normal.

In January the Telephone Department introduced quarterly billing instead of the previous annual billing. All the while, however, the war was beginning to encroach on daily life. A National Service Office was opened in Conway Street (allocated the number Central 1151) ready for recruiting under the recently introduced UK National Service Act. This law did not extend to Jersey although there was no shortage of volunteers for all the services. Already at this time there was a steady stream of people leaving the island for the UK as the fighting crept nearer to the islands. By May the war was just across the water in France and further emergency measures were introduced, including the introduction of low levels of lighting in telephone boxes. This was unsuccessful and so blacking out of the boxes' windows was carried out instead.

By June the situation was becoming desperate. A curfew was introduced because of the possibility of air raids and the Telephone Department appealed for the return of retired operating staff to replace the many who had joined the evacuation. A good response to the appeal was reported. A request for the public to use the phone only in emergencies was made and trunk calls were strictly limited to three minutes. A desperate scramble to leave the island resulted in many of the engineering staff leaving the Telephone Department too and thus its skilled human resources would be depleted during the war years.

On 14 June the trunk office in Minden Place was closed to the public thus leaving the remaining UK bound telephone circuits fully available for official business. An appeal was made for extra staff at St Peter's exchange to cope with the increased traffic from the airport. The last Telephone Committee meeting was held on 21 June when it was reported that over one third of the staff had opted to evacuate including the Telephone Manager.

The inevitable invasion took place on 1 July and the island braced itself for the occupation. In the absence of the Telephone Committee the Department would be run directly by the German military.

Chapter 15 - Telecommunications in the Occupation 1940 – 1945

Just prior to the German invasion, the submarine communications cables with France were cut by the Post Office engineers. Communication northward continued until the 2 July when the occupation forces disconnected the cables off Guernsey. The cables to France were quickly restored by the German military engineers and taken into the control of the occupation forces who immediately took over the Post Office repeater station at Trinity Gardens. Access to these cables by the civilian population was eventually restored but only through a licensing scheme, whereby subscribers were approved by the military authorities to communicate with French partners for the purposes of trade. Special permission was required to make personal calls to relations in France. The majority of cable traffic was from the occupation forces themselves.

Communication between the islands was less contentious but nevertheless also monitored and controlled by the occupying forces. Communications were reopened from the 12 July 1940 and continued throughout the war.

Immediately after the occupation began, the use of wireless sets was banned but this restriction was lifted on 15 July. However, on 15 October, all amateur wireless transmitters were confiscated by order of the Field Commandant. This effectively cut off communications to the outside world for the civilian population.

Although the submarine cables to both Guernsey and France continued in use and were controlled by the occupation forces, it was still possible for traders to communicate to the immediate environs of France. This was set out in a German order published on 18 July 1940 whereby special permits were given to designated traders and States officials for the purpose of procuring supplies for the island. This order also covered the distribution of post which, for the time being, was still being received from the UK via France. The reception of UK mail (via France) was stopped at the end of October and at the same time restrictions on the use of photography were imposed.

Communications are vital to the military and thus on arriving in Jersey the German forces immediately recognized the importance of the established telephone network for its own purposes. The army thus requisitioned significant numbers of lines in order to establish its command network. At first this was not too difficult for the Telephone Department to meet, since a large number of the telephone-owing middle class had evacuated during the few weeks immediately prior to the invasion. Thus in 1940 the Telephone Department provided a considerable number of lines for the use of the German military and ancillary services. This requirement meant that a large quantity of its stores was consumed in providing the additional connectivity necessary to interconnect German camps and other military posts around the coast and elsewhere.

At the end of September a proclamation from the German Commandant created a military zone around the coast of the island. This meant that a substantial number of telephone lines became 'off-limits' and this necessitated the recovery of upwards of three miles of overhead telephone routes in order to clear space for coastal defences and anti-aircraft batteries. About the same time guards were permanently assigned to the larger island exchanges.

The depletion of its stores was made much worse by a great storm in mid November 1940 which brought down a considerable number of overhead lines and it also inflicted further damage on the overhead electricity distribution system. Over 300 lines were affected in and around St Helier while more than 1,000 lines in rural areas were damaged. This amounted to almost a quarter of the total number of connected lines. Many poles were brought down and thus restoring service was a difficult and lengthy business. This was not made any easier by the insistence that military lines were restored first.

In January 1941 fuel shortages began to interrupt electricity supplies. By February The occupation authorities restricted the hours of electricity generation, closing the system between 11PM and 7AM each day. These fuel shortages had a knock-on impact to the telephone system and the States Department of Essential Facilities limited telephone calls to 3 minutes and advised that only urgent calls be made. Telephone operators were order to strictly apply these rules.

In March a new order on post and telecommunications was issued, further restricting the use of the telephone service. Special permission would be needed to call outside the immediate area (which included the coastal areas of Normandy and Brittany) and all post would be carried, subject to conditions, by the German military post office. Telegram delivery via the telephone was prohibited, although to what extent any telegraph traffic to the general population still existed is unknown. Contravention of this order would result in punishment by hard labour.

While at this time the population was allowed to continue listening to the BBC, the occupation forces occasionally confiscated wireless sets as punishment in areas where anti-German slogans or other resistance activity was exercised. A typical defiant act would be as in July 1941 when sets were confiscated from homes in the Rouge Bouillion/St Saviour's Road area for the daubing of 'V' signs on walls. The 'V' sign defiance was a subversive resistance action encouraged by the BBC, typified by its call sign using the opening bars of Beethoven's Fifth Symphony which represented three dots and a dash - the Morse code for the letter 'V'. The German authorities further required the civil population to provide patrols to ensure that this activity was deterred. Confiscated radio sets were later allowed to be collected from Woolworth's store in King Street.

In April 1942 the news was received that the former Engineer Manager J H

Stanhope died in Harrogate, Yorkshire. Stanhope had left the island during the last evacuations before the occupation.

The ARP was still active under occupation and the telephone system was still a key component. The German command issued orders that telephones were not to be used during air raids or tests. The German forces evidently relied on the pre-war siren system for their own purposes.

During this period the German signals engineers were also engaged in providing a military communications network. Much of the communications system would be provided in the military exclusion zone around the coast and thus it would need to be secure against bomb blast. The German engineers used armoured cable buried in the ground between signalling centres around the coastal areas and suspended cables from existing telegraph poles inland. In some cases this additional load had a detrimental effect on the stability of the overhead routes.

The military central switchboard was installed in Lyric Hall, Cattle Street, opposite Central exchange. This enabled circuits to be connected to the existing telephone network via a tie cable between the exchanges. A States Telephone Department technician was permanently assigned for this purpose. Ultimately the military system would extend to some three further switchboards and 15 cable diversion posts around the island[86].

In July 1942 the Field Commander ordered the confiscation of all wireless sets, no doubt as a consequence of the turn of events in the war. Collection posts were set up at all Parish Halls around the island. Instructions were given to ensure that a label with the name and address of the set's owner was pasted to the equipment. Later, in December, Bel Royal Radio Limited was given permission to purchase some of the confiscated sets, this may have led to the post-war confusion when a large number of sets went 'missing' when owners tried to reclaim them. In the end, the States decided that too much time was being spent looking for missing sets and terminated the repatriation leaving many angry claimants. This could have also been exacerbated by the German authority's decision to allow the return of the

---
86 See the Channel Islands Occupation Society https://www.cios.org.je/ (last accessed 23 May 2017)

gramophone section of some confiscated equipment in January 1943.

The resources of the Telephone Department were again stretched in October 1942 when a severe storm brought down many overhead lines, some weakened by the additional load of military signalling cables. Spare parts were by now in short supply and therefore the Department relied on refurbishing equipment. As many as 500 telephone instruments had been requisitioned by the occupying forces by the end of the war and thus the telephone workshops at 34 Parade and Minden Place relied on recovering usable parts from scrapped equipment to maintain the system. Even recovered overhead wire was recycled and reused in order to eke out the remaining stores. This process led to a side industry in providing a source of parts for the construction of illegal wireless sets. Parts such as telephone receivers, induction coils, capacitors and wires enabled the construction of crude but effective radio tuners, capable of receiving BBC broadcasts.

In December 1942 the public telephone box located at Green Street slipway was removed to make way for the German military railway.

The years 1943 and 1944 were particularly difficult as fuel shortages gradually reduced the ability of the JEC to produce electricity for any length of time. The number of mains operating hours per day gradually became less and less. The exchanges were all equipped with their own batteries for normal operation, but the CB exchanges relied on the mains to keep the secondary cells charged. Consequently, calls were strictly time limited in order to conserve the batteries. However, this did not prevent the acting Engineer Manager, Percy Luxon, from setting up a broadcast system between the exchanges to enable the distribution of local drama and music programmes to telephone subscribers over the exchange network. He developed a distribution system that enabled several hundred subscribers at a time to listen-in to locally produced plays and concerts during the darkest years of the occupation. This provided a much needed fillip for the austere days of continuing shortages and power cuts.

In June 1944 immediately following the D-Day landings on the Normandy coast, the German command order the closure of all telephone exchanges. In some cases at the minor magneto exchanges, German engineers actually short-circuited the operator batteries to prevent service. This blanket closure was relaxed from the 1 August when the exchanges were again allowed to open between the hours of 8AM to 8PM, although because of the damage to batteries at St Lawrence and Sion exchanges, these switchboards were not returned to service until the 9 September.

The reason for the closure of all the exchanges may have been to prevent communication with France, although what strategic purpose this may have had is questionable. Nevertheless, all cable communications with France ceased on 8 August when St Lo in Normandy, where the cables connected to the French network, fell to the American forces and the cable was cut at the French terminal. Communications with Berlin now relied entirely on wireless.

The reliability of the public mains supply continued to get worse as fuel supplies diminished. The Allied blockade of the islands was almost complete and the electricity supply failed entirely on 3 January 1945. Severe restrictions on the use of telephones were imposed via the Department of Essential Services by the Platzkommandantur. This order published in the JEP on 6 January advised users to use the telephone only in emergency and to keep call time to a minimum. The same day the telephone service at the minor CB exchanges of St Aubin, Trinity, Five Oaks, Millbrook and Gorey was reduced to just a few essential lines, such as doctors, police and lines to German military offices. The available lines were published in the Jersey Evening Post during January (see Table below). Batteries at these exchanges were probably occasionally recharged by portable generators provided by the German military. The other country exchanges were largely unaffected as magneto switchboards do not rely on mains power and have their own primary cell batteries. Central exchange remained relatively unaffected as it had been possible to keep the batteries recharged by coupling a steamroller to the exchange battery charging dynamo. Presumably there were sufficient remaining supplies of fuel to keep the steamroller in steam for at least part of the day. This situation continued until a few days after the liberation on 9 May.

St Aubin		Five Oaks		Gorey	
No	Subscriber	No	Subscriber	No	Subscriber
6	Dr J Hanna	5	J A Perrée	5	Home for Girls
21	P Biard	9	G J Mourant	24	Mental Hospital
22	Dr JE Lewis	13	W E Le V dit Durell	27	Home for Boys
40	C S Le Gros	35	J Le Sueur, Clairvail Farm	30	A Webley
41	St Brelade's Hospital	42	T P Mourant	48	R Jackson
46	H Le Rossignol	57	Mal Assis Mill	50	R Lawrence
62	C S Fisk	75	Fairview Farm	52	Faldouet Dairy
78	A M Coutanche	78	E J Mourant	61	O P Journeaux
102	Caretaker, German Cemetery	83	Army of Occupation, Colege House	73	Rev J Valpy
104	W Benest	84	H C Garden	84	E A Dorey
128	La Moye Nursing Home	85	P E Le cuirot	111	P Briard
155	F M Burrell	98	Dr J Gallagher	120	J Norman
194	Lilywhite Laundry	102	Army of Occupation, Bagatelle	163	F Abrahams
247	F A Laxen	153	St Martin's RC Presbytery	191	C Bilot
256	E Le Quesne	188	Army of Occupation, Goodlands Cottage	196	A S Perchard
349	R Burrow	201	Canon J R Wilford	205	Dr P G Bentliff
358	St Brelade's Reservoir	202	Dr D Coutts	224	C Pallot
362	Mrs Skillett	210	Field Police	231	Rev S R Knapp
465	Rev G R Balleine	223	W Perchard	271	E Messervy
		238	E Abraham	278	Miss M Bruford
		255	G Sutherland	280	E Labey
		257	R Andrews		
		270	P Le Gros		
		289	R Kilmister		
		303	Dr T Warrington		
		311	Bon Air Nursing Home		
		320	Army of Occupation, Linden court		
		321	Army of Occupation, Newlands		
		335	R Mollet		
		349	C W Rice		
		364	St Saviour's Parish Hall		

Table 26 Exchange lines with preferential service January 1945

Chapter 16 - The Post War Years 1945 – 1959

The island was liberated on 9 May 1945, one day later than the official end of the war in the rest of Europe. A few days later a team of GPO of engineers arrived from Guernsey, where they had already restored wireless communications services. The GPO quickly re-established the Jersey telegraph communications using a temporary wireless system. By the end of May, the pre-war cables had been repaired and temporary terminal equipment was installed to enable the reopening of telephone services with the UK, some 4 circuits being installed and two circuits were provided between the islands.

At the beginning of June the press service telegraph was restored to both Guernsey and Jersey over the repaired cables, enabling full reporting of UK and international news to return to local papers. At the end of June the States appointed Percy Luxon as Engineer Manager after having acted in that role throughout the war in place of the late Stanhope. A few days later, Rumfit resigned after 22 years to take up a post in the UK. In August Luxon was presented with a gold watch by the Bailiff, Alexander M Coutanche, on behalf of the members of the Jersey musical and drama societies in appreciation for his efforts in providing the broadcast network during the war. The presentation had been organized by the well known local musician, Walter Larbalestier.

In October the first telephone subscriber list since the occupation was published, consisting of 73½ pages of subscribers. At the end of the year at the elections in December, the Telephone Department arranged its first public telephone information service by arranging for a dedicated operator to continually read out the latest results. Any subscriber could request the service and be connected to listen to the recital.

Over the Christmas period the newly liberated telephone system enjoyed a resurgence in subscriber activity. During the period 44,290 calls were handled by the island's switchboards of which 6,000 were made on Christmas day. The Saturday before Christmas 12,966 calls were made which was more than the whole of the same period under occupation conditions. Outgoing trunk calls totaled 680 with 733 inbound.

During the war the telephone system had continued to operate using much the same equipment that had been installed almost 50 years earlier by the NTCo. It had become apparent that with the liberation a fresh approach to the telephone system was required. The pre-war upgrades to some of the country exchanges had resulted in more efficiencies but there were still too many exchanges in the network. In April 1946 La Rocque magneto exchange was closed and subscribers transferred to Gorey. Numbers were changed by adding 400 to the original La Rocque number. Around the same time La Moye exchange was also closed and subscribers transferred to St Aubin. This reduced the number of operational exchanges on the island to 11, five of which were still magneto. At the same time Luxon announced that plans were afoot to reduce the number of country exchanges to four, covering the western, northern, eastern and southern areas of the island. Clearly, the experience of the occupation followed by the increased demand for telephone service from the returning evacuees and resurgent economy had focused the Departments mind and it became clear that the present network was not of a suitable standard for economic development. This announcement led to questions in the States asking if the proposed development would include automatic exchanges.

At the beginning of May the first post-war Muratti Cup tie was again broadcast by the JEP. The commentator was the newspaper's sports reporter, Bill Custard and the services of the Telephone Department's engineers C J Syvret and R Perrin were acknowledged for their assistance. Jersey won the contest 1 - 0. It is worthwhile noting that the JEP appealed to its readers not to call the office during these broadcasts as the incoming calls interfered with the quality of the commentary.

In September 1946 an unusually ferocious storm hit the island; much damage was caused to the telephone system by the high winds gusting at up to 90mph. However, more unusually, a different phenomenon caused more problems. The storm resulted in salt water being deposited on much of the open wire routes near the sea and the Telephone Department's line crews were faced with the task of washing the salt deposits off the porcelain insulators with scrubbing brushes and fresh water in order to restore the circuit insulation. Hundreds of subscriber and trunk lines were affected and the task of restoration took some time.

The Telephone Department decided to introduce a full annual telephone directory instead of frequently issued updates. The printing of this was assigned local company Bigwoods. In October the publication of the new telephone directory upset the local business community because a decision to sort names alphabetically disregarding the prefixes 'Le' or 'De' was a deviation from former practice. The Department was accused of being 'out of touch with business'. (The surname practice in Jersey is to capitalize the French prefix).

Another storm rounded off the year with 1,017 lines reported down. Luxon claimed that the Department had the situation under control, but faced criticism in the States that it was not as efficient as under Stanhope. This did nevertheless illustrate the fragility of the overhead distribution network, and the extent of the damage caused by the German occupation army's indiscriminate use of the overhead routes for its own cables.

Further woes continued into 1947 as first heavy snow in January brought down many lines then a lightning storm in early summer affected over 500 lines. In the spring a new new Secretary was appointed to replace retiring P G Cabot who had held the post since 1940. George J Le Cornu from the States Treasury was appointed in March. In May the Telephone Committee commissioned the engineering department of the GPO to conduct a survey of the island's telephone system. The GPO was to report back to the committee and make recommendations for the improvement of telephone services in the island. This more comprehensive survey followed on from an earlier report undertaken by Luxon, who was aware of the fragile state of the network following the occupation.

Meanwhile, the debate over the format of the telephone directory continued. In September the Chamber of Commerce canvassed its members and sent out 636 questionnaires from which it received 350 replies but only five of the responses were in favour of new system. Nevertheless the directory continued in the new format.

By now the Department was managing to get to grips with the increasing demand for telephones, although there was an extensive waiting list. In September figures published showed that there were now 7,672 telephone instruments[87] in operation compared to 5,960 in 1939. These were connected to total of 5,936 subscriber lines, there having been just 4,826 in 1939. Altogether these lines generated an average of 124,322 local and 5,936 trunk calls per week against averages of 96,426 and 2,767 respectively per week before the war. This represented only about a 5% increase in the average calls per line for local calls but a more than double the number of trunk calls. This may have been an indicator of the increasing prosperity of the island through both commerce and tourism, even though at this time the number of available trunk circuits was still very limited.

In November 1947 the first post-war Lieutenant Governor Lt-Gen. Sir Arthur Edward Grasett, visited Central telephone exchange. At this time the design capacity of the original exchange had been greatly exceeded and more remedial work would soon become necessary in order to enable further subscriber connections. The Governor was particularly impressed with the exchange power equipment, according to a report in the JEP, taking great interest in the ringing machines.

---

87 Telephone authorities of the era had a preference to express the number of instruments in operation rather than the number of lines as the former was a larger number. This statistic was easily realizable when the telephone authorities had an absolute monopoly on the supply of telephones and other premises equipment.

At the same time the States approved a Telephone Committee request for a grant of £63,000 for urgent remedial work. Questioned on the poor state of the telephone system the committee president, Constable SG Crill of St Clement, said it was likely that an estimate of £300,000 to £400,000 would be required to overhaul and update the system and that the committee would be reporting back to the States later.

At the first States' sitting in 1948 the Telephone Committee president reported on the current states of the network in more detail, drawing on both the Luxon and the GPO reports. Luxon had noted that the unprecedented demand following the war had put enormous strain on the existing network and exchanges. He noted that the only solution to increasing capacity on Central exchange, which was the fasted growing, would be to resort to unorthodox solutions which would reduce efficiency and service. Further consideration of options included the building of an automatic exchange at an estimated cost of £120,000, but the committee decided to defer such a scheme until there was more certainty of the Postmaster General's intentions with regard to the renewal of the current telephone licence which was due to expire in 1953. The Post Office Engineering Department conducted a survey of the island's telephone network during the summer of 1947 and provided a '... comprehensive report consisting of 12 foolscap pages and 8 appendices...' which considered more ambitious options than that of Luxon.

The first recommendations were already being acted upon. The committee had already placed an order for equipment to extend Central exchange by 600 lines and a search was on for a site for a new Western exchange to replace the ageing magneto switchboards at St Ouen and St Peter which had already been extended well beyond their original design capacity. In addition, an extensive new cabling programme was beginning and this was being carried out as the materials arrived. Such materials were in short supply following the war as the demand for telephone network infrastructure was quite general. The GPO was also to provide an upgraded trunk switchboard at Central exchange which was due for completion before the tourist season began.

However, the committee president stressed that the measures put in place for Central exchange could only be viewed as temporary as ultimately conversion to automatic would be necessary. The GPO report which suggested a single island exchange estimated this to be in the order of £500,000 which would include the necessary upgrade of the underground cable and overhead line plant to enable automatic telephones to operate. In addition, the overall increased demand for new equipment following the war meant that the proposal would probably need to be delayed to a later date.

The development of the network statistics as of the end of 1947 were also announced to the States:

	1939	1946	1947
Subscribers	4,626	5,375	6,052
Local calls	4,049,542	4,160,505	5,063,885
Trunk calls	63,391	157,441	230,418
Operators	50	68	75

Table 27

The number of subscribers was less than in Guernsey at the same time (6,818) although the average calling rate per subscriber was higher (836 against 623).

The proposed extension to Central exchange would provide a separate island suite of switchboards which would require tie circuits to be used to enable calls to be passed between the two separate switchboards as there was no capacity on the original board for the replication of the subscriber line multiple from the new extension. This would inevitably slow down connection times as in some cases two operators would be required to complete a call. At the same time it was decided to introduce subscriber meters onto the existing exchange to improve the efficiency of the now overburdened call ticketing system. Quotes were received from Ericsson and Siemens which would also incorporate an upgrade to the exchange to

Image by the author

A recent picture of the West exchange building which has remained basically the same since construction apart from a few cosmetic changes. There is a separate building to the rear which housed the later automatic switchroom.

bring it up to CB1 standards with automatic 'double beat' ringing. The contract was awarded to Ericsson.

After a brief consideration of sharing the new Harbours and Airport radio communications site at La Chasse, in May the States agreed to the purchase of a plot of land at St Ouen for the new Western telephone exchange for the sum of £480 together with a £50 sum paid to the tennant to break the lease. The site, which still houses the modern day equipment, on Grande Route de St Pierre (directly opposite the Farmer's Inn) was purchased from Mrs Bowditch the owner of the existing adjacent St Ouen's exchange at 'Sans Ennui'. Building work by the contractor commenced on 9 July when at a special ceremony the foundation stone was laid by the Telephone Committee president S G Crill (Constable of St Clement) with other committee members Jurat G P Billot, H le F Grant (Constable of St Helier) and Deputies W Kruschefski and Venables, in the presence of F Le Boutillier (Constable of St Ouen). Also in attendance were the senior managers of the Department; Luxon, Syvret and Le Cornu. The architects were Queree and Swain and the appointed builder Messrs Le Selleur Ltd. The building design was thoroughly utility, being of square construction with a flat roof was competed at a cost of £7,569-12-0d.

While the building work progressed, the Department set about updating its cable distribution network for the new exchange. A special American-made John Hodge and Co Ltd cable plough was hired from Messrs Clough, Smith and Co of Redhill, Surrey and used to install cable along roads and soft verges throughout the proposed exchange area with the addition of some preparatory work for the proposed new Northern exchange. In addition, the department also laid cables for the airport telecommunications department which was in the process of setting up new wireless communications sites at La Chasse, St Ouen and Les Platons, Trinity. The contractors were able to lay up to a mile of cable per week in 3ft to 3ft 6in deep trenches. The lead sheathed paper insulated cable was laid directly into the ground without protection.

The equipment for the new exchange was ordered from Messrs Ericsson of Beeston, Nottingham. It was not a standard CB10 exchange of the type favoured by the GPO but a proprietary switchboard which had originally been destined for Abassynia (Ethoipia) and was therefore released from the factory in a tropicalized version. It is likely that the Department purchased this exchange at an advantageous price since it was part of an order cancellation. It is also likely that it took advantage of this offer as telephone equipment manufacturers were fully stretched by the post war demand and long waiting lists were common. It would also appear that the equipment was acquired before the commencement of the building work as the exchange accommodation was tailored to the equipment dimensions.

The equipment for the new exchange was unusual compared to the standard CB10 equipment used elsewhere. The equipment room was based on an 8ft 6in standard height for both the line relay equipment racks and the combined MDF/IDF cross connection frame; whereas CB10 used the PO standard 10ft 6½in rack profile. The exchange multiple was also different, having a repeat of 80 rather than the standard 100. This was reflected on the MDF and IDF layouts where the exchange side lightning arrestors were mounted in verticals of 80 instead of 100. The switch used a standard 40 volt central battery which was provided with only a single battery with a mains rectifier that could be configured for both charging and floating operations. The exchange main ringing supply was provided by a static inverter which produced a $16_{2/3}$ Hz 70V ringing source using the public mains as an exciter current. Under mains failure conditions a small 40V battery driven ringing machine took over. This was also non-standard compared to other exchanges which normally provided ringing using a mains-powered dynamotor with a battery powered machine for use under power failure conditions. Interestingly, all the exchange circuit diagram blueprints, which were printed on cloth, were provided in the attached contact format favoured by continental manufacturers rather than the detached contact standard used in the UK.

The switch room was provided on the first floor with a single suite consisting of a cable turning section positioned above the MDF which contained the line test meter, two 'B' junction positions and eight 'A'

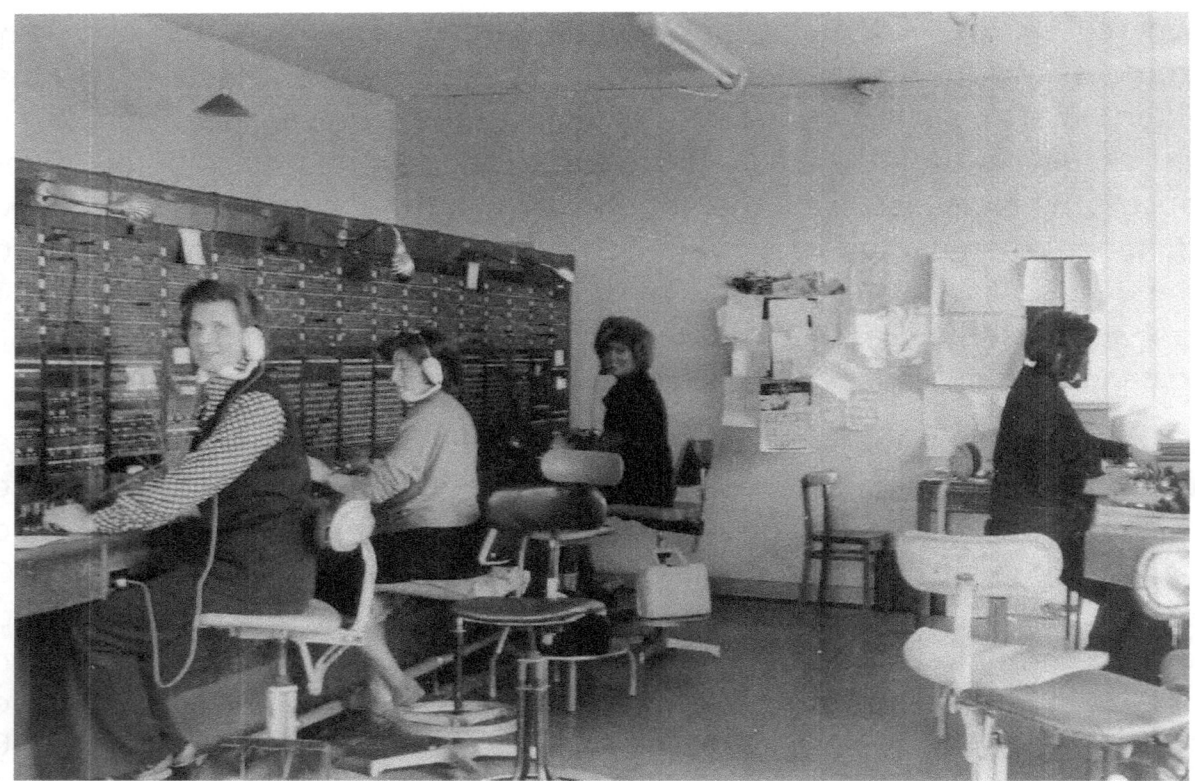
Image by the author

The photo above shows a general view of the switchroom at West CB exchange taken in the late 1960s.

positions. The original design size of the exchange was 800 with expansion to 1,200 planned. The other accommodation consisted of an operator's rest room and toilets on the first floor with a lineman room, toilet and two store rooms on the ground floor. At the rear of the building a caretaker's flat was provided. The entrance was fitted with double hardwood glazed doors for equipment access. The downstairs floor surface was painted rubber compound while the upper floor was lino tiled. This building still exists today although now contains no switching equipment.

The exchange was officially opened on 5 August 1949. Initially subscriber lines were still equipped with the old magneto telephones which had to be modified to work correctly on the new system. Since this work could not be carried out to any great extent before change over, it meant a busy day for Department engineers on opening. The old telephones would not work efficiently on the central battery system and therefore a programme of replacement was undertaken which took some time as supplies of modern telephones were still in short supply.

A site was needed for the new Northern exchange and in March 1949 a location conveneintly near the existing Sion exchange was offered by J E Renouf to the Committee at a sum of £600. However, this was turned down by the Beauties Naturelles Committee as it was considered prime farm land. In October the States approved the purchase, also in the sum of £600, of a parcel of land at a property known as 'L'Abri', Chemin de Hérupé, St John from a Mr Bichard for the site of the new Northern exchange that would eventually replace the old exchanges at St John, Sion, St Lawrence and Trinity. This site was on land that had recently been rezoned from farming to residential use. The new exchange building was radically different from that used to house the recently opened Western exchange and had been designed by the States' Public Works Department. The contract for construction was won by Messrs Farley and Company Limited. The States had acceded to a request of £35,000 made by the Telephone Committee for the construction of the new exchange.

Meanwhile, the debate over the future of Central exchange continued in the States. A new site for a building to house the new exchange was proposed on either the fish market or the old market, both sites being adjacent to the existing Telephone Department buildings. It was recognized that a larger exchange would be necessary and the debate extended to the question of whether this new exchange should be automatic. The Telephone Committee president, Wilfred Kruchefski, pointed out to the assembly that such an exchange was likely to cost in excess of £300,000, this figure having been ratified by GPO experts in their recent report. Kruchefski was also concerned that the GPO licence was close to renewal and was, therefore, reluctant to proceed with such a venture until this matter had been settled. The States approved a total expenditure of £109,000 for the Telephone committee at its November sitting which included the annual £8,000 GPO royalty payment and £45,000 for network construction, which did not include the building of new exchanges.

At the March 1950 States sitting, further requests were made for the purchase of a site for Eastern exchange. A suitable location had been identified in a disused quarry on Grouville Hill (Rue à Don). The Telephone Committee had offered the Parish of Grouville £100 for its purchase; however, at a parish hall meeting, the fathers of the parish rejected the offer as '...derisory for a site of about half a vergee'. The Telephone Committee made a further request at the July sitting of the States for £30,000 for a new Southern exchange on a site identified at Red Houses. Deputy Morrison noted that sites for Eastern and Northern had been agreed, while the exchange for the western parishes was already working. St Aubin's exchange was now almost at full capacity with about 1000 lines working. The request was lodged *Au Greffe* until 13 September when it was debated in the house. At this sitting the request for the new Southern exchange was approved. It was reported that the Northern exchange building was now well under way.

Lower call charges to the smaller islands were introduced at the end of October, reducing a 3 minute call to Sark or Alderney from 9d to 6d. The Telephone Department introduced a special temporary exchange for the trade's exhibition at Springfield in order to relieve the main exchange operators during this busy period. The revised budget plans were presented

Image by the author

A recent picture of South exchange at Red Houses. This style of building was also adopted for North and East exchanges. The building remains much the same as when it was build save a few cosmetic exchanges. The extension to the rear was constructed for the automatic exchange in the 1960s. The 'chimney' in fact houses mobile telephone antennae.

to the States at the end of November. A sum of £101,500 was requested of which £37,000 was for a Central relief exchange and £35,000 for Eastern exchange. However, these last two items were withdrawn before the debate on the request of the Finance Committee.

The Northern exchange building was completed during January 1951 and installation of the CB10 switchboard was under way. The exchange equipment was supplied by the General Electric Company Limited (GEC) of Coventry and cost about £10,000 including labour. Meanwhile, the Telephone Department engineers were busy re-routeing underground cables from the old exchanges ready for connection to the new switchboard. This was quite a complex task as the new exchange would replace four existing exchanges although the new building was about equidistant from them all. Earlier civil works had partly prepared for the task, nevertheless considerable work would have been necessary. The exchange was opened at 1.00PM on 3 April with the official cut-over being performed by T G Le Marinel, Constable of St John. Also in attendance at the low-key ceremony were the president of the Telephone Committee, the Constable of St Lawrence, Senator C H B Avarne of the Public Works Committee, G C Law the States' engineer, and representatives from Messrs Farleys and GEC.

The annual Muratti match was once again broadcast by the JEP to thronging crowds in Bath Street. The commentary on the 1951 game was undertaken by the famous sports broadcaster John Arlott. The highlights of the match were also rebroadcast by the BBC on its west of England Home Service between 4.30PM and 5.00PM the same day. The JEP broadcast facilities were once again supported by the Telephone Department and this year, for the first time, the public address system was provided by Rediffusion Limited, which had just entered the radio relay market in the early part of the year. The JEP continued its broadcast service until 1960 when the service was effectively replaced by improved coverage on BBC television and radio.

The potato exporting and the holiday seasons brought about extra strains on the Departments switchboards, particularly the trunk boards located at Central exchange. The operators handled enormous amounts of traffic which was largely conducted through a trunk booking system, whereby a call had to be ordered in advance if required during busy periods. The Department reported 11, 687 calls made in June, with a record of 1,018 calls during a single day. The trunk traffic demand was rapidly outstripping the submarine cable capacity and thus representations were made to the GPO to improve the situation.

At the same time, Central exchange was also nearing its maximum capacity of subscribers. The original exchange had been designed for a maximum capacity of 3,000 lines, but had been extended through the use of non-standard means to accommodate 4000 lines. This put extra strain on the operators who had individually to deal with more subscribers that the switchboard design capacity. The Telephone Committee therefore made a request from the States for an advance of £24,000 to lease and convert Lyric Hall in Cattle Street (opposite the Department's offices) for the installation of a relief switchboard. The committee was once again quizzed in the States on why there were no plans for an automatic exchange but again pointed to the uncertainty of the GPO licence renewal. However, Senator Kruchefski, deputy president of the Telephone Committee, assured the house that this issue was still being studied by the Department in conjunction with GPO experts. However, he commented that while a replacement for Central could be envisaged, it would be many years before an automatic exchange could be built. Deputy Venables voiced his opinion on the 'interference' by the GPO in island affairs, but accepted that the current licensing arrangements were unavoidable. Finally, the States were asked to grant £105,000 to the Telephone Committee for the necessary development work of which £24,000 was for the switchboard equipment while £9,500 was by way of the annual Post Office royalty.

In the spring of 1952 Southern exchange was opened. This was another low key affair like that of the Northern. Only a select few dignitaries were present as the exchange was cut over. The CB10 equipment was once again provided by Messrs GEC of Coventry and the exchange building was a mirror of but otherwise substantially the same as that of Northern exchange designed by the States architects. The new switchboard connected all the

Image by permission of Société Jersiaise

The photo above shows the States Telephone Department administration offices as they appeared in 1952 shortly after the addition of the Jersey Crest above the door. At this time Central CB10 exchange was housed in the granite building to the left and the garage door to the right in Cattle Street was used by the Department's mechanic for vehicle repairs. The Fish Market is seen to the right of the picture.

subscribers from St Aubin's exchange and a few from Western in cases where line length had been a problem for adequate working. Shortly after the opening, the premature death on 27 July 1952 of P K Luxon the Engineering Manager was announced at the age of 54. Luxon had only been in the post since June 1945 and had been ill for some time, although he attempted to conceal this fact from his colleagues. He was a former GPO engineer and had served in the Royal Engineers during the First World War, returning in 1919 as an Inspector. He transferred to the Telephone Department on takeover at the same rank, which he held until taking over as acting manager in 1940. His funeral attracted many mourners as he had been a popular and active member of society, assisting with many charitable works. S G Syvret took over the reins as acting manager at a salary of £800pa.

In the autumn the works for the provision of the relief exchange in Lyric Hall were now well under way. Lyric Hall was a practical choice for the exchange since it had been used by the German occupying forces and thus probably had existing cable access and other features allowing its adaptation as a telephone exchange. The 3000 line CB10 switchboard and power equipment, which was again supplied by Messrs GEC Limited, was to be installed by Departmental engineers. The old exchange was now completely full, in fact more than 1,000 more subscribers than the design capacity, a number which had been reached during 1952, and the waiting list for new connections was growing. There were a total of 6,414 lines in operation island-wide, of which nearly 4,000 were connected to Central, the remainder being spread among the six country exchanges. Over 6.5 million local calls were being handled annually. Syvret said that an automatic exchange had been considered but rejected as no suitable site could be found within a reasonable distance of Central. This, he claimed, was necessary because of the number of cables involved. This would seem to be a weak argument, given the recent underground plant engineering effort put into the establishment of the new country exchanges.

In the meantime the GPO had undertaken survey work to improve the trunk service to the UK. In October *CS Alert* was on station adding submarine repeaters to the existing cables between Dartmouth and Guernsey. In order to accommodate this extra capacity a new trunk switchboard was required. The present board in Central exchange could not handle the additional circuits and there was insufficient room for expansion. Negotiations between the GPO and the Department had started the previous October and a number of accommodation options had been considered including Lyric Hall and additional building work at Northern exchange, but finally it was agreed to provide a second floor extension above the Department's office building on the corner of Minden Place and Cattle Street for the sum of £8,826. The Department proposed to carry out the equipment installation work on this project too. Around the same time the PMG renewed the telephone licence for a further 20 years.

It was calculated that the floor area necessary for the trunk exchange equipment and manual board would require the whole of the available space on the extension, therefore it was decided to provide access to the switchroom via an external metal fire-escape rather than the use of an internal staircase. Once the contract between the States and the Post Office Telephone Department was signed, all haste was made to complete the building works. The new cable upgrade would increase the number of trunk circuits between Jersey and Guernsey and onward to Dartmouth from 24 to 120, although not all would be available for telephone traffic as some were required as bearer circuits for Post Office and private telegraph lines and the capacity would be shared with Guernsey. This additional work extended the budget proposals before the States to £117,300 to complete the work in hand.

The relief exchange in Lyric Hall was completed during July 1953. Arrangements were made to move some subscribers from the existing switchboard in order to spread the workload for the operating staff. It was decided to relocate subscribers with numbers between 2000 and 4100. The cut over of the subscribers was carried out at 12.15AM on Sunday 26 July. This would also reduce the workload on the existing switchboard and allow controlled expansion across the two locations. While the load was now spread between two boards, the operating procedure was more complex as it was now necessary to employ transfer circuit working between the two boards to complete calls on the same exchange. Initially 40 circuits had been installed, but it quickly became apparent that that was insufficient, so a further 60 circuits were added shortly after opening. The workload for same exchange call completion across these circuits would have vastly eclipsed the junction traffic from the other island exchanges.

The new exchange had been installed using the labour of 5 States Telephone Department engineers and one apprentice. The complete installation of 18 positions, which were arranged in a 'U' shape around the hall, had taken five and a half months. It was reported that over 4.5 miles of cable had been used and over 400,000 soldered terminations made. The current workload of the whole system was high, over 7 million calls having been made during the previous year and all indications that this would exceed 10 million by the end of 1953. Two new enquiries positions positions were installed in Lyric Hall to relieve the trunk positions in the old exchange. The new relief exchange would allow the waiting list, which had stood at 495 before its opening, to be substantially reduced, however, despite the fact that 400 new lines were connected to the new switchboard, the waiting list was still at 109 and demand for new lines was growing. It was estimated by the Engineer Manager, C J Syvret, that the new exchange would only last about 7 years after which it would be necessary to introduce automatic working as he estimated that the number of lines on Central would by then be approaching 10,000. He also requested that the States allocate the recently vacated old fire station at the top of James Street (now called Rue Funchal - a road off Minden Place), to the Department so as a new exchange could be built on the vacated site. This, he avowed, would be the most economic site available as it would substantially reduce engineering work and cabling requirements.

Traffic was still growing rapidly, local calls increased by almost 2 million during the year and trunk calls by almost 15% during the same period. During 1953 1,100 new lines were connected and at the year end the waiting list for new lines had again grown to 300.

In April 1954, the States approved the construction of a new Eastern exchange at Hambye, St Saviour. This site had been selected after an exhaustive search and failed negotiations over the site on Rue à Don, Grouville. The site at Hambye was far from ideal as it was isolated from the main populated areas that it would serve, save that it was about equidistant from both Five Oaks and Gorey exchanges which it would replace. The building contract was awarded to Jersey Construction Ltd at a cost of £9,938 for a building similar to the Northern exchange building but slightly larger to accommodate a larger switchboard.

This was followed by the official opening of the new trunk exchange on 16 June conducted by the Lt Governor Randolph Stewart Gresham Nicholson, who made an inaugural call to the PMG, Earl De La Warr. The new exchange which was again supplied by Messrs GEC[88] was of the latest GPO specified Sleeve Control[89] design. The switchboard consisted of a number of standard height operating positions and a separate low level enquiries suite, for the provision of directory information for all international, UK and local numbers. The signalling apparatus was of the GPO standard AC1[90] type junction equipment which provided 2 to 4-wire speech and signalling facilities to enable semi-automatic trunk working inland to the UK as well as services to the GPO international exchanges. The direct circuits to France at Rennes were also incorporated onto the switchboard as well as the local trunk circuits to Guernsey. These changes released space on the existing Central exchange trunk positions which could now concentrate on local junction working only. Although the switchboard and trunk signalling equipment was owned by the GPO, the staffing and maintenance was provided by the States Telephone Department under a contractual agreement.

In November 1954 the Telephone Committee requested a loan from the States of £350,000 for the construction of an automatic exchange which by

---

88 The contract for the trunk exchange had been awarded by the GPO. Under the Bulk Supply Agreement (BSA) between the GPO and the UK manufacturers, Jersey was considered GEC territory. The BSA was effectively a price fixing cartel. It was finally outlawed in 1974 under the newly introduced Restrictive Trade Practices Act.
89 Telephony : A Detailed Exposition of the Telephone Systems of the British Post Office : Volume II : Automatic Exchange Systems – Atkinson J
90 POEEJ Vol 44 Part 2 1951 Signalling System A.C. No.1 (2 V.F.)

now was becoming increasingly necessary. However, the request was withdrawn pending a further review of costs which would be provided to the States at a later stage. At the end of December the operation of the inter-island trunk service was handed over to the local telephone committees by the GPO, thus removing the GPO's premium from the cost of calls. This prompted the Guernsey States to offer free inter-island trunk calls over the New Year period. The Jersey Telephone Committee did not follow this example.

In early 1955 the construction of the new Eastern exchange building was now complete and engineers from GEC began the installation of the new £49,547 switchboard contract. Under questioning in the States the Telephone Committee revealed that contemporary cost of installing a single subscriber line was in the region of £15 to £20 on average. The cost was somewhat less that that charged to the subscriber for the installation but, the Committee asserted, the difference was quickly recovered from rental and call charges. In May the States approved the request for the upgrade of Central exchange to automatic working, and in June approved the £350,000 loan pending the provision of further cost details.

On 27 October 1955 Eastern exchange was officially opened at a ceremony attended by the President of the Telephone Committee Senator G Crill, the Engineer Manager Mr C J Syvret, the Telephone Department secretary Mr J A Norman and a number of other dignitaries including Sir Alexander Coutanche, Mr Riley and Mr Suggars representing GEC, together with their wives, Mr Walton of Jersey Construction, Mr G C Law the States engineer and Mr G Le Cornu the States Treasurer. At opening the exchange had 1,500 subscribers connected and had a design capacity of 3,000. During the construction over 12.5 miles of cable and over 10,000 miles of copper wire were used by the Departments own engineers under the direction of the Department inspector R Perrin. The first exchange supervisor was Miss Foster. Eastern exchange was again of the CB10 design supplied previously by GEC for both Northern and Southern exchanges. It was the largest of the country exchanges and installed in a spacious building finished to a high standard with parquet wood floors on both levels.

Budget details revealed at the December States sitting showed that the Telephone Department was making a good profit on its operation, receipts during the previous year being £170,000 against expenditure of £154,000. Despite this profit, it was announced in January 1956 that new line installation charges would be raised by £1. Network cabling and overhead construction improvement estimates for 1956 were listed as Southern: £6,375; Eastern: £4,500 and Central: £13,125. The newly renewed GPO licence royalty remained at £1,200. Management and office staff wages were stated as £9,250, of which the Engineer Manager was awarded £1,075pa; Operators and Engineers were allocated £85,000 and Pensions £2,000.

The number of subscribers on the system now represented a penetration of 20% of the population. Of the 300 distribution points on Central exchange over 90% were now full and some cables had been in service for over 45 years (that is to say, some cables that had originally been installed by the National Telephone Company were still connecting subscribers and exchanges). The Committee revealed that about 150 new distribution points would be required for Central exchange for both additional connections and to replace existing cables to the required insulation standard necessary for automatic working. It was also announced that part of the Old Market, adjacent to the Minden Place exchange would be transferred to the Telephone Committee as a site for the new automatic exchange building.

Four tenders were received from suppliers for the new automatic exchange, the prices were reported to the States by Senator Venables, the Telephone Committee president as follows: Siemens Brothers: £286,993, Ericssons of Beeston: £287,700 Standard Telephones and Cables: £288,500 while GEC submitted the lowest tender of £284,494. However, as can be seen, the prices were very similar and this prompted a suggestion from Senator Le Marquand that the matter should be lodged *au Greffe*. However, Venables explained that this was no great surprise as the manufacturers worked from a pricing standard set by the GPO, (at that time the UK telecommunications manufacturers operated a GPO sponsored cartel (the BSA), which maintained a fixed price for the most common telephone exchange parts[91]) and that the tenders had been checked by the GPO at the Committee's request and found to be in order. It is also no surprise that GEC provided the lowest tender as Jersey was considered to be a 'GEC territory' under the same rules. The GEC tender was accepted.

Automatic exchanges were already in operation on the network but these were confined to private branch exchanges (PABX), for the internal use of large organizations. Three 200 line Pre-2000 type Strowger[92] PABXs had been installed at the States Offices in the Royal Square, the General Hospital and at the Hotel de France. These exchanges required special accommodation as they were constructed using standard (10' 6½") exchange size racking. These exchanges were of the semi-automatic type which required an operator to complete outgoing calls to the public exchange, a corded switchboard being provided to manage incoming and outgoing calls. Smaller PABXs were usually able to be installed in standard office accommodation. Many other private exchanges were in operation but these were various types of manually controlled switchboards (PMBX). The Pomme D'Or Hotel switchboard which was installed in the early 1950s is now on exhibition at the Jersey War Tunnels.

With continued strong growth in trunk traffic the GPO set out plans for a new submarine cable that would be landed at Le Dicq. The Assistant PMG, K Thompson, in a Commons statement confirmed that various new European cables would be laid following completion of the new transatlantic cable, TAT1[93]; this would include the Channel Islands.

In October the Department announced plans to fit over 8,000 dials to existing telephones on Central and Millbrook exchanges. This project was scheduled to begin the following Easter when the necessary stores had been procured. At the time of the announcement over 100 telephones had already been converted as upgrading work or line transfer became

---

91 Telecommunications in Europe - Eli M Noam. New York : Oxford University Press, 1992. ISBN-13: 978-0195070521
92 For details of the Strowger system see: Telephony : A Detailed Exposition of the Telephone Systems of the British Post Office : Volume II; Automatic Exchange Systems – Atkinson J
93 Mattingley, F (1957). "Manufacture of Submarine Cable at Ocean Works, Erith". Post Office Electrical Engineers' Journal 49 (4), p. 308, January 1957

necessary. The plan was to test subscriber premises wiring concurrently and make any upgrades if necessary. The quality of insulation required for automatic working was much greater than that necessary for manual exchanges since a higher exchange voltage, 50V compared to 40V, was employed and distortion-less dialled information over the connecting wires was necessary.

Over the following few months the resources of the Department were greatly tested. In November gales brought down over 1,000 lines and some poles. In December, road works at West Park damaged a trunk cable and disconnected Millbrook, Southern and Western junction circuits. This necessitated re-routing traffic via Northern exchange which greatly restricted the number of local junction calls that could be handled.

This was quickly followed by another great storm in late winter 1958 which probably caused the greatest damage to the Jersey telephone network in all its history. On the night of the 7 and 8 March, gales brought down more than 60 poles and over 4,700 lines, this being over 40% of all the island lines. Such was the quantity of faulty lines that the effect on Central exchange was drastic. The number of lines in the 'permanent glow' state, that is, calling the exchange and thus making the calling lamp on the switchboard to glow, brought about the rupture of the main battery fuse, thereby causing total exchange failure. The main Central switchboard was out of service for over an hour while engineering staff disconnected the faulty lines at the exchange Main Distribution Frame. Meanwhile the same problem also affected the Lyric Hall relief exchange which was out of service for over two hours, finally being returned to service at 10.30AM.

The management authorized unlimited overtime and the Department staff rallied to the call to return service to subscribers as soon as possible. A priority list was drawn up placing key subscribers such as doctors, honorary police and other essential workers at the top. The situation was further exacerbated on 10 March when a major 800 pair underground distribution cable failed at the corner of La Motte Street and St James Street, affecting subscribers in the Georgetown and Samares areas. However, by 13 March 450 subscribers on the cable and 1,800 subscribers on overhead distribution had been returned to service. The Telephone Committee president revealed to the States on 19 March that it was costing around £200 to £300 per week in overtime costs, but the service was 'almost back to normal' with 3,400 subscribers reconnected. On 25 March the Telephone Committee Vice-president presented a final report to the States detailing the extent of the damage. Wire was procured from the GPO South West Region, GEC Limited of Coventry, Henley Cables, the States of Guernsey Telephone Department, The Jersey Electricity Company Limited and Rediffusion Limited. Four tons of overhead wire, which equated to 224 miles, had been used at a cost of £4000. The statement went on to say that it would take months to restore the network properly as many temporary repairs had had to be made. Cables were laid around fields which would be replaced with overhead wires when the potato crops were dug. He recorded a tribute to the Telephone Department's staff and management, particularly mentioning the Engineer Manager C J Syvret, the Assistant Engineering Manager, H W Coppock, Mr H St George, Mr R Perrin and Mr J le Vouguer. Compensation claims from business and residential subscribers affected by the damage would covered by the States insurers.

Meanwhile, engineering work for the new automatic exchange continued. The building work, contracted to Charles Le Quesne Limited, on the new two storey exchange building in Minden Place was progressing well and it was scheduled for completion by the end of March. New 6-way earthenware cables ducts with 11 joint boxes were being laid west from Central exchange, Minden Place via Halkett Place, Hill Street, across the weighbridge to the Esplanade and thence along Victoria Avenue to Millbrook and Beaumont. To the east via Belmont Road, St Saviour's Road, Don Road and St Clement's road to the Dicq. This latter cable duct would also serve the new submarine cable, due to be laid at the end of June, which would be continued along St Saviour's Road, Springfield Road and into the GPO Repeater Station at Trinity Gardens. The cables to Beaumont would consist of two 1,400 pair, local trunk cables of lead with paper core insulation which would run to West Park whereupon it would be sub-divided to provide 2,000 connections to the Millbrook exchange area while a further 750 pairs would be drawn to Beaumont. In total, 16 miles of new cables would be deployed.

In May 1958 the equipment for the new exchange started to arrive from the GEC Coventry factory. Mr I O Hodge of the GPO Telephones South West Region, Bristol, was seconded for a period of up to two years as the Clerk of Works overseeing the installation of the new exchange. By the beginning of June, 17 of the first of the eventual 40 installation engineers from GEC were on site and equipment installation was well under way. It was estimated that it would take approximately 18 months for the installation to be completed. Installation of the step by step exchange of the GEC SE50 type purchased by the Department was not a complex task, but it was a long project. Each rack, which could weigh upward of half a tonne each, had to be erected together with all the overhead ironwork to support both the racks themselves and the interconnecting cable and power distribution. Each rack was wired individually, either directly to other racks or via an Intermediation Distribution Frame to increase equipment utility. Each connection had to be made by hand and soldered, hundreds of thousands of connections would be required by the time the exchange was complete. In addition, each switching element of the exchange would need to be individually installed, adjusted and tested, which required expert installation engineers.

Early in 1959 the GPO commissioned its new submarine cable and expanded the number of trunk circuits to the UK. This new cable quadrupled the potential number of trunk circuits.

The work on the new exchange absorbed much of the Department's resources over the next 18 months. Apart from the construction of the exchange itself, which was mostly undertaken by the contractors, the maintenance staff eventually destined to work the exchange pursued training courses provided both by the manufacturer and the GPO training centres. The external line plant and subscriber installations required a considerable amount of reconstruction in order to reach the standards necessary for automatic working. Telephone boxes in and around St Helier required updating to the pre-payment Button A and B[94] type. This work was

---

94 Telephony : A Deatailed Exposition of the Telephone Systems of the British Post Office : Volume I : General Principles and Manual Exchange Systems – Atkinson J

accompanied by a general refurbishment of all the kiosks at the same time. New notice frames were to be provided in each kiosk to instruct users on the method of operation and the correct codes to dial for the remaining manual exchanges and emergency, enquiry and operator services.

Notices were also provided to all subscribers on the use of the new system, including the information that the country exchanges would have their names truncated in order to avoid confusion, particularly when the exchange name was passed over poor quality connections to operators Thus Southern would become South, after removal of the suffix '-ern' and so on. The new codes to dial for individual exchanges would be:

North	4
South	5
East	6
West	7
Millbrook	8

Additional services were also allocated dialling codes, for trunk calls '0' was allocated; for the telegram service '90'; for faults, enquiries and the time '91'. The existing services for weather reports ('92') and mail boat arrival times ('93') were now allocated separate dialling codes. Emergency services would be accessed using the UK standard '999' code; this would enable subscribers' free calls to the Fire, Police, Ambulance and Life Boat services. In practice only '99' was needed as the last digit was redundant, pulsing directly on the manual board connecting relay set. The faultsman ring-back facility was initially allocated to '97'.

The exchange was cut into service on Sunday 1 November 1959. At 8.13AM the Engineer Manager Mr C J Syvret gave the order to start the cut-over and 2 minutes and 4 seconds later, it was reported, technicians confirmed that the process was complete. At 8.30 Deputy S J Venables

Image by the author

This picture shows a telephone bought in ready for the changeover of Central exchange to automatic working in 1959. The telephone is a GPO Type 706L in black fitted with a braid cord and a metal dial plate. The more familiar curly cord and plastic dial plate was introduced in the 1960s. Here the dial number ring shows the codes needed to access the remaining manual exchanges, North, South, East and West.

switched off the battery to the old manual exchange completing the process. The official opening took place on 6 November by the Lt Governor, General Sir George Erskine who made an inaugural call to HM Postmaster General, John R Bevins. The ceremony was also attended by the Bailiff, Sir Alexander Coutanche, Deputy Venables, President of the Telephones Committee, and S J Syvret, Engineer Manager. A plaque was unveiled in the entrance to the exchange followed by a reception at the Omaroo Hotel for various island dignitaries, the GEC Engineer Mr C Riley, the GPO Clerk of Works and a number of visitors including L G Semple Regional Director of the GPO South West Region, Mr P Lintell the Guernsey telephone Engineer Manager, members of the Guernsey States Telephone Committee and the Telephone Department senior management.

Initially the exchange was dimensioned for a multiple of 9,000 lines. On cut-over 5,300 subscribers were connected to the new switch. The exchange was of the latest GEC step by step type, employing the newly developed SE50 selector mechanism. It was the first public exchange to use this new equipment type which would be given the designation 4000 Type[95] when adopted by the GPO. The exchange was accommodated in the new building erected next to the old Central manual exchange on part of the Old Market site in Minden Place. The building had been designed by the States Engineer Mr G Law from a specification prepared by GEC. The building which was over 2 floors was constructed by building contractors Charles Le Quesne Limited. The First floor was 'L' shaped with an extension bridging the Old Market to the rear of the equipment room with a room partitioned off with a glass screen accommodating the test suite at the front. The ground floor housed the power room, the battery room, the subscriber meter room, toilets, kitchen and 4 offices, the largest at the front for the Engineer Manager. Windows overlooked the market and Minden Place on the ground floor and also the old exchange building on the first floor. The decoration was finished in cream with light oak parquet flooring throughout.

The equipment room contained the Main Distribution Frame, which was

---

95 POEEJ Vol 51 Part 3 October - December 1958 The 4000 Type Selector

Image by the author

A view of the Central Main Distribution Frame (right) and the Subscriber Intermediate Distribution Frame (left) taken in the early 1970s after fluorescent lighting had replaced the original incandescent lamps which were fitted to runways. An access ladder is seen clipped to the equipment floor rail.

finished in Battleship Grey, the Equipment Intermeditate Distribution Frame (EIDF) and Subscriber IDF (SIDF) were both finished in a cream finish (Light Straw) while the switching equipment was also finished in Battleship Grey. The ultimate design size of the exchange was about 16,000 lines, which meant that there was sufficient space left between the ranks of equipment for future expansion. The original numbering scheme was set at 5 digits utilizing the ranges 20000 to 24999 and 30000 to 33999.

The decision to continue with manual exchanges after the war was an opportunity missed. While it could be argued that the general shortage of equipment immediately following the cessation of hostilities may have influenced the decision for West exchange, the same cannot be said for the remaining country exchanges. The authorities in Guernsey were able to purchase two small automatic exchanges in the early 1950s. The arguments on cost were also erroneous, while the original GPO report did suggest a considerable investment, it reflected the costs of automating the whole island network into a single exchange which was a sensible if not altogether practical suggestion. The approach adopted in Guernsey of conversion one area at a time (which indeed, was also the approach used in Jersey after automation was rejected as too costly) was a far more pragmatic solution. The additional expense that may have been incurred initially by the introduction of automatic working would have been recouped in smaller exchange building and consequential lower costs as no switchboard and operator accommodation would have been needed and long-term savings on operator staff wages. The other argument regarding the uncertainty of the GPO licence was also overplayed. Guernsey faced the same dilemma, but it did not prevent the development of a sensible approach to the question of modernization.

Chapter 17 - The Automatic Era 1960 – 1972

After the cut-over of Central exchange, Syvret announced his retirement. As expected, recent ex-GPO recruit Harry W Coppock was appointed by the Committee as the new Engineer Manager which was confirmed by the States on 6 January 1960. However, Syvret remained as a consultant to the Committee until the cut-over of Millbrook exchange had been completed.

In order to reduce the workload of the available staff, Millbrook exchange would remain in service for about 4 months after the opening of the new Central automatic while engineering work was completed to make the transfer. The exchange was finally closed down on 31 March 1960 when just under 1,000 subscribers transferred onto the new automatic exchange. A small ceremony was held at the old exchange building as many of the operators were either made redundant or retired. The exchange caretaker and night operator, Mr and Mrs Alfred Le Marquand, were transferred to North exchange. The exchange was dismantled and the building eventually sold as a private dwelling to Department employee Alfred Le Mettais.

At this point the island seemed to be recovering from a mini-recession. Over the past few years the number of calls both on and off island had declined from a peak in 1956/1957 but after the opening of the exchange calling rates were again on the increase. This was further stimulated by an extension of the off-peak lower charge band introduced by the GPO in July 1960. Coppock announced that more trunk operators were required to service the increased demand. Also during the early part of the year the

Department procured three experimental recorded announcement machines[96] from the Post Office research establishment at Dollis Hill, London. These machines would provide an automatic repeated announcement service for the Mail boat Arrivals and Weather services. Two machines were put into service on the codes '92' and '93', replacing the services previously provided by the manual board operators. The third machine was kept as a spare.

The new machines quickly proved their worth, being used not only for the original services but also during the States elections later in the year provided a platform for the candidates election addresses. This latter use would seem quite at odds with a government owned and managed telephone system. Later in the year a service was offered for children to call Father Christmas, in this case normal call charges were suspended for the period. The service attracted 889 calls by Christmas Eve 1960. In the first year over 45,000 calls were made to the new automated services. After this initial burst of seasonal generosity and while the Father Christmas service continued for many years, the charges were not suspended thereafter.

During 1961 the GPO entered into negotiations with the States Telephone Committee with regard to the automation of the trunk service. The Post Office had started to introduce Subscriber Trunk Dialling (STD)[97] beginning with the first experimental exchange open by HM the Queen Elizabeth on 5 December 1958 at Bristol. The STD system was based on the existing switching technologies and required a number of modifications to existing exchanges as well as the provision of a special automated trunk switch on the island. The switch would be provided by the Post Office under their powers of telephone licensing monopoly in the Channel Islands, empowered by the Telephone (Channel Islands) Order, 1952. In order to facilitate the new services, switching equipment accommodation would be required next to Central exchange and changes to Central exchange itself would be necessary. It was announced in the States in December that it would likely be at least 4 years before the service would be opened. The Department also worked in conjunction with the Post Office on the building of new ducts

---

96 Post Office Electrical Engineering Journal Volume 52 p231
97 POEEJ Vol 50 Part 4 Jan – Mar 1958 The General Plan for Subscriber Trunk Dialling

between the site of the new Frémont Point television transmitter that was to be provided for the new ITV service to be operated by the successful bidder Channel Islands Communications (Television) Limited trading as Channel Television. The ducts would be provisioned from the transmitter to the GPO repeater station at Trinity Gardens passing by North exchange, thus a joint operation was desirable as the duct work could be shared by both. When questioned in the States the Telephone Committee chairman justified the £27,000 cost because of the net benefits to the service, customers had been complaining of poor service and in particular wireless interference on junction calls. (The latter was probably a combination of the poor cable quality and crosstalk from the adjacent Rediffusion cable service).

On the 28 February 1962 a heavy snow storm hit the island and over 4,000 telephone lines and more than 100 telegraph poles were brought down. This was a similar scale to the 1958 storm, although this time the faults seem to have been confined to mainly overhead plant. The Department was seemingly better prepared too, since all the repairs were carried out within a few weeks, using over 200 miles of open wire, and on this occasion temporary repairs seemed to have been largely avoided. The repair cost reported to the States was £10,000.

The Telephone Departments attention was now turned towards the development of the telephone network. It had become clear that the demand for telephone services had increased, the waiting list for connections had remained around 600 for the past few years and the capacity of some exchanges was approaching maximum, in particular South exchange, as a result of States' urban development policies. Its current multiple of 2,240 lines (the original design maximum was 2,000) would be exceeded by 1965 and thus conversion to automatic became urgent. Plans were therefore made to make the necessary network changes automatic working and planning the changeover of South exchange would require a building extension as well as new junction cables and improvements to the subscriber line plant. In addition, the capacity of the new Central automatic exchange was also being exhausted and soon an extension would be necessary to cater for demand. To exacerbate matters, the Post Office plans for conversion to STD would require yet more work by the Department.

The Budget was presented to the Committee late in 1962 and the outline budget places before the States. The total estimated costs over the period up to the conversion of South and the introduction of STD by 1965 were £383,000, this being the augmented total of the original loan agreed two years earlier. This was debated before the house in April of 1963. The president of the Telephone Committee, Senator Venables presented the figures. The total would be required over 3 years, already approved in principle at the supply day debate in November 1962. The expenditure would be broken down as follows:

Construction of building extension to South exchange	£20,000
Extension of 3,000 lines at Central exchange	£40,000
Junction cables Central-South-West	£28,000
Civil works for cable ducts, new overhead construction and upgrading existing line plant	£25,000

The president stated that this work was a matter of urgency if the number of subscribers on the waiting list was to be reduced. In addition, the committee recognized the urgent need to upgrade the network. At this time much of the subscriber network was of overhead open wire construction and as the network expanded this was increasingly susceptible to weather damage as had recently been experienced.

In the meantime, day to day work continued. A small extension to the trunk exchange was carried out in time for the visitor season and civil works were carried out in preparation for the construction of a large new housing development at Quennevais that was forecast to increase demand for lines on South exchange, which itself would now require a temporary 200 line extension prior to conversion to automatic.

The budget statement before the States in November 1963 highlighted the areas of development:

Wage rises and general maintenance	£16,000
Underground cable	£28,000
South exchange equipment	£79,000
South exchange labour	£13,000
South to Central junction cable	£12,000
Construct new trunk exchange Central	£48,000
STD equipment for South exchange	£3,500
STD equipment for Central exchange	£8,000

The Telephone budget in the States debate attracted little attention, however the Constable of St Ouen did question the wisdom of the STD conversion before the entire island's exchanges were automatic and Senator J Le Marquand asked if the building work was really needed. The committee president, Senator Venables, replied that without the building there was no space to install sufficient capacity for the growing trunk traffic on the existing manual board and thus they had little option but to install the STD equipment Moreover, the STD equipment would be necessary as this was required by the GPO and therefore some of the investment could be recouped directly from it. It was impossible to wait until all exchanges were automatic, things in the telephone world were changing rapidly. He reminded the house that it had rejected an island wide automatic system in 1946 because it was thought too expensive, but that the house had not foreseen the enormous growth in demand for the telephone. The sums were agreed without further debate.

During the early part of 1964 changes were afoot for the preparation of automated trunk dialling. The telephone workshop and internal equipment stores were relocated from the old Central manual exchange building to a site known as the Red Barn at St Aubin on which a lease had been obtained. This made the way clear for the demolition of the building ready for the construction of new accommodation for the STD exchange and additional office space. Around the same time the external equipment stores were relocated from Gullivers Yard on the Parade, where they had been since the days of the NTCo (the site now occupied in part by the States administration offices, Cyril Le Marquand House) to the German tunnels at St Aubin, within a short walking distance from the Red Barn along the former railway line. The heavier items, such as poles and manhole covers, were moved to the tunnel from Biles' Field, First Tower. The German tunnels were constructed off the tunnel originally used by the Jersey Western Railway and Tramway Company Limited. They extend deep under the cliff towards the west and provide a constant temperature for the storage of external telecommunications equipment. The complex consists of a main tunnel with a number of branch tunnels off to each side. The tunnel storage was shared by a number of other enterprises including, at one time, the storage of hire cars. The Departments garage and vehicle maintenance was moved from the Cattle Street side of the Departments offices to a store in Ingouville Lane where it remained for about one year before being moved to Springfield in July 1965.

The telephone workshop had been in existence from the beginnings of the Department and had been responsible for the refurbishment of telephone instruments for reuse on subscriber lines. Technologies had developed and over time the instruments had become more complex. The introduction of the A-B telephone box had meant more work for the workshop in keeping equipment in good order and in preparation of subscriber and public installations, by pre-mounting equipment in the appropriate layout on backboards.

With the introduction of the Post Office 700 series telephones in 1959[98], and its subsequent adoption by the Department in the early '60s, there was even more onus on the workshop in preparing the complex installations of subscriber equipment that the new telephone allowed, using the GPO 'N'

---

98 New 700 type telephone, Post Office Electrical Engineering Journal Vol 49 p69

series telephone plan diagrams. Consequently, the move to the larger premises at the Red Barn was a welcome relief for the workshop staff. The 700 series telephone came in a number of variants for various functions and it was also available in a number of colours, breaking away from the 4 available with the 300 series (black, red, green and ivory). The 700 series was available in black, but also a number of other more exotic colours including two-tone variants. The Department adopted two-tone grey as it standard. It also purchased a cheaper compatible alternative from Pye TMC Limited, the telephone number 1806, which was virtually identical to the 700 series apart from the type of dial and certain cosmetic features. At this time the standard telephone generally supplied to subscriber lines was usually a refurbished black Bakelite type. Other colours attracted a premium on rental charges. The new two-tone grey was now also available, but at first these tended to be reserved for connections on automatic exchanges.

The contract for the new trunk exchange building construction was awarded to Charles Le Quesne Limited. It would occupy the space between the Telephone Department offices and the Central exchange building. The building, the frontage of which was designed to match the Central exchange building was over two floors. The ground floor would initially be equipped with offices and the first floor would contain office space for senior management at the front with an equipment floor over the remainder of the building for the trunk exchange.

The frailty of the overhead distribution network was tested during 1964 as the following graph illustrates:

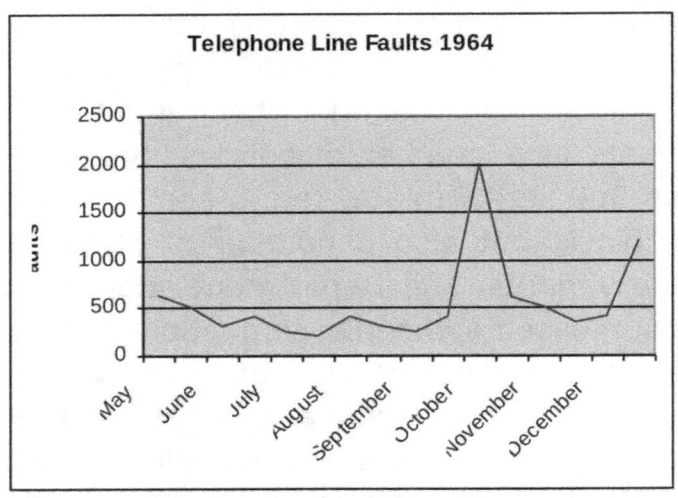

Figure 5

The mean monthly fault level for the year was around the 500 per month mark. This had been caused by the susceptibility of the overhead distribution network to vagaries of the weather. In May a lightning storm accounted for over 600 faults in one day. In October a gale and stormy weather caused nearly 2000 faults over a few days and again on New Year's Eve nearly 1,200 lines were brought down by snow accumulating on overhead wires. A spokesman for the Telephone Department told a JEP reporter that the level of faults was mainly due to the overhead distribution system, but that planned network upgrades over the next few years should mitigate the problem. The level of faults was well below the numbers experienced in 1958 or 1962 which is perhaps as a result of the gradual improvements in the network during that time. There was now a concerted drive to deploy more underground cable and to reduce the length and size of overhead distribution routes, although many long pole routes remained, especially in rural areas. Here too, the improvements in the financing of the Department meant that this crisis was able to be met out of planned financial resources, unlike in previous times.

During the spring of 1964 work began on an extension to the rear of South exchange at Red houses. It was a simple flat roofed construction measuring 65' by 34' (19.8m x 10.4m) with windows down the long sides

Image by permission of Charles Watts

This photo shows one of the large distribution poles that were prevalent around St Helier in the 1960s. There is a mix of insulator types visible ncluding some Bennett insulators on the third and fourth arms counting from the bottom, These were introduced just after the States' takeover of the telephone system.

and a double equipment door at the rear. The building contract was awarded to Messrs Regal Construction Limited. Three separate rooms were constructed within the basic outline to provide a new battery room and a separate room for the MDF and test desk equipment. The whole being finished in cream emulsion and grey vinyl tiles. The new accommodation was connected to the existing exchange by creating a door between the old equipment room and the new MDF room. Thus the new exchange was able to share the existing domestic facilities.

In October 1964 the 3,000 line extension to Central exchange was commissioned and immediately the number of the Jersey Police HQ benefited from a number change to a more memorable 25511 from 21906 in order to provide more access lines to the Private Branch Exchange. The newly available number ranges were 25XXX, 34XXX and 35XXX. At the same time changes in preparation for STD were also introduced. The GPO STD system was unable to deal with single digit access codes and thus it was necessary to add the digit '8' in front of of the existing minor exchange codes, thus North became '84', East '86' and West '87'. However, as South exchange would be converted to automatic before the need for these changes came into effect, its code remained '5'. At this time a further 200 subscribers in the Beaumont area were transferred from South exchange to Central to relieve numbering capacity. The normal process of placing numbers in quarantine until the publication of the next directory was waived because of the continuing demand for new lines.

The automated announcement services were also brought into the '8' number range, becoming '82' and '83' and were also augmented by the addition of a time announcement service (TIM) which was allocated the code '81'. The voice of the TIM service was provided by the winner of a local competition held in January 1964, Mrs Pauline Le Conte. The TIM equipment was provided by Messrs Ericsson of Beeston, Nottingham[99]. The time service received over 225,000 calls during its first full year of operation, proving its worth as a service and revenue earner – the total calls were worth approximately £2,800, equivalent to more than two senior technicians

---

99 Ericsson Bulletin No 45, July 1962.

wages at that time. Changes were also made to Operator access. To come into line with the current GPO standard, operator access was changed from '0' to '100' and Directory Information from '91' to '191. The code '190' was added for telegrams and '197' for fault reporting. A rudimentary Subscriber Automatic Line Tester (SALT) was also installed at around this time. It was manufactured by Ericsson of Beeston and installed at Central exchange. The SALT device allowed linemen (and others!) to automatically test lines while tracing faults. The tester was accessed using the code '14' and it would do a basic test of line insulation and earth faults on both the A and B wires of the line. After dialling, the normal calling party release feature was overridden and the connection held while the test proceeded. At the end of the test the caller would be rung and a series of tones would indicate the results. Since the device was thermionic valve based it required daily alignment to counteract the usual valve ageing process.

Equipment for the new automatic exchange at South began to arrive towards the end of the year from Messrs GEC of Coventry as a follow-on contract from the recent Central extension. The exchange equipment was of the latest version of the GEC SE50 type, this time the whole exchange being finished in Light Straw (cream). This version of the switching equipment followed the latest design standard currently used by the GPO which was termed the Rack IDF (Intermediate Distribution Frame) standard. This idea attempted to construct telephone exchanges in convenient units, theoretically each 1,000 lines could occupy a certain footprint within the exchange, ideally one rank of racks. In practice this proved to be too idealistic. The GEC engineer in charge of the project was Mr Royston C Vaughan.

Connection between South exchange and the other exchanges was greatly simplified by the relatively short junctions required on the island. Signalling for automatic connections could be carried out using simple loop-disconnect signalling. This was supplemented with additional features for various types of traffic to trunk operators, local manual exchanges and trunk dialling. The latter type facilitated the extension of metering facilities over the junction to enable future direct dialled trunk calls to be charged correctly. Around the same time the signalling between the local manual exchanges and the automatic exchanges was modified from battery dialling to loop disconnect. Battery dialling was a simplified system that could be employed from manual exchanges but required additional equipment at the automatic exchange. This method was also thought to have poorer transmission characteristics than loop disconnect relay sets and thus they were provided at North, East and West exchanges ready for the introduction of South automatic and to upgrade junctions to Central. The equipment was provided by GEC in pre-2000 type practice relay sets mounted on the manual exchange 2' 6" racks.

Meanwhile the upgrading of the telephone network continued. Particular attention was given to the South exchange area as the changeover to automatic working would require higher quality line plant. The development strategy chosen was to reduce the number of overhead distribution routes and to replace, where economically feasible, long routes with underground cable. The last drop to the subscriber was still to be above ground but the Department selected the GPO ring pole distribution method. Ring poles were designed to be used in urban environments where the distribution to the subscriber could be completed in a single span, although in some cases this may involve the addition of an intermediary span branched from a neighbouring house. The GPO had introduced the ring pole principle in 1932 but had used it more widely after the war where refurbishment and new development was required.

Open wire was still the order of the day, although at this time plastic insulated 'drop wire', first adopted by the GPO[100] around the mid 1950s, was also available and was widely used in other jurisdictions, particularly the USA for the final span to the home. It was, however, regarded by the Department as only a temporary solution for subscriber distribution. Open wire was still economic at this time as labour costs were comparatively low compared to distribution assets. Open wire was also less obtrusive in the environment but nevertheless still had the potential for higher fault rates than insulated connections. The Department also adopted the GPO

---

100 POEEJ Vol 51 Part 3 p242 - Light-Weight Drop-Wire Construction.

Image by the author

A picture of a surviving ring pole taken in 2010. The pole still seems to have active open wire connections at this time. The Telephone Department installed hundreds of these poles during the network upgrade of the 1960s and early 1970s.

technique of open wire suspension for new work in preference to the line wire wrap technique introduced by Bennett which was, however, still utilized throughout the rest of the network. The GPO had formalized the methodology of open wire erection[101] before WWII. Each wire was regulated to a specified tension determined from tables that took into account a number of parameters including the span length, wire diameter and air temperature at the time of erection. Special clamps called ratchet and tongs were used to grip each wire and when a complete new pole was being established, each wire would be tensioned equally in order to maintain the pole as vertical as possible given the various loads in different directions. The wires were then looped around the insulator using a copper sleeve and a special tool to insert the correct number of equally balanced twists.

Around this time the Department also abandoned the use of pole climbing spikes for linemen, these were metal spikes that were strapped to the boot and around the the calf using leather straps. The lineman was then able to climb a pole up to the working steps by digging the spikes into the surface of the wood. This was a precarious exercise as the whole body weight of the lineman was held at times on a single strip of metal dug only about 20mm (¾") into the surface of the pole. As the pole aged and the number of climbs increased the surface became worn and flaky and the whole operation was quite dangerous. Climbers were very convenient for the Department as it meant that linemen could use motorcycles to attend faults, carrying the bare minimum of spares in panniers on the back of the machine. With the withdrawal of climbers the Department also withdrew motorcycles and equipped the linemen with Morris Minor vans with ladder racks. The Morris Minor was the favoured vehicle, possibly because it was used by the GPO. In fact, the entire vehicle fleet of the Department was based on the Morris range of commercial vehicles. Later, other local vehicle suppliers objected to this apparent monopoly and the Telephone Committee opted to buy alternative marques. Vans enabled the carrying of more spares and offered the lineman some protection from the elements, but were less manoeuvrable than bikes. The Department did not entirely abandon two-

---

[101] POEEJ Vol 30 Part 1 p56 - The Maintenance Aspects of Overhead Constrction

wheel transport, however, as it briefly equipped subscriber apparatus fitters with specially adapted scooters in the latter part of the 1960s.

On the large new housing development at Quennevais, recently approved by the States, the department, along with the other utility companies, laid underground cables into each property. This and other new developments in the South exchange area added extra burden to the upgrade process. New underground cabling was completed largely in polythene cable, although stocks of lead covered paper insulated cable was still utilized on repair work elsewhere. All-polythene cable had been trialed by the GPO[102] as early as the late 1940s but prices had dropped and the materials improved thus its popularity had increased. The new junction cable between West, South and Central exchanges was paper insulated but polythene sheathed throughout. The Department also adopted the GPO practice of introducing further line plant flexibility into the network by installing street cabinets and pillars[103]. Main local distribution cables were terminated on the cabinet or pillar and then the local distribution cables were also brought to these points. This allowed further optimization of the cabling at a more local level by cross connecting the main cables to the local cables when new lines were connected. The Department also adopted GPO plastic terminal blocks for new work while continuing to use the proprietary Standard Telephones and Cables metal covered terminal boxes in areas where lead cable was still employed.

At this stage the Department was faced with an increasing demand for new connections, not only because of the new housing developments in the South exchange area, but also because of the States Housing Committee's other developments elsewhere. Of particular concern was the sudden increase in demand from development in the Havre des Pas area from both new private and public apartment construction. Resources and manpower were limited for the civil works necessary to provide additional line plant and so a novel solution to the problem was implemented. The Department bought from ATE Telecommunications Limited of Liverpool three line

---

102 POEEJ Vol 50 Part 4 p219 - Recent Developments in the Use of Polythene Cables for Subscribers' Lines.
103 POEEJ Vol 39 Part 3 p100 - Flexibility Units for Local Line Networks - Part 1

Image by the author

This picture show a street cross-connection cabinet of the type introduced in the 1960 when the external network was being upgraded. These street side cabinets enabled optimum usage of underground cabling schemes by introducing flexibility at various points in the network. Subscriber distribution cables were terminated on one side of the cabinet while the main cable to the exchange was terminated on the other side. This meant that more subscriber connections could be provided than connections to the exchange taking advantage of the fact that at that time not every home had or wanted a telephone.

concentrators[104]. This electromechanical solution based on a light crossbar switch manufactured in Switzerland by Gfeller AG, would enable the better usage of existing line plant by enabling 49 subscribers to share just 11 ordinary subscriber connections. These units required equipment at both the remote site and at the exchange. While this was an expedient measure of some use, the complexity of the equipment introduced an element of unreliability into the system. These concentrators were originally installed at Marett Court, St Helier but later moved to the States social housing development at La Collette flats in Green Street, St Helier. Elsewhere other types of line connectors enabling 10 subscribers to share 2 normal lines, manufactured by the Telephone Manufacturing Company (TMC)[105] of St mary Cray were deployed as temporary solutions to shortages on various exchanges as circumstances required.

At the same time, the Department also introduced shared service. Shared service is considered in many circles as the choice of last resort since it significantly reduces the quality of service as a consequence of two subscribers sharing the same connection. However, when line plant is in short supply any service is better than none at all. Shared service was possible after the first extension on Central exchange that included the installation of 50 point linefinders instead of uniselectors[106]. The 50 point linefinder[107] was fully equipped for shared service operation provided the appropriate configuration was carried out.

The additional workload brought about by the network upgrades had seen the Departments engineering workforce increased from 110 in 1959 to 160 by the end of 1964, although the number of operators had remained stable at 170 during the same period, despite the large numbers released by the Central automatic exchange conversion. Notwithstanding this increase in staff and investment, the Department was still making good returns to the

---

104 See The Telecommunication Journal of Australia, October 1964 p408 for a full description
105 POEEJ Vol 52 Part 4 Jan- Mar 1960 Subscribers' Line Connectors
106 Telephony: A Detailed Exposition of the Telephone Exchange Systems of the British Post Office (Volume 2) Atkinson J
107 The Ericsson Buletin No 37 p65

States' Treasury and requests for additional funding easily passed through the States more or less 'on the nod'.

In the spring of 1965 installation staff from Messrs GEC Telecommunications arrived to start construction of the new STD exchange at Minden Place under the supervision of the Post Office engineering department Bournemouth area, this contract again being awarded under the BSA. The equipment to be installed in the new building was primarily for the trunk switching and signalling functions, but at the same time it was also necessary to install additional equipment in the existing automatic exchange in order to extend the automated trunk switching functions to subscribers. The trunk equipment was again based on the GEC SE50 switch but equipment also included Motor Uniselector type group selectors for 4-wire switching and the Type 4 electromechanical register-translator equipment. The earlier changes in access codes had made available the code '0' which would be used for access to the trunk switching equipment, the standard access code throughout the UK. The introduction of STD meant that additional equipment would also be required at South exchange to enable the repeating of trunk meter units over the junctions to Central. The GPO system also proposed changing public call boxes to enable STD, but at this point the Telephone Committee decided that the considerable expense of this upgrade was not necessary for the local market and decided to continue routing call box originated trunk calls via the trunk operator.

Work continued throughout the year on the new local and trunk exchanges. The Department decided to install dust-proofing, offered as an optional extra by Messrs GEC, on the equipment racks at South exchange and also experimentally on some existing equipment racks at Central exchange.

During the early summer the overhead network again suffered from inclement weather. Lighting storms struck seven major cables in addition to overhead routes and around 400 subscriber lines were affected. This indicated the improving condition of overhead line plant as work to reduce the number of open wire routes continued. Trunk traffic was also increasing and particularly during the summer visitor season the limitation of available trunks caused many delays. The Department again had to request the States for emergency funding of £12,000 to cover the extraordinary expenditure caused by this storm and from one earlier in the year. The Department reported that it was 'pleased' that only 90 faults were reported after a storm late in November.

The autumn also saw additional funding for improvements to the local junction network with a new scheme for East exchange receiving States budget approval. Rental rises were also approved at the December States sitting with residential line rental increasing from £6 to £8 and business lines from £8 to £10. The Committee chairman was able to report to the States that the: '[Telephone Department] is a profitable well run States department'[108].

In January 1966 the death of the former Engineer Manager S G Syvret was recorded just 6 years after his retirement. Meanwhile work progressed on the new South exchange and trunk exchange. In May the Department demonstrated the Subscriber Trunk Dialling system in a special display, firstly at Red Houses, St Brelade then at the department store of F Le Gallais and Sons, St Helier. Interested parties were shown how the system worked by two switchboard supervisors Mrs P Moody and Mrs D Coutanche where free calls could be made by the lucky subscribers.

The new trunk exchange opened on 2 June, although the official opening following on the 9 June when it was combined with the opening of South exchange which had been cut into service the previous day. The Lt Governor Vice Admiral Sir Michael Vickers KBG OBE made the inaugural STD trunk call to Mr R J Chappe at the Home Office. The ceremony was attended by the Bailiff, members of the Telephones Committee and the Telephone management as well as various dignitaries and representatives from both the GPO and Messrs GEC Limited. Interesting facts revealed during the speeches were that the new South exchange conversion had included 750 poles, 8,000 insulators, 75,000 yards of cable, 53,000 yards of underground duct, 350 jointing chambers and 750 loading coils for the junction cables. Working parties had installed 3,070 telephone sets of which 1,100 had been

---
108 Jersey Evening Post States sitting report 2 December 1965

replacements, 1,500 dials fitted to existing instruments, 630 extension bells, 80 coin boxes and 27 private switchboards. On opening the exchange had a multiple of 3,000 lines and a final design capacity of 7,000. It was integrated into the Central exchange numbering scheme having numbers in the range 41111 to 43999. The original manual exchange had cost £12,000 when installed but now only commanded £26 as scrap. South exchange was indeed scrapped but not before sufficient plant had been recovered to install a 200 extension at West exchange later in the year using both components and recovered cable. Other recovered parts were retained for further work at North and East exchanges as the need arose.

The transition to STD was surprisingly smooth, with just a few problems reported, these being mainly from subscribers failing to wait long enough after dialling the number to receive supervisory tones. Consumers, it would seem, were ready for STD and this acceptance had a surprising knock-on effect to call box trunk traffic. The original intention was to retain the existing pre-pay coin boxes for the foreseeable future thus avoiding what the Committee saw as an unnecessary large investment in equipment. However increases in demand required the introduction of a new operator access code (90), for call box users to complete calls to the local manual exchanges. This was because the trunk manual board operators were then able to distinguish trunk calls which remained via the operator access code '0' from those calls intended to be terminated locally. This differentiation prevented overload of the trunk operator positions. It was not technically possible for direct connection to the local manual exchanges since until the 'A' button was pressed the called party could not hear the caller, thus if the caller pressed the button to speak when the operator answered and the desired number was then found to be engaged, then the caller lost the call fee. Inevitably, the increase in trunk traffic led to congestion and delays on the calls that still needed to be manually connected. The growth in trunk traffic far exceeded that forecast and thus the need for more off-island trunk capacity became urgent. The GPO trunk call tariff reductions announced early in 1967 only served to stimulate demand.

At the end of the 1966 it was announced that the telephone directory would be printed locally, but that the traditional 'thumb index' would be abandoned to reduce costs. This proved to be an unpopular move among subscribers, but the Department did not bow to user pressure and went ahead with a format that is still largely in use today. In the States it was revealed that a yearly 'profit' of at least £50,000 could be expected from the telephone system (this would represent about 10% of turnover). For the first time the salaries and wages bill was more than half of the annual turnover at £284,430.

With the conversion of South exchange completed, the Department turned its attention to the general upgrade of the distribution network. The underground plant in some areas was pre-war and a fair amount of cable which had been laid by the German occupying forces was still in service, particularly in rural areas. There were also many overhead distribution routes, some of them quite lengthy. These were a constant source of faults and a maintenance problem since they had grown to be large and over time the quality of the poles and other structural components had deteriorated. Around St Helier there were a number of large poles some in excess of 18 metres (60 feet) which distributed telephone lines from 8-way cross arms across roads and rooftops. These lines were notoriously difficult to maintain. Many of these poles had been erected by the NTCo and had gradually increased in capacity. Some of the most well known of these were at Snow Hill, Garden Lane and Tunnell Street. Replacing these large distribution poles with alternative distribution required the provision of new underground plant. In the centre of St Helier this was achieved by providing new duct work and polythene cables along the streets and distributing from terminal blocks attached to buildings. This improved not only the aesthetic appearance of the town but also the quality of the connection, greatly reducing the risk of faults. This work continued throughout the network for the next few years, with the realization that the line plant would have to be generally improved in the country exchange areas for automation.

In 1967 the Wilson government's review of the General Post Office, under the leadership of the Post Master General, Anthony Wedgewood-Benn, proposed the division of the GPO into two separate corporations: Posts and

Telecommunications. This would be finally realized under the Post Office Act 1969 but part of the recommendations of the White Paper was that the postal and telecommunications services on the offshore islands (the Channel Islands and the Isle of Man) should be offered to the local governments to operate. After a full States debate an investigation was delegated to a committee including the Engineer Manager, H W Coppock. The delegation met with the PMG in June 1967. Discussions continued through the rest of the year including joint meetings with the States of Guernsey. While there was little comment from the GPO staff involved with the telecommunications operations, the postal workers and their union vehemently opposed the takeover. Nevertheless, after a further debate in January 1968, the States approved the purchase of the GPO assets and went on to form the Posts Committee which would take over the running of mail services. The takeover of the telecommunications division was somewhat more complex as currently there was a further new submarine cable system under construction and the finer details of the interconnection and control of these cables would require careful consideration.

The Department finally decided to convert call boxes to STD as the growth in trunk traffic had proved to be burdensome on the trunk operators. Consequently, approval was given to install the necessary exchange equipment to control the new boxes at both Central and South exchanges. The Coin and Fee Checking (CFC) equipment was expensive and complex. It used a sophisticated tariff management system required by the British Board of Trade that was constructed using relay technology. The CFC relay sets were so large that together with their counting uniselectors only 15 could be mounted on a standard exchange equipment 4' 6" rack. Consequently, their floor footprint within the exchanges was disproportionately large given the numbers required to service the coin boxes then in use.

The conversion of the remaining manual exchanges to automatic was approved in principle by the States at the end of 1967. Estimates for the work would need to be submitted before the States for the 1968 budget. It had also become clear that the location of the existing East exchange was not ideal for the future growth of the system. Since it had been built the distribution of housing within its service area had shifted to the east. The Assistant Engineer Manager, Howard James, stated that a new site was then actively being sought in the Grouville area and that 'alternative systems' were also being considered for automatic working. This implied that the existing Strowger technology would not be considered for future development. This perhaps reflected the Department's general disappointment in the GEC supplied SE50 Strowger switch. Although this equipment was the latest development of the Strowger step by step switch that had been invented at the end of the nineteenth century, the technology itself had hardly changed since it original conception. The SE50 had proved to be less than robust in practice, despite its much vaunted ease of maintenance and its marginally faster switching speed than its predecessors. The dust proofing system too had not lived up to its expectations. While dust-proofed racks had worked successfully on private switches, the increased traffic load of public exchanges had produced unexpected results. The air contained within the sealed rack units quickly ionized under heavy switch use and caused oxidization of the relay and switch contacts, particularly in tone paths which did not have the higher currents that assisted contact self cleaning. This resulted in many 'no tone' switch faults causing degradation of the service; subscribers would hang up early if no signal tone signal were heard.

The pressure for new connections on East exchange was also relieved early in the year with a transfer of 550 subscribers onto Central exchange. These subscribers were largely from the St Clement's area. Presumably, subscribers did not object strongly to the inconvenience of a number change given the considerable improvement in service offered by transferring to an automatic system. Central exchange was, in turn, in the process of further extension work that would add another 2,000 subscriber lines in the 26XXX and 36XXX ranges. There was also a large 'regrading' programme under way to reconfigure and extend the switching equipment to cope with the ever rising demand for both local and trunk calls.

The late 1960s was a time of much innovation in the telecommunications industry. The proposed alteration to the GPO structure heralded a new era of change. The reliance on the traditional systems and practices was now under review. New technologies were being more rapidly accepted, such as the introduction of the Standard Telephones and Cables revolutionary Deltaphone (called the Trimphone by the GPO), which for the first time offered subscribers an alternative from the traditional telephone bell and presented a radical new design. Chic did not come cheap, however, as the quarterly rental was an additional £3 for a Trimphone. Around the same time, the Department also offered business subscribers the new Pye TMC repertory dialler. This was perhaps the first telephone product to use integrated circuit technology purchased by the Department. It offered busy executives access to up to 32 pre-programmed numbers at the touch of a button. The equipment by modern standards was quite large and required expert programming using dozens of wire jumper connections. But it was a huge advance in convenience and the forerunner of the modern telephone. Again, advance came at a price of £30 per quarter rental.

In July 1968 the long awaited additional submarine capacity was commissioned by the GPO and a group of 12 new circuits to Salisbury was opened relieving some of the congestion on trunk calls. At the same time the Department was completing the process of upgrading its public telephones and private user coin boxes to STD working by replacing the old Button A-B boxes with the Pay-on-Answer type. South exchange was completed first as there were less coin boxes to change. Central exchange took a little longer as there were about 600 to convert. Because of the configuration of the exchange equipment, the conversion had to be managed in groups of 25 at a time. Hence a long programme was necessary to complete all the coin box locations.

The budget for the conversion of North, East and West exchanges was presented to the States in October 1968. It was estimated that the total cost of the project, including switching equipment, new buildings and line plant would be £1,390,000. It was planned that North and East should be completed by 1973 and West in 1975. The budget was approved with little comment by the States at the November budget sitting. At the same time an estimated budget of £800,000 for the takeover of the trunk telephone network from the GPO was also approved as was a further request for £236,500 for general network development. The Telephone Committee also approved a 5½% wage rise for operators and engineers that would, for the first time, bring them on par with their UK equivalents. This was the beginning of the major spending days for telecommunications and also probably marked the beginning of the rise in profit margins from which the telephone system would continue to benefit for many years into the future.

The telephone system was growing rapidly, the demand for new lines was continuous as a result of the expansion of the finance industry and consequentially the switched traffic also grew. The increase in traffic was particularly noticeable on Central exchange as the centre of the Jersey business district. The GEC SE50 switches proved to be mechanically fragile in use and much routine maintenance and repair effort was required. The Department decided to recruit additional exchange maintenance technicians from Post Office Telecommunications to bolster the exchange maintenance staff. Elsewhere staff was increased to service the reconstruction of the network in readiness for conversion of the remaining manual exchanges to automatic.

Around this time the GPO had recently introduced its improved data transmission Datel 200 service[109]. This enabled data communications over telephone lines using a modem[110] – a familiar piece of equipment in these days of the Internet but at that time something quite new and expensive. Many of the local financial business requested Datel services for the interconnection of their offices to their new mini computer systems, and with their main frame computers in the UK. The now familiar Automatic Teller Machine (ATM) was also an early user of this service. The first ATMs were deployed in the UK in 1967 and arrived in Jersey in 1969. While nowadays setting up modems is a simple 'plug-and-play' exercise, in 1969 it required expert attention and much complex test equipment.

---

109 An Introduction to the Post Office Datel Services. Shows modem 1A and 2A POEEJ v59 pt1 pg1
110 Modem is derived from the words Modulator and Demodulator the two functions of a data transmission device that converts digital 1 and 0 signals into analogue frequencies for sending over a standard voice frequency telephone circuit.

Planning for full automatic working was now well under way. New buildings would be required at all the exchanges as the existing ones were too small. At North exchange there was a suitable site available adjacent to the existing building. East exchange was more difficult as the existing building was not in the ideal location. However, a new site was identified on Public Health Committee land at Sandybrook, Grouville. Agreement was made between the committees to transfer the land and the States architect was instructed to draw up a suitable building plan. Howard James confirmed that the MDF equipment was already on order but the switching equipment tender documentation was still in preparation.

In July 1969 the original TIM machine finally failed after 5 years of reasonably accurate if somewhat erratic service. In this time it had received over a million calls, generating about £12,000 of revenue for the Department. The equipment was replaced by a pair of machines costing £2,500 from Assmann GmbH of Germany working in automatic fail-over mode. The original recording was again utilized and the service was inaugurated on 16 August by Senator S J Venables. The familiar 'pip' denoting the time was replaced by a single electronically generated 'gong' sound. Around the same time the ageing information services machines were also replaced by new machines from the Plessey Australian subsidiary company, Rola. These Variable Message Repeater[111] (VMR) machines were purpose built for the telecommunications industry and although similar in concept to the old machines they were horizontally mounted recorders and used return springs rather than gravity for the reset function. The additional VMRs enabled the Department to split its weather and mail boat services so that the weather was now available using the access code '85'. Later, further services were introduced; '88' for share information and '89' for lottery results (on the day of the draw). Overall, the information services were financially successful with 161,889 calls made during 1971 with more than half a million calls to TIM during the same year.

At the end of the year the Department was hit by a serious influenza epidemic which affected up to 50% of operators on the minor exchanges.

---

111 See http://www.oneillassociates.com.au/~poneill/pdf/VMR%20Pub%20PV53.pdf last accessed 08/08/2016

The Telephone Manager, Harry Coppock, appealed to the public to be patient until the staff had recovered. This reduction in service unfortunately coincided with the New Year celebrations.

The States approved the building of the new North exchange in February 1970. The building contract was awarded to Jersey Contractors Limited for the sum of £19, 133. The building architects were W V Jelley and Associates, who had been instructed by the States Public Buildings and Works Committee. In June the States were asked to agree the Telephone Committee request to take over the fish market site in Cattle Street from the Public Buildings and Works Committee, as it would become free after the reorganization of the markets. The fish market was to be condensed into the Beresford Street market as the number of stall holders had decreased over the previous few years and the building itself was in some disrepair. Coppock stated that the present buildings had reached the limit of their development and that further space would soon be required for further expansion of the existing exchanges. It was envisaged at this time that expansion of the existing switched network would be based on an electromechanical system.

At this time the Department was also preparing for the impending Decimalization Day (D Day) which had been scheduled in line with the UK on 14 February 1971. D Day would make fundamental changes to the call charging system of the Department. A strategic plan needed to be developed to ensure the rapid conversion of public and private rented call boxes to accept the new coinage. New coin slots, guides and screening components would need to be installed in all of the 600 or so coin boxes within the period allowed for the changeover to the new denominations. The Committee also used this opportunity to adjust tariffs. At the end of the year it announced the changes that would come into effect. Calls from ordinary lines would be increased from 3d to 1.5p (3.6d) an increase of 20% due to rounding error, a significant windfall increase in revenues. Call box calls on minor exchanges would rise from 4d to 2p (4.8d) while automatic call box calls would fall from 6d (2.5p) to 2p (4.8d). Other changes included new line installation charge up £2 to £10; extension telephone rental up £1 to £3 per quarter and Private Branch Exchange installation and rental charges increased by 25%. For the first time the Telephone Committee revenues exceeded £1M during 1970. The demand for new lines was unabated and a further 2,000 line extension was ordered for Central exchange together with expansion of call box equipment to satisfy increased demand and traffic.

At the January 1971 sitting of the States approval was given for the award of the East exchange building contract to A C Mauger Limited in the sum of £25,141 14s 2d. It was explained that while this was not the lowest tender, the contractor promised a completion date some 12 weeks earlier than that of the cheapest. The Committee also reported that a contract for the exchange MDF had also been placed with Messrs GEC Limited, while the exchange equipment was subject to a separate tender. The equipment tender for both North and East exchanges had been submitted to five suppliers: GEC Telecommunications Ltd; Plessey Telecommunications Ltd; Pye TMC Ltd; Standard Telephones and Cables Ltd and the Swedish company L M Ericsson. The responses had been assessed and the choice of supplier made by March. GEC, Plessey and Pye TMC had all offered step by step Strowger systems. Ericsson offered a crossbar switch but was expensive, given the current exchange rate and general weakness of the Pound Sterling at the time. STC was unable to bid because of its commitments to the Post Office, but a solution was offered from its fellow International Telegraph and Telephone (ITT) subsidiary, Fabbrica Apparecchiature per Comunicazioni Elettriche (FACE) of Milan, Italy. FACE manufactured the Pentaconta crossbar system developed during the 1940s and 1950s by the French Compagnie Générale de Construction Téléphonique (also part of ITT) and had been installed in some quantities in France and elsewhere on the continent. The version offered by FACE was the latest development of the switch, using semi-electronic control systems. Pentaconta[112] was a centralized control system (as opposed to the Strowger step-by-step system) which enabled more sophisticated facilities to be offered. The FACE bid was accepted by the States and the approximate cost of the two exchanges was £250,000.

---
112 100 Years of Telephone Switching by Robert J. Chapuis, A. E., Jr. Joel. IOC Press

The crossbar system was a little more compact in design than the equivalent sized Strowger exchange, thus the buildings at both North and East would be totally suitable as they had been designed for a Strowger exchange footprint. Ultimately, it would mean that the buildings would be able to accommodate more subscriber lines than the equivalent Strowger exchange.

Engineering staff at the end of 1970 numbered 165, while operators employed full time were 100 with a further 52 part-time staff. The number of operators employed on the trunk switchboard had declined as STD callboxes had enabled more direct dialled trunk traffic. The additional staff employed over the last few years enabled the Department to retain its level of maintenance as well as devote manpower to network upgrade and development. Indeed, the improvements in line plant quality had ensured that the maintenance effort was greatly reduced. With the buildings for both new exchanges completed and the MDFs in place, the new underground line plant necessary for automatic working could be rolled out.

Of the two exchanges, East would be the most challenging. The exchange would need to be moved from its existing location in St Saviour to Grouville, some two miles distant. In order to achieve this end, the existing MDF would need to be connected to the new MDF during the transfer period to enable subscribers to access the current working exchange. A temporary tie cable was therefore installed between the two sites and as new line plant was introduced onto the new MDF the subscriber would be connected back to the existing number on the old exchange. The same process would be adopted at North but here the new MDF was in the adjacent building and thus the transmission losses experienced by those subscribers on East as a result of the additional cable length would be avoided.

Meanwhile the changeover to the new decimal coinage passed without major incident on the 14 February 1971. All public call boxes were changed over before the month end and private rented boxes completed in time for the withdrawal of the old coinage.

During the remainder of the year negotiations continued between the Telephone Committee and the Post Office regarding the takeover of the

Image by the author

A view of the Trunk Exchange at Minden Place showing the patch frames and trunk test desk with the special faults telephone and MDF to the left. The patch frame enabled the re-routing of faulty circuits as a temporary solution during fault finding or maintenance.

trunk network. These talks were conducted in tandem with Guernsey as the takeover had to be synchronized to come into effect at the same time. The law draughtsmen were instructed to produce a suitable legislative framework. The costs of the network had been fixed early on by Post Office Telecommunications and probably did not reflect the actual cost of the component parts and consequently bargaining on the value of the assets did not seem to form part of the discussions.

In view of the added costs to the Department following the takeover the costs of calls were again increased on the 1 January 1972 in order to ensure that the costs of running the larger network could be covered. Calls increased by 33% from 1.5p to 2p per unit while other installation and rental charges were also increased. This passed through the States with little debate, as it seemed now becoming generally accepted the the Telephone Committee operated its network 'as a commercial business'. In addition the annual rate of inflation during this period was almost 10%. The average cost of installing a subscriber line was stated by Deputy Tanguy, the committee president, to be £30.

In May the States agreed to fund the purchase and construction of a new telephone garage, stores and workshop at Le Geyt farm, Five Oaks. This would provide a permanent site for much of the department's external support activities and consolidate the current three leased sites at Springfield, St Aubin's tunnel and the Red Barn, St Aubin.

Work continued at both North and East exchanges on the external network upgrade and new construction activities for automation. In the meantime, the demand for telephone lines continued and to overcome the shortage of numbers at these exchanges, immediate release of ceased lines was authorized with numbers prefixed by the digit '7' to provide differentiation between ceased and new lines. Thus the ceased number East 1234 would be reissued as East 71234. Suitable marking was made on the switchboard to alert operators to the reissue.

Around this time the specification for the replacement automatic exchange at West was sent to potential suppliers. The request to tender also included a relief unit for Central exchange. The same list of suppliers as for North and East exchanges were invited to tender for these two projects.

The Telecommunications draft law was presented to the States on 27 June 1972 by the Telephone Committee president Deputy Tanguy who summarized the history of the telephone system. The GPO had assumed the running of the island network on the takeover of the NTCo in 1912 and had subsequently, in 1923, licensed the States of Jersey to operate the system under the Telegraph Act 1869 as extended by Order in Council to the island in 1872. HM Government under a review of the General Post Office in 1967 had offered the States the opportunity to purchase the remaining GPO operations in Jersey including the postal and trunk telephone system. While the postal operation had been taken over by the States in 1969, the negotiations on the trunk system had been more complex and had taken somewhat longer to conclude. Once the transfer of assets had taken place the total telecommunications holding of the States would be £1,771,000 and the telephone operation would no longer be subject to a licence. It was proposed to change the name of the Telephone Committee to the Telecommunications Board. This change was challenged by Deputy Thomas who wanted to ensure that any such board should have competent business leaders, possibly non-States members in the same way at the Jersey Electricity Company. However, the States agreed that the committee could be replaced by a board consisting of States members as many of them had business experience in any case. The name change to the Telecommunications Board also necessitated a change in the air traffic control department which had until know been known as the Telecommunications Department. To avoid confusion this States airport activity was renamed the States Department of Electronics.

The resulting board would have joint control of the interconnecting submarine cables with the Post Office Telecommunications in the UK, the French authorities and with the Guernsey Telecommunications Council. The States also approved the adoption of the ownership and leases on properties transferred with the cable systems, these comprising the cable huts at Fliquet and Plémont and the repeater station at Trinity gardens together with sundry leases for equipment space at Broad Street Post Office, Channel TV, Frémont point and Les Platons. The majority of the engineering staff employed by Post Office Telecommunications were transferred into States employment, retaining their previous pension rights and some of the terms of employment, including a half day's holiday on the Queen's birthday.

The new Telecommunications (Jersey) Law 1972 came into full vigour at midnight on 31 December 1972, thereafter the States Telephone Department ceased to exist and the business activities were be transferred in name to the States of Jersey Telecommunications Board.

Chapter 18 - The States of Jersey Telecommunications Board 1973 – 1985

As of the 1 January the whole of the telecommunications systems on the island were taken under the control of the newly formed Telecommunications Board which has an absolute monopoly over all telecommunications activities on the island, with the exception of spectrum licensing which remained with the Post Office. The immediate effect was not apparent as it was largely business as usual, but slowly changes were introduced. The first publicly visible sign was the introduction of a new logo which started the first tentative moves towards a more commercial approach towards the business of providing telephone and associated services. The new logo was a somewhat amateurish affair and firmly based in the old world of telephony. It was based on the telephone dial with the words 'Jersey Tels' integrated into the finger holes.

Early in the New Year the information services were augmented with the addition of the religious based daily prayer (accessed by code 80). This was made possible with the purchase of further recorders which were now installed on a purpose built equipment rack in Central exchange. At the same time the Lottery Results were given a dedicated announcement machine and access code (89).

On 1 April the Telecommunications Board celebrated the 50th anniversary of the States' ownership of the telephone system. A celebratory dinner was held for members of the board and senior staff and other guests at the Old Court House hotel at Gorey on 12 May. Guests included the Board president Deputy Cyril Tanguy, the longest serving committee member Senator Wilfred Kruchefski, other members and former members of the committee, some 50 of the long serving staff racking up over 1,200 years of service including Doug Le Maistre who had completed 41 years at the Department, and senior managers including Harry Coppock the Telephone Manager, Howard James and Roderick Perrin. Other guests included the branch representative of the Electrical Trades Union, Mark Young and the local union representative William McKain, the managing director of Channel TV, Ken Killip, senior operators and Stella Williamson – Miss Telephone Personality of the Year. Later that summer during August, telephone exchanges were opened to the public for guided tours, the Board's training officer Brian Fells and senior operator Mary Moody leading the tour parties.

The reality of the separated network was that a new commercial agreement was now required with the GPO for the connection of off-island private circuits. Senior management, sales and planning staff held a joint meeting with counterparts from Guernsey in London to finalize arrangements for connection and invoicing of such circuits.

Early in 1973 the Department made a break with the traditional Strowger step-by-step private automatic switching systems that it had been installing since the 1950s. It installed a Thorn-Ericsson ARD 561 crossbar Private Automatic Branch Exchange (PABX) in its own offices at Minden Place to take over from an ageing GEC step by step Strowger system. The crossbar switch offered many advantages over its predecessor in both reliability and features. It also occupied a smaller footprint than the equivalent size Strowger system and it was much quieter in operation, enabling it to be installed in normal office environments. At around the same time following internal reorganization the Exchange Telegraph Company[113] (Extel) upgraded its 'blower' service that provided racing results and commentary to betting shops. The original service had been via a relay system of

---

113 Extel 100 - The Centenary History of the Exchange Telegraph Company, J M Scott, Benn 1972 ISBN 0 510-27951-1

exchanges, the results and occasional commentaries being relayed by a series of local offices connected to the main switchboard in Manchester. After the reorganization, the local offices no longer relayed the information but instead the central commentary was broadcast over the telephone system. Bookies dialled into the local system and the service was extended by the operator. A distribution amplifier and test system was installed in the Central exchange and connected to the Extel office in Minden Street where it was distributed to local betting shops.

A new data transmission network was also installed for the Resources Recovery Board's new telemetry system designed and supplied by GEC Elliott Automation Limited. This new system would enable the centralized monitoring of sewerage pumping stations around the island. A network of branching panels mounted on standard open 19" racking was installed at each exchange and connected to each remote station over private circuits. Considerable difficulty was encountered on the setting up of the network as the overall return losses at some stations caused instability in the network. Although the system was initially designed to work as a branched 2-wire to 4-wire system, some outlying stations had to be connected as 4-wire circuits in order to meet transmission standards.

During the summer the switching equipment for North and East exchanges began to arrive from the FACE factory in Milan, Italy. The deliveries were supposed to begin in the early spring, but had been delayed by a combination of the effects of the prevailing OPEC oil crisis[114] and its effect on the plastic industry and industrial action in the Italian factories. Large plywood clad boxes marked 'Yersi' were delivered to each exchange. The FACE engineers and technicians followed in September under the direction of their project manager signore Paolo Pangrazzi. Despite the late delivery of the equipment the Italians were typically optimistic and assured the Board that the exchanges would still be installed on time.

The Italian engineers were highly enterprising and quickly converted the equipment packaging into shelves for the storage of components, wire, cable, etc and the remaining materials were then quickly snapped up by

---
114 Opec: Twenty-Five Years of Prices and Politics Ian Skeet 1991 isbn: 9780521405720

local engineering staff for home projects.

It quickly transpired that the floor at North exchange was of sub-standard quality. When attempting to drill holes to secure the racking, the screed crumbled. A rapid expedient solution was to overlay the entire floor with ¼" steel plate over which grey plastic tiles were laid. This ensured that the equipment was probably the most securely fixed anywhere in the world!

Crossbar equipment of the ITT type required an almost complete framework for racking to be put into place over the entire floor area. The equipment bays were then secured to the framework where required, leaving gaps for future expansion with the floor and other steel frame components already present. This was contrary to Strowger practice which only completed the overhead racking throughout the exchange leaving clear space beneath for future expansion.

During the spring the responses to the tender for West and Central exchanges were received and evaluated. Responses were received from GEC, Plessey, Pye TMC, Standard Telephones and Cables (STC) had expected to get a follow-on contract for crossbar via its Italian partner company) and L M Ericsson. The most interesting was from Pyeb TMC which was offering a new product from its parent company Philips of the Netherlands. Pye TMC offered the PRX 205[115] processor controlled reed switch exchange. This was a new innovative solution that Philips was trying to sell in the UK in competition to the Post Office designed TXE exchanges then being made by the British telecommunications manufactures. Jersey was seen as a possible jumping-off point and thus was offered the switch at a very competitive price. The PRX system was already being deployed in the Netherlands. Negotiations were conducted through the late spring and summer with the Pye TMC commercial director J V (Jimmy) Greenfield. A contract for £350,000 was signed with the company on 17 September 1973 for two switches, one for West and the other to be installed as a relief exchange for Central, which was expanding rapidly and had outgrown the floorspace available for further extensions to the existing Strowger

---
115 Philips Telecommunication Review, Vol 31, No 2, September 1973. General Description of PRX-system – A W van r'Slot, M J Laarakker.

equipment. The award of this contract to Pye TMC greatly disappointed the board's traditional supplier, GEC, who belatedly tried to rescue the tender by making an offer of its version of the Post Office electronic exchange systemTXE2[116] in place of the TXK1[117] crossbar system in its original response to the tender, but this system was not as advanced as the PRX. In Guernsey, however, the States Telephone council later did opt for a version of TXE2 supplied by Plessey.

The selection of the PRX had other benefits to the Board. The proposed extension in the former fish market now became unnecessary since the PRX exchange's overall footprint was considerably smaller than equivalent sized Strowger or crossbar switches, or indeed, TXE2. Consequently, the new relief switch for Central could be installed in the existing carpenter's shop and the carpenters would be accommodated at the Board's garage at Springfield until a more permanent location could be found. Similarly, the new building at West exchange could also be made smaller and thus save on construction costs. Therefore the plans for the new building were revised before the building tender was advertised.

Demand for more lines at the minor exchanges was also a problem. At West exchange this demand was able to be met with a small extension of 100 lines, however at the other manual exchanges the capacity of the existing switchboards had already been exceeded. The interim solution of reallocating ceased numbers immediately with the addition of a prefix was not sufficient. Two further temporary extensions were added at both North and East manual exchanges, this time using an automatic switching solution. Specially modified Private Automatic Branch Exchanges (PABXs) were installed at both exchanges. Subscribers were connected through these switches so that they appeared to be connected on Central exchange. At North subscribers were allocated numbers in the range 382XX while at East numbers were 390XX. Twenty five lines were supplied at East exchange where the GEC PABX was installed at the new exchange building next to the MDF. At North fifty lines were installed with an STC Stepmaster PABX installed in the old exchange equipment room. Power supply at East was

---

116 100 Years of Telephone Switching by Robert J. Chapuis, A. E., Jr. Joel. IOC Press
117 ibid

supplied by a floated battery consisting of 4 spare car batteries from the Board's garage, while at North a single car battery floated across a mains 12V power supply was inserted between the switch and the local 40V battery. For all local calls the exchanges behaved as normal automatic lines, but because there was no provision for metering, trunk calls had to be completed via the operator. Subscribers were offered line rental at a rate that included unlimited local calls.

The OPEC-led oil crisis had its effects on Jersey. The restriction of oil supplies had consequential effects on the electricity supply for the coming winter, and cuts in service were forecast. The Board took action to lower its power consumption by reducing the number of electric lamps in use throughout its offices and exchanges, substituting fluorescent lamps for incandescent lamps where appropriate. At the same time, it was also clear that there could be grave consequences for the Board if mains supply to exchanges were cut for any length of time. The burgeoning finance industry now relied heavily on the telephone and in particular on international communications to conduct its day-to-day business. It was therefore decided to purchase a mobile generator that could be sent to any of the island's exchanges in cases of prolonged mains outage. Each exchange was therefore upgraded to enable this facility; the upgrade ensured that power services within the buildings were distributed as equally as possible across the three phases of the mains supply. As the state of the single battery at West exchange was now judged to be poor, it was decided to station the mobile generator permanently at this site.

Similarly the provision of power for the forthcoming International Subscriber Dialling (ISD) extension to the trunk exchange and the Central PRX was also critical. After evaluation it was determined that the existing Central exchange batteries were of sufficient capacity and reliability to manage the complete power requirement of all the present and future switching equipment, but this was contingent on a continuous mains power supply. Thus it was decided to build a new power room on the old fish market site that had originally be earmarked for the Central local exchange extension building that was now not required.

Meanwhile, demand for new lines continued apace. It became clear that it would not be possible to wait for the new Central relief exchange to satisfy demand. An interim solution was arrived at by the provision of a small 400 line extension using specially modified final selectors that could absorb and discriminate the third digit of the number. This enabled the extension to be completed without the need for a third numerical group selector rack for which there was no space available. The special extension was designed by the author and implemented in short time using number ranges 280XX/281XX and 290XX/291XX. This modification was recovered in 1976 and subscribers transferred to the PRX after its first extension reallocating to 7XXXX numbers. The line plant reconstruction effort was now concentrated in the North and East exchange areas in preparation for the new automatic equipment. Therefore in Central the shortage of line plant now became critical. As an interim solution the Board deployed the newly available Subscriber Carrier 1 + 1[118] system in Central exchange. This system, also known as WB 900, effectively permitted the doubling of capacity on each subscriber line by adding a virtual circuit using an amplitude modulated high frequency carrier. This was a great improvement over the simple shared service option as it ensured privacy between parties using the same line. Problems with the recently introduced Group 1 fax machines meant that where the sharing party was a business it was necessary to ensure that the physical circuit was used rather than the carrier circuit. Various other modifications to the existing line concentrator equipment at Central exchange released a number of subscriber line circuits to help alleviate the demand for new lines.

At this time a large part of the remaining floorspace in Central exchange was required for interconnecting equipment for the new automatic exchanges, changes for ISD and the provision of more call box equipment as demand grew from the tourist industry. Changes to the Central test suite were completed to enable access to the Pentaconta 'supernumerary' numbers. These were telephone lines that had no associated directory number, such as outgoing PABX lines, and were accessed by dialling '11'.

---

118 Post Office Electrical Engineering Journal Vol 64 Part 4 January 1972 - A Subscribers' Carrier System for the Local Network

FACE had provided special dials but as these could not easily be installed on the existing test switchboard a modification to the test dial circuit enabled the sharing of a locally designed circuit that produced 11 dial pulses on demand from the test clerk. Also during this period a new SALT/ALIT (combined Subscriber Automatic Line Tester and Automatic Line Insulation Tester)[119] system was commissioned. This was a new version of the original SALT device and incorporated an overnight subscriber line test to pre-empt potential line faults by routinely measuring line insulation. The tester would automatically work its way through the subscriber ranges and produce a printout of potential or actual line faults that could be passed to the exchange test clerks for further investigation.

By the end of the year it was becoming apparent that the new crossbar exchanges would not be completed within the original time scale. This was clearly an embarrassment for both the management and the Board. Early in the New Year the Board announced that the new proposed opening date would be 1 May 1974. At the same time subscribers were advised to retain the 1973 directory until the change over as the new 1974 directory currently under distribution showed only the new automatic numbers. Recorded announcements were connected to number levels '5' and '6' to advise callers to refer to the 1973 directory or call directory enquiries.

Further problems were being experienced by the supplier. Ongoing industrial action in Italy accompanied by continued shortages of plastics as a result of the oil crisis had meant that equipment deliveries were being continually delayed. Further prolonged strikes in the spring resulted in the president of the Board, Deputy Tanguy, and Coppock travelling to Italy to renegotiate the contract. However, the meeting failed to determine any clear date for completion of the project and they were only able to report to the States that senior FACE managers would be visiting the island at the end of the month. Following that visit all parties declared their belief that the exchanges would be brought into service 'before the end of the year'. Clearly there had been a breakdown of the trust between the parties. FACE had expected to receive follow-on contracts for both West and Central relief,

---

[119] The principles of this tester are described in the POEEJ Jul 1958 (Vol. 51 Part 2) although the installed equipment was of a later series of the design.

thus the award of the contract to Pye TMC must certainly have soured relations. Notwithstanding the industrial action, there was now little incentive for FACE to complete what was, after all, a minor contract.

The building contract for the new West exchange building was awarded to Messrs Charles Le Quesne at the States' sitting on 1 May 1974 in the sum of £29,969. The new flat roofed building was to be situated immediately behind the existing building. At the same sitting approval was also given for the construction of the power room on the old fish market site. The demand for new lines was unceasing, and it was decided to lease a Mobile Non-Director automatic Exchange (MNDX)[120] from Post Office Telecommunications to provide further relief at Central until the PRX exchange was installed. The exchange in a trailer 22' long by 7'6" (6.7m x 2.3m) arrived from Birmingham at the end of June and was set in position on the fish market site, but not before part of the historic wall had been dismantled and subsequently rebuilt to facilitate entry. Howard James, the deputy director, told the media that the decision to request the 400 line 2000 type MNDX was taken as there was otherwise a distinct possibility that Central exchange would run out of numbers before the new relief exchange was commissioned.

The numbering ranges for the whole of Jersey were now set. The new PRX relief exchange was to be prefixed with '7' while West would use '8'. This meant that the island would have a fully linked numbering scheme based on 5 digits with no further capacity to expand except within the exchanges themselves. As a consequence of the arrival of the MNDX, the '7' prefix was allocated to it temporarily. The use of '8' as the first digit of subscriber numbers for West also meant that the existing information services required renumbering and so it was decided to allocate these to the prefix '18' thus making TIM '181' etc. Work to permit this change was carried out in time for the 1975 telephone directory.

The ongoing delays and the embarrassment of the directory fiasco had proved too much for the Board, and in July Harry Coppock announced his

---

[120] POEEJ Vol 42 July 1949 - A 200-Line Mobile Automatic Exchange by E. Siddall and A.A. Page

early retirement. This was followed shortly after by the sudden and mysterious resignation of Howard James. The vacant role of Assistant Telephone Manager was filled by Roderick Perrin after a shuffling of senior management. At the same time there was a public row as the Board decided to levy a 15% surcharge on telephone bills as a consequence of the extraordinary inflationary trend over the previous few months as a consequence of the oil crisis. Deputy Tanguy, the Board chairman, justified the surcharge as an expedient measure before resetting tariffs in the budget for the following January. The matter was debated further at the October States sitting when the new tariffs were proposed for approval but there was little opposition to the Board's position. It was clear that the Telephone Board was now given extraordinary latitude in determining its own finances as it delivered a regular and increasing income into the States treasury. This was despite the further delays to both North and East exchanges that were announced shortly before the debate. It was evident that the exchanges would now not be commissioned before the year end. Further delays in components and development had been experienced in Italy, particularly the design of the coin-box control equipment which had proved a challenge to the Italian design engineers. This and other problems did not prevent the completion of contracts for two large Pentaconta urban exchanges (PABXs) which were installed at the General Hospital and Highlands College.

At the November sitting of the States Liverpool-born Leslie May was appointed as the new Director of Telecommunications. May had been with the Board since 1973 as the Chief Accountant. This was the first time that a non-engineer had been appointed to this post.

The Board rented a house at St Brelade to house the Pye TMC project manager, Mr T B (Tom) Renouf, an ex-patriot Jerseyman, who was to supervise the delivery and construction of the two PRX exchanges. The building at West would be completed in January and the former carpenter's shop was now in the course of refurbishment. The first task of the project was to install the power supply for the exchange. The building on the fish market site was now nearing completion and the interconnecting overhead

Image by permission of Charles Watts

The Pentaconta PABX being tested during installation at the General Hospital, St Helier. As can be see, the racks barely fit beneath the cellar ceiling.

racking was being installed to connect the new power equipment with the existing batteries and exchange equipment throughout the building. A Dormond Diesel 50 kVA standby generator set ensured that power breakdowns would not affect telephone services, at least maintaining the trunk and subscriber switches on Central exchange. Dutch engineers from Philips Telecommunicatie BV, Pye TMC's parent company, installed the cables between the power room and the PRX suite ready for equipment deliveries in the New Year. At about the same time, the long delayed second floor extension adjacent to the trunk exchange manual suite was completed. The building had been delayed as the contract had to be stopped when it was realized that there were no windows on the North elevation, which was undesirable as the spare floor space was earmarked for offices.

In February 1975 the finance for the long awaited Telephone Engineering Centre was approved by the States. The new facility finally opened towards the end of the year. The original proposal had been agreed in 1967 and planning permission granted at the end of 1974, however the scale of the project had been reduced in line with the available budget. The contract in the sum of £209,976.12 was awarded to Messrs A C Mauger & Son (trading as Sunwin Ltd) being the lowest tender. This project would bring together all the external engineering and administration activities of the Board which were currently spread over a number of locations. More staff from Pye TMC arrived on the island as the buildings became ready and equipment began to arrive. The Dutch technicians quickly installed the PRX processor and memory suite which enabled the software developers from Pye TMC, St Mary Cray, under the supervision of its project manager Harry Constantine, to begin testing. The developers began testing and evaluating the TOS v5 telephony operation system software that had been developed specifically for the UK market using the Telecommunications Board's internally developed specification. The MDF was installed at West exchange, although strangely it was not fitted with lightning protectors despite a poor record of lightning faults within the West exchange area. This meant that after experiencing several service problem later in the exchange's operation, an

Image by the author

North exchange the day after cutover. The telephones on the switchboard were for advising important subscribers of the changeover. The island suite installed during an extension to the exchange in the 1960s can be seen on the right. The box on the chair in the foreground is full of discarded headsets.

expensive retro-fit was necessary.

The following month the long awaited opening of North and East exchanges was announced to be 1 May. As a consequence the new telephone directory would not be distributed until 1 April, having traditionally been delivered from the beginning of the year. This change of date remains to this day. The directory would leap from 1973 to 1975, the 1974 directory having been largely useless on account of the 18 month delay in the crossbar exchange project. An additional £125,000 order for a further 2,500 lines at Central was also announced, even though the PRX exchanges were not yet even fully installed.

North and East exchanges were finally cut over on 1 May 1975, 1,935 lines out of a total installed capacity of 2,400 at North and 2,652 out of 4,000 subscriber lines at East. The official opening of North exchange was carried out by the Bailiff H F C Ereaut and that of East by the Lt Governor General Sir Desmond Fitzpatrick on 17 July. North exchange had floor space available for up to 4,500 lines while East could ultimately cater for 10,000. The new exchanges provided new facilities for both subscribers and maintenance staff. Automatic fault reporting on teletypes was available and displays on the emergency switchboard immediately gave the calling number identity. For subscriber's Dual Tone Multi-Frequency (DTMF) press-button dialling became possible together with automated call transfer. For engineers automated fault detection and reporting enabled easier and quicker detection of faults. At the same time the Board introduced meter recording cameras that enabled staff to take photographs of subscribers' mechanical meters 200 at a time. These photos were then passed to the accounts department for billing, greatly speeding the process and reducing the cost and time of manual reading. As a consequence of the opening, a number of subscribers were changed from Central exchange to East exchange in the Le Hocq area of St Clement, freeing up some spare capacity on the Central switch.

Engineers and equipment from Messrs GEC of Coventry started to arrive for the installation of the ISD exchange. Power services for the new switch room had been provided by the Board's own staff. The introduction of ISD also required extensive modification to the existing trunk switching equipment to accommodate the added complexity. AC11 Signalling band SSMF2[121] (similar to ITU R2 DTMF signalling[122]) equipment had to be incorporated into the existing trunk register and translation equipment. International dialling registers were to be accessed from the existing equipment in order to store the extra digits necessary for dialling international numbers. Additional charging rates were required for call metering purposes. This requirement had been realized earlier, when the frequent tariff charges demanded by the Post Office had resulted in expensive changes to the existing electromechanical pulse generator equipment. Former Internal Planning Director Howard James had commissioned an internal development of an electronic pulse generator which in principle would be easier to change. The design was completed but a decision to use a solid state relay driver output to the metering equipment proved to be less than successful. However, this equipment was now installed and there was no opportunity to step back as the electromechanical equipment was now obsolete. This equipment nevertheless continued in service till the end of the electromechanical trunk exchanges, although it was later modified to incorporate heavy duty relays for pulse distribution.

During the spring, new information machines were purchased from Assmann GmbH of Germany to replace the ageing and increasingly unreliable Rola machines. The Assmann machines were in principle practically identical to the TIM machines purchased in 1969 which had proven extremely reliable in service. The new machines used transistor amplifiers instead of the thermionic valve amplifier version in the TIM machines. The opportunity was taken to relocate the existing TIM and new announcement machines to a smaller 19" rack in the trunk exchange thus freeing more space in the overcrowded Central switch room. A semi-electronic version of the access relay sets was later developed locally and installed in January 1978 replacing the existing electromechanical sets thus

---

[121] POEEJ Vol 63 part 2 July – September 1970 Transit-Trunk-Network Signalling Systems Part 2 – Multi-Frequency Signalling Equipment
[122] ITU Q.400-Q.490 : Specifications of Signalling System R2

reducing further the equipment footprint.

In July a submarine cable fault restricted dialling availability and calls via the operator were limited to 6 minutes. Fortunately the fault was rectified within 3 days and services returned to normal. This was the first submarine cable fault since the States' takeover of the trunk system.

In October 1975 telephone call unit charges were raised from 2p per unit to 2.5p, except for call boxes which were at the time set to take 2p and 10p coins only. An investigation had been made to consider the raising of call box charges to 4p, but this required the user to deposit 2 coins. Although a technical solution was designed, the Board felt that this was a step too far for users and abandonned the plan. At the end of the month the Central PRX relief exchange was cut into service at 7.00AM. The subscribers connected to the MNDX would be transferred to the new switch, but because of the need to provision the incoming junctions to the new switch which was using the same 7XXXX number ranges, a delay of some 15 minutes was experienced before the subscribers could be contacted. The PRX was one of the most advanced switches in production at that time. It offered a range of services and facilities considerably in advance of the other automatic exchanges on the island. It automatically retained all subscriber call data and was able to produce computer readable output that simplified billing tasks and other administration. Call transfer and other diversion services were completely within the control of the subscriber and remote connections meant that the exchange could be administered without attendance. The system was also self testing and provided automatic reports of errors and potential problems. The exchange was also fully compatible with the existing network, special interfaces having been developed by Charles Holden in the Pye TMC laboratories at St Mary Cray. Because of the delicate nature of the electronic switching system it had been necessary to modify some of the services on the existing Central exchange, particular attention being paid to the 'howler' service. The howler was a tool used by telephone administrations to alert subscribers that have accidentally left phones 'off the hook'. In its simple form it applied a loud noise to the line to attract the subscriber's attention. In the

Image by the author

A general view of West PRX exchange circa 1975. The exchange processors can be seen in the middle of the picture and system control panel can be seen on cabinet at the front of the first row. The input/output control equipment is in the foreground with an ASR35 teleprinter.

existing system this noise was generated by a simple vibrating relay contact driving a transformer with a simple stepped amplitude control. The resulting signal was not only very loud but also contained large voltage spikes which could be detrimental to the electronics of the PRX (and, indeed, some modern subscriber apparatus). The new howler used a form of frequency shift modulation that provided a loud audible warbling sound, this principle was carried forward into other modern exchanges.

From the engineering viewpoint the PRX marked the end of the old style switching technology. This began with the allocation of directory numbers which up to this time had been directly associated with the subscribers line. In the PRX system subscribers were identified only by their equipment number, the directory number was associated with the equipment by means of the switch software. Thus contiguous line circuits no longer had contiguous numbers. This added a new flexibility to the telephone network which meant that in some cases when subscribers moved premises, their number could be associated with their new location without any physical wiring changes at the premise or the exchange, offering further savings in time and labour. Nevertheless, the PRX was back engineered to provide existing Strowger services such as Service Interception and C&FC.

Early in 1976 the lowest bid from a tender process was accepted from Messers Le Vesconte and Coutanche in the sum of £4,700 for the reconfiguration of office partitions on the ground floor of the trunk exchange building to make space for the 2,500 line extension to the PRX exchange. On 6 March West PRX exchange was opened in a low key ceremony and the automation of the Jersey telephone system was finally completed. The move to full automatic switching signalled a significant reduction in the number of staff employed by the board. Operators at three manual exchanges had been made redundant or relocated over the previous 10 months resulting in staff levels dropping from 437 full-time staff in December 1974 to 310 full-time with 78 part-time staff in December 1975, further reductions resulted from the cut-over of West and more would result from the introduction of ISD which opened on 2 May. By the year end the number of full-time staff had dropped to 276 with 51 part-time.

The Jersey telephone system was now a fully linked 5 digit numbering plan:

Exchange	Type	Number Range
Central	Strowger	2XXXX 3XXXX
South	Strowger	4XXXX
East	Pentaconta	5XXXX
North	Pentaconta	6XXXX
Central Relief	PRX	7XXXX
West	PRX	8XXXX

Table 28

Thus all numbers became simply 'Jersey' under the UK National Dialling Code 0534.

With the completion of the ISD trunk exchange extension, the board had over a period of just 11 years completed the changeover of four manual exchanges to automatic, a relief exchange for Central, the introduction of both STD and ISD and upgrades to the power systems at Central inclusive of new buildings in each case. Aside from the actual exchanges themselves, the Board had also made substantial upgrades to the external line plant, reducing the amount of open wire overhead line distribution substantially with a programme of civil works to enhance and extend the underground distribution network and to re-plan distribution poles to ring pole distribution and in some cases the use of insulated drop-wire to replace open wire. Much of the exchange extension work had been carried out by its own engineers, while new work was installed by contractors. In total almost £5M had been invested in these major projects and it was with some relief that the Board was now able to reap the benefits of those changes and enjoy a period of relative stability for the next few years. Now the focus would be on innovation and improving facilities and services rather than a strict programme of necessary change.

Chapter 19 - From Analogue to Digital 1976 – 1993

Technology was now rapidly changing and push button telephones were becoming available not only in DTMF format, which could only be used on the crossbar and PRX exchanges, but now large scale integration (LSI) made available the same functionality for Strowger. Pye TMC introduced the first range of push button to dial pulse telephones based on the Post Office telephone 746 instrument, the latest iteration of the 700 series standard PO design. Some versions of this telephone later incorporated number storage facilities, a feature which up until this time had only been available to businesses or on certain telephone exchange types (such as PRX). At the same time new ranges of subscriber equipment started to be available including new loudspeaking telephones, more sophisticated PABX's and key systems and new lower cost ranges of facsimile (Fax) machines from Japanese manufacturers.

On the evening of 25 January 1977 the unthinkable happened; all three submarine cables connecting the Channel Islands to the UK were cut during stormy weather, probably by dragging anchors. The odds of this happening simultaneously were quite long since the cables were separated by large distances and landed at different locations in the islands and in the UK. However, the consequences were grave. The loss of the cables was signalled to the UK by the Guernsey harbour radio. The cables between Jersey and Guernsey and Jersey and France remained intact so very limited communication with the outside world was possible via the operator. Calls via France were limited to 3 minutes and had to be booked in advance. The effects on the finance industry were obviously severe. Some companies sent representatives off the island so that business could continue, updated paperwork being sent via air charter. The airport was able to continue operating under difficult circumstances; communications normally transferred by telex were being conducted via ground to air wireless telegraphy. The Post Office cable ship *CS Monarch* and sent to the area to search for the cables which could have drifted some distance from their original course following the cut. The first cable, the largest carrying some 1,380 circuits from Guernsey to Bournemouth, was located about 20 miles north of Guernsey. This cable was returned to service on the morning of the 28 January with a temporary repair, which enabled a near normal service to be resumed. The two Dartmouth cables were repaired by 30 January. Final permanent repairs were carried out during the weekend of 25 February.

The introduction of ISD was not without problems. Exposed to the possibilities of dialling direct to distant places, subscribers suddenly found greatly increased bills. Complaints were made directly to the Board and also aired in the local press. The inability to differentiate international and UK calls from the total on the bill caused some difficulties. The Board chairman explained in the press that all calls were accumulated on a single subscriber meter and that unlike other jurisdictions such as the USA, no means of separately identifying trunk calls was possible on Jersey exchanges. (This was not strictly true since the PRX had this ability if required.) The only solution for disgruntled subscribers was to request a Printed Meter Check (PMC)[123] which only provided a record of calls *after* the complaint, which would not be particularly useful. Adding PMC to a subscriber's line was a complex and temporary solution, since only limited numbers of this equipment was available at each exchange. An alternative solution offered to subscribers was the provision (as rental) of a Subscriber Private Meter[124] which displayed in real time the number of charge units per call. Proprietary solutions for use with telephones were also available from office

---

123 Post Office Telecommunications Journal (vol. 13, no. 3, Summer 1961, page 139)
124 Subscriber private Meter Equipment POEEJ Vol 51 Part 4 Page 338

equipment suppliers that relied on a modified digital clock that could be programmed with dialling codes and costs using a card reader or other input system.

By this stage the Telecommunications Board was beginning to run the operation as if were a business rather than a States public service. Historically the committee had been given considerable latitude within the States budgetary system and rarely was there any debate over proposed investment. This was reflected in its attitude to the pricing of services. Some controversy was created when charges for the Night Service Transfer service were increased from £3 to £5 per quarter. This service was frequently used by doctors that wanted the normal surgery number diverted to the duty doctor's home after hours. This had to be effected via the operator only for Central and South exchange subscribers. The doctors protested that a state owned and highly profitable service should make such increases. The Board justified the increase based on the higher costs of providing this service manually and that the fees would help to subsidize other widely used services such as directory inquiries. On the other exchanges this service could be set up by the subscriber, but a quarterly enabling fee was charged. This tension between public service and commercial operation was again highlighted the following year when in March 1978 Deputy Norman Le Brocq, who had earlier in the month accused States trading departments as 'taxing by stealth', proposed free telephones for the aged. Although it was not until November 1979 that this proposal was finally debated by the States, the suggestion was roundly rejected. By now the Telecommunications Board was returning well over £1M to the States treasury annually.

Technological developments continued apace, but at this stage were largely confined to customer apparatus where the risks of leading edge developments were generally considered more acceptable than in public network equipment. In particular, equipment was becoming more specialized and this was illustrate by a much publicized opening of the Grand Hotel's new Ericsson ARD 561H[125] switchboard. This was an adaption

---

125 Ericsson Review. 1962. V.39 Part 3

of the standard crossbar switched PABX to incorporate features specific to the hotel and leisure industry. Meanwhile at the Trinity Gardens repeater station a programme of upgrading saw the replacement of the the remaining valve based transmission channel equipment with transistorized versions in 62 type construction practice.

Early in 1979 the Jersey Post Office philatelic bureau published a set of stamps celebrating Europa 1979 which featured a manual switchboard and the PRX exchange. This gave the Philips exchange some unexpected publicity. At the time the philatelic department of the Jersey post office was a profitable adjunct to its regular business. A 1,000 line extension was completed at East exchange by Italian contractors; while at South exchange automatic routine test equipment was installed to replace the laborious manual testers and to enable the continued single engineer maintenance operation. Obsolescent equipment at the Trinity Gardens repeater station was replaced where necessary with more modern equivalents in an effort to reduce maintenance effort and improve efficiency of the transmission network.

The Telex system, which had come under the auspices of the Telecommunications Board as a result of the takeover of the Post Office monopoly, was still growing in popularity among businesses. This was as a result of its *de facto* position as a written conformation of instructions, and in this role was considered a legal document[126]. Telex was used by many businesses that required certainty in transactions such as banks, importers and travel agents. The advantage of Telex was its 'answer back' facility. When another Telex machine was called, its address code would be transmitted back to the sender, giving immediate confirmation that the connection was made correctly. When the Board took over the service all Telex connections were backhauled over 12 channel (52 type equipment practice) or 24 channel (62 type equipment practice) multi-channel voice frequency (MCVF) circuits to the Telex Bristol Zone exchange at Bournemouth. The number of Telex lines had expanded rapidly from 86 at

---

126 Acceptance of a contract see *Entores Ltd v Miles Far East Corp.*, [1955] 2 QB 327

Image by permission of Charles Watts

This picture shows the Ericsson ARD 561H crossbar PABX being installed at the Grand Hotel, St Helier. The switch could be scaled from 60 to 270 extension lines and from 6 to 40 main exchange lines. While the switch appears large with modern electronic systems, it was nevertheless considerably smaller than the Strowger step-by-step system it replaced.

the takeover in 1973 to 286 by mid 1979. The UK Telex[127] network had been developed during the 1950s and the country had been divided into eight zones and 50 charging areas with a mixed 5 and 6 digit numbering plan. The Bournemouth area was in the Bristol zone (Zone 4) and had numbers allocated in the range 41XXX. Thus the Jersey Telex population occupied more than 25% of the available numbers on the exchange.

The Board bought a Telex concentrator switch from Northampton based Databit Limited, a subsidiary of Databit Inc. founded by an American, Peter Cohen, who had patented[128] a method of multiplexing telegraph signals. The core of the switch was electronic, controlled by a processor unit. The interfaces to the connected teleprinters and to the Bournemouth main exchange were based on the standard ± 80 volt telegraph signalling system. Power supplies within the equipment converted the standard -50V exchange battery to the required signalling and logic voltages. The introduction of this concentrator enabled a considerable reduction (probably in the order of 90%) in the number of backhaul circuits required to Bournemouth and also allowed the Board to manage its own Telex lines. Further advantages accrued to the Post Office which was able to recoup a large number of Telex lines on its exchange. A new numbering range 4192XXX was allocated to Jersey Telex subscribers after modifications at the Bournemouth exchange. Existing users had their number amended by removing the '41' prefix and substituting '4192' instead. It is interesting to note that the Board had entered into discussions with Pye TMC Limited during the installation of the PRX exchanges regarding the possibility of integrating Telex switching. An experimental model of a Telex line card[129] was developed in the St Mary Cray laboratories but the project was never pursued.

The PRX exchange at West was badly affected by a lightning storm in September. The lack of lightning protection on the MDF damaged several subscribe line circuits and the resulting surge current caused a processor

---

[127] POEEJ Vol 48 Part 1 - Automatic Sub-Centres for the New Telex Service.
[128] Adaptive Sampling Rate Time Division Multiplexer and Method, United States Patent 3862373, Peter Cohen, Databit Inc.
[129] See www.prx205.org last accessad 08/08/2016

fault which required a complete reload of the system software. Engineers from Pye TMC's laboratories in Malmesbury, Wiltshire were called to assist in the system rebuild.

Later in 1979 radio paging was also introduced. The Motorola system[130] was an electronic switch and was connected to the main public exchange via a 6 digit number in the range 28XXXX. Each pager had two distinctive tones available activated by separate numbers, thus enabling the user to have a rudimentary signalling system. The equipment also provided a recording feature that enabled the caller to leave a message and the pager user to recover the message at a later time. The alternative number would alert the pager user to the need to call a specific number in order to act upon the alert. Although Motorola marketed its product as a campus system (producing a wide area system later), the relatively small size of the island enabled the system to operate satisfactorily. Transmitters were originally located at Les Platons and Five Oaks with a third site located at Trinity Gardens brought into service in 1983. This service was particularly popular with the medical profession. By the year end 264 pagers were in service.

Towards the end of the year the Board began to experience the beginning of the liberalization of the telecommunications market. Local retailers began to introduce ranges of imported Asian 'novelty' telephones. This was part of a trend elsewhere to deregulate the supply of telephone instruments but at this stage the Board were not ready to accept this change. This led the Director of Telecommunications, Leslie May, to issue a statement on the Board's position disapproving the connection of these telephones and warning of the potential to degrade the quality of the network. This was a reaffirmation of the monopoly position adopted on the introduction of the 1972 Law. Admittedly, the provision of subscriber apparatus at that time was not straightforward. The methodology then in use was cumbersome and for the most part connection of additional telephones to an existing line required fixed wiring changes. The Board's view was that it owned the cabling, even on subscriber premises, and therefore connection of these

---
130 Introduction to Paging Systems, One-Way, Two-Way, Pocsag, Ermes, Flex, Reflex, & Inflexion Lawrence Harte (ISBN: 0974694371 / 0-9746943-7-1)

instruments was strictly speaking breaking the Law. The standard defence was that these phones could damage the network. This was clearly nonsense, since these imported telephones were widely used elsewhere without detrimental effect. However, it was clear that this was a problem that was not going away and in any event the Board's policy would be difficult to police.

On 18 December 1979 there was a repeat of the 'million to one' event experienced in 1977 when once again all three submarine cables to the UK were cut during stormy weather in the English Channel. Although the previous experience had been an embarrassment the view prevailed that the odds of it happening again were extremely long. The new event was, by consequence of the growth in the telephone network and the finance industry's increasing reliance on telecommunications, even more disruptive. The break severed all but the most rudimentary voice links with the outside world, the remaining connections being again via the old pre-war 12 channel cable to Rennes, of which only 5 voice circuits were available for Jersey and two for Guernsey. Urgent calls were limited to 3 minutes and queues of up to 5 hours were experienced. Severe weather delayed the arrival of the cable repair ships and again the finance industry was thrown into chaos. Paradoxically, only the previous week the States had approved capital for the installation of a new cable to Dartmouth, but planning, manufacturing and installation would mean that this would not be available for two to three years. The cables were out of service for 3½ days until the weather abated sufficiently for the Post Office Cable Ship *CS Monarch* to start repairs. The first cable was temporarily repaired shortly after midnight on 22 of December followed shortly after by the other two cables.

The repeat of this catastrophic failure focused the minds of the Board and proposals began to be put forward on ways to circumvent or mitigate future problems. One of the possible options was a satellite link, but the alternatives of microwave links to France and perhaps to the UK were somewhat more feasible. The president of the Board, Deputy Ellis, told the press that all options would be considered. Talks were already under way with the UK authorities for a new submarine cable and a delegation was to go to Paris on 23 January. In the event it was decided to construct a 960 channel microwave link to the Isle of Wight from Alderney, which would then be linked back to Guernsey over a further microwave route and then on to Jersey via the existing inter-island cable. Such a link was at the very limit of line of sight radio links, especially given the tidal range over which it was sited. Each island would have access to half the circuits under fault conditions.

At the end of January 1980 work began on the construction of the new £4M submarine cable that would be laid from Dartmouth to Greve de Lecq. Construction work began on the underground ducts which would stretch along Grande Route de St Ouen, Rue de la Fontaine , St Peter's Valley to Bel Royal and then along Victoria Avenue and Rouge Boullion to the repeater station at Trinity Gardens. Meetings were also held with the French telecommunications authority (PTT), to discuss a new microwave link to Normandy. The French were sympathetic but wanted the link to be a 'live' service carrying normal traffic and not just for emergencies. Further talks would involve the British Post Office in order to continue the link onward to the UK via France. In April the representatives from both Jersey and Guernsey and Post Office Telecommunications, signed an £8M contract with Standard Telephones and Cables Limited for the manufacture and laying of a new cable (designated CI No. 6) from Dartmouth to Greve de Lecq.

Not all submarine cable was destined for the UK, as in February the cable to Elizabeth Castle was renewed, the original having been installed by the German occupation forces circa 1942/43. The cable was laid in the sand adjacent to the causeway from West Park by the Board's external engineering department and local building contractors.

In June the Board announced increases in the cost of telephone calls from 2.5p to 3p per unit. The cost for call box users would also rise from 2p per call to 5p. The Board president, Deputy Rumbold, justified the increases as necessary to fund future improvements to the system. The increases would add approximately £600,000pa to the revenue which at this time was delivering a hefty 14% annual return on turnover to the States treasury. At the same time, however, it was announced that a programme to convert public telephones from pay-on-answer to a more modern pre-payment type that would allow more international dialling for the first time, would be started in the autumn. The existing telephone boxes would be converted to accept 5p and 10p coins during this period. Approximately 1,250 public and private user coin boxes would be converted during the project which would start in 1981 would require modifications to the exchange equipment that would provide suitable controls for the new coin boxes. Additional 50Hz subscriber private meter equipment (SPM) adapters, already in use on standard and switchboard lines, would be installed. A beneficial trade-off of this move enabled the freeing-up of the considerable exchange rack space occupied by the existing coin box control equipment.

In June the States approved the budget of £450,000 towards the joint construction of the Jersey-France microwave link. The contract for the supply of the microwave equipment was awarded to Telettra SpA of Italy. The link that spanned the English Channel between the Guernsey Telecom radio tower at the Alderney exchange building and the ITA 405 line TV transmitter mast at Chillerton Down[131] was supplied and installed by Post Office Telecommunications and jointly funded between the parties. The Chillerton Down site is some 166m above sea level and the mast measures 229m. The diversity-switching microwave dish array was located close to the top of the mast (it was actually moved closer to the summit after the closure of the 405 line TV service in 1985). Work on this system started at the end of 1980 and it was commissioned at the end of 1981. The system was subject to the vagaries of the weather and the tides since it was barely 'line of sight'. Fortunately, it was only called upon once in earnest and was abandonned later when circuits via France were increased.

The Minden Place buildings were home to a number of exchanges and switching systems. The original buildings had been designed to house a manual trunk switchboard and a single Strowger exchange. By 1980 the buildings contained a 16,000 line Strowger switch, a trunk exchange with an international switch extension, a manual trunk switchboard, directory inquiries, the PRX exchange and a Telex switch plus a number of other

---

[131] The Transmission Gallery website http://tx.mb21.co.uk/gallery/chillerton-down.php last accessed 08062017

subsystems. At this time the reliability of the Strowger exchanges was becoming critical. The maintenance effort required increased continuously as the equipment aged. Routine test equipment was used intensively in order to detect faults before they caused service problems. The maintenance group developed in-house a 'fault-trap' device to assist in the identification of faulty equipment during hours when the exchanges were unattended and this had a significant impact on the requirements for out of hours working, resulting in the lowest number of engineer call-outs for many years. Maintenance effort on the other crossbar and PRX exchanges remained low, as these systems were less prone to wear and mechanical failure.

Although trunk calls were now increasingly being direct dialled by subscribers, there were now 112 destinations available which had necessitated a further extension of the trunk switch, the number of operator enquiries calls remained static at over 2,000 per day. This included subscriber services such as call diversion and interception which still relied on manual intervention on the old Strowger switches.

Because of these maintenance issues and the continually rising traffic during the year the Board also started to investigate the future strategy of the telephone system. It was becoming clear that the existing technologies were increasingly more difficult to upgrade and maintain. At this time the telecommunications industry was in flux; the long promised digital exchanges were still not in full production but it was also clear that existing space-switched analogue systems were coming to the end of that technology's ability to provide innovation. Consumer demand for new services was also beginning to develop, particularly within the finance industry. Now that more international destinations could now be dialled direct there was less demand for operator intervention and thus the reduction of operating staff was in prospect. The Board therefore presented a budget estimate of £4.125M for the proposed replacement of the existing trunk exchange with a new 'electronic' switch, but at this stage no clear picture of what form this would take was apparent. Although the PRX switching system could have been adapted to trunk switching, no development of British Post Office type interfaces had been carried out by the manufacturer as the exchanges in Jersey were, in the end, the only two examples in service in a UK environment. Indeed, the Board had not developed the existing PRX system software in line with other administrations and had adopted a deliberate policy of minimizing costs of maintenance which later proved false economy.

A new yellow and white colour scheme was adopted for the Board's telephone kiosks and van liveries. Another development was the introduction of a new logo, expressing the further commercialization of the Board. The new image replaced the rather amateurish internally developed earlier version with a new design commissioned from a marketing company. It is also interesting to note that during this year the Board experimented with bicycles instead of vans around St Helier in order to improve subscriber service response times.

1981 began with an industrial dispute over pay that threatened strike action. For a time a work to rule was called which affected overtime and other out of hours working. Although this action was resolved after negotiations, further union disputes arose in November over the States sponsored job creation programme that was a feature of the early 1980s recession. The union shop steward, Bill McKain, objected to plans to introduce work creation staff into areas normally operated by skilled technicians. Again, after a threat of industrial action, a negotiated resolution was reached to enable 8 placements. McKain himself became embroiled in dispute with the Board following his appointment as full-time local EEPTU representative. The Board president, Deputy Ellis, accused McKain of alleged fraudulent returns on time sheets. After a public spat between the union and the Board the matter was put to rest.

During the year upgrades to the MDFs in Central and West exchanges were carried out, retrofitting lightning protection to prevent further damage to the PRX subscriber line cards. The third extension to the Central PRX was competed adding a further 2,048 lines bringing the switch up to over 8,000 line capacity. The first extension of the Telex concentrator was also completed, confirming the continuing popularity of the Telex for business.

The first large additions to the Resources Recovery Board's telemetry system was carried out, revamping the entire branching network under a major extension of its sewerage and surface water monitoring system. Tests during 1979 on the main exchange battery, which was installed at the time of the first automatic exchange in 1959, had shown that it was now in need of replacement. A new battery was ordered from Tungstone Batteries Limited of Market Harborough. At the same time the mains rectifier plant was also increased by the addition of a 1,000A 50V rectifier unit bringing up the total power capacity to 2,500A.

Preparatory work for the planned 120 channel microwave system to France which would replace the pre-war submarine cable link was carried out at La Chasse, St Ouen where a transmitter tower was sited. The rise in popularity of dial-up modem data connections between businesses on the island led the Board to consider the application of local call metering equipment and an investigation into the practicality of this was also completed although this idea was later abandoned following the ordering of new exchange equipment. Finally, a computerized line plant management system that was developed internally and running on Commodore Pet computers was commissioned, improving efficiency of allocation and management of orders.

Also early in 1981 the Board announced that after long negotiations with the newly formed British Telecom (BT) that it was able to bring the new Prestel[132] service to Jersey. Prestel had been developed by BT at its Martlesham laboratories and in principle it was similar to the video text systems used by the BBC (Ceefax) and ITV (Teletext), the fundamental difference being that the Prestel service was connected via a telephone line modem to the Prestel computer from which data could be downloaded. The service first introduced by BT in 1979 still required a television to display the data in simple text format, its main advantage being that more data could be distributed than via the broadcasters' systems which relied on the limited spare capacity in the video channel. Prestel can be considered as an early version of a bulletin board, it was rather less functional than either

---

132 Fedida, S. and Malik, R. (1979). The Viewdata Revolution. London, UK, Associated Business Press, ISBN 0-85227-214-6

email or the Internet. Prestel's earliest users were news agencies and travel companies, indeed this remained the case throughout its life. This service was relatively expensive since it required both a subscription at £17 per quarter (for residential users £7 per quarter) and in addition a £20 connection charge and a fee for each call made to the system. The calls were especially expensive from Jersey as users needed to dial a UK number at trunk call rates, compared to BT users that largely had access to local numbers at local connection costs. The Jersey Evening Post became a Prestel information provider in January 1982.

In the spring of 1981, the Board installed an experimental telephone kiosk in Ruette Haguais, St Helier. This kiosk was constructed in aluminium and was specially designed for easy wheelchair access, coinciding with the International Year of Disabled People. This heralded a 20 year search for a replacement of the traditional Jersey kiosk which had been previously designed and built by island building companies and the Board's own carpenters.

Preparatory work for the new submarine cable to Dartmouth began at Greve de Lecq in February 1982 when the cable ship *CS Libation* arrived to lay the 3.5 nautical mile shore end. The shore work had been carried out earlier to provide a ducted route to the Trinity Gardens repeater station. Forty engineers and labourers floated the cable ashore on over 200 special buoys to where it was drawn into the ducts at the foot of the slipway. Thereafter the cable shore end was entrenched in the sand to prevent damage from shipping. In April a new sea-earth system was laid in the Peoples' Park, St Helier and connected back to the repeater station. An efficient system earth is a necessary requirement for the reliable operation of submarine systems and the park was the nearest open space that could be used to provide sufficient surface area for the earth conductor.

In July the programme of changing all public call boxes to pre-payment began. The new BT CT22B coin collecting telephones were selected for the programme which accepted 5p, 10p and 50p coins and thus enabled direct dialled international calls. At the same time the cost of ordinary telephone calls rose from 3p to 3.5p. This increase was blamed by the Board on the

Finance Committee's insistence that States trading committees were self funding, and this was especially relevant in the case of the proposed budget for the upgrade of the telephone trunk and switching network. A number of other charges were increased at the same time, especially operator connected calls. Also announced that month was the ending of the overnight telegram service. The growth of the telephone had been eroding the market for telegrams for many years and now the economics of the telegraph system were called into question. A new facility that partially replaced the telegram was the Royal Mail Intelpost public Fax service that was introduced the same month by Jersey Post.

The specification for the new electronic telephone exchanges was now in the course of preparation. Over the last 12 months a clearer picture of the type of exchanges that would become available from manufactures had formed. Many manufacturers were now confidently touting their digital switching technologies. As usual subscriber apparatus was ahead of public switching and the Board started to deploy the BT Business Systems Monarch PABX[133] which was one of the first digital systems to market and twenty five installations were completed during the year.

A more controversial move came in late August 1982 when local politician, Senator Dick Shenton, proposed the part privatization of the telecommunications system. A proposition was lodged *au Greffe* for debate in the States assembly. This proposal envisioned the floating of 49% of the interest in the business through a public share offer. Shenton was clearly inspired by the proposed share offer for BT following its incorporation by the Thatcher government. There was some disquiet among both States members and the public regarding the operation of States trading departments, as profits were returned to the Treasury and not used for future development. This created a more costly arrangement of bidding for funds in the annual budget which were then subject to interest payments to the Treasurer. Shenton further claimed that a part privatized operation would be more efficient and profitable.

In an attempt to head off the debate the president of the

---

[133] POEEJ Vol 73 Part 1 1980 - Monarch 120 – A New Digital PABX

Telecommunications Board, Bill Morvan, as reported by the JEP the day before the States sitting of 6 October 1982, called for an independent inquiry in the the proposal. While he agreed that the Board was in negotiations with the Finance and Economics Committee over repayment plans he was not convinced that a sell-off was the right answer. Under the current arrangements the Board was permanently in debt to the States and, he agreed, this was inefficient. There was continued need for investment in infrastructure in order to keep the system up-to-date. However, he cautioned a rash decision and insisted that a through investigation of all the States trading departments should be undertaken. However, the president of the Finance and Economic Committee was more forthright. Senator Ralph Vibert condemned the idea as 'complete nonsense' and that any investigation would be a 'waste of time and money'. He was convinced that the present arrangement was adequate and that the States could provide the necessary estimated £20M investment over the next few years. The Board was well run by the States, he claimed, and that there was no need for outsiders to advise the States on how to run their businesses. He further accused supporters of the proposal of living in 'cloud cuckoo land'.

The subsequent debate in the States was very much polarized between two opposing camps. Senator Vibert of the Finance and Economics Committee led the assault against the proposition which was firmly based in the *status quo*. He reasoned that while there may be inefficiencies in the present system of finance, the States owned the operation and that if it were sold it could become an uncontrolled monopoly. This view was backed by the States only communist member, Deputy Norman Le Brocq, who observed that the current position of the Jersey Electricity Company (JEC) demonstrated that fact admirably. Senator Piere Horsfall (president of the Policy and Advisory Committee (PAC)), pointed out that in the case of the JEC the States had taken up shares in a private company which was the reverse position. However, he agreed that a review of the Telecommunications Board's position in light of the fast-changing industry should be conducted to inform house. In his opinion the current structure was unsuited to the task but he also agreed that an injection of private capital could benefit the States. Bill Morvan still preferred the option of referring the matter to the PAC. At the end of the debate the house divided and Shenton's proposition was lost by the narrowest of margins: 20 votes for; 22 against. Clearly, the States were somewhat open to the idea of a privatized operation but in what form this should take was at this point unclear, however the seeds had been sown for future debate and policy. Had the vote gone the other way, the eventual opening of competition in the telecommunications market may have taken a different course.

November 1982 saw third occurrence of the 'one-in-a-million' event of all four of the UK submarine cables failing at the same time. Embarrassingly, the failure included the recently completed cable from Dartmouth (CI No 6) one week into service. This time, however, the Board was prepared. The newly installed microwave link between Guernsey and Jersey was hurriedly commissioned enabling the full sharing of the 960 circuits between the islands of the Isle of Wight to Alderney microwave link commissioned the previous year and telephone and telex traffic could be rapidly restored even if to a lower grade of service. The available capacity was sufficient to cater for about 60% of the Jersey traffic during that time of year. The Alderney to IOW microwave link was operating to the very limits of the distance across open sea and the quality of the connection was somewhat indifferent. The BT cable ship *CS Monarch* was already in Southampton and would start repairs as soon as the weather improved, although soon after starting repairs the CS Monarch had to leave the area for other contractual commitments and so the CS *Alert* took over the operations and repairs were completed within a week. It was decided that one of the 120 circuit Dartmouth cables, the CI No 3, (which had failed prior to the complete outage in October) was beyond economic repair and should be abandoned. Later that year the Board announced plans to bury approximately 27 nautical miles of the most vulnerable parts of the new cable in the main Channel shipping lanes to mitigate the problem. This task would be carried out by a recently developed submersible cable plough called *Seadog*. Work continued with the French authorities on the microwave link to replace the ageing pre-war cable from Fliquet to Pirou.

The Board's financial department introduced Apple computers into its operations to support financial modelling and payroll functions. A counter terminal was introduced for bill payment. At the same time investigation of the possible integration of the Board's requirements into the new States mainframe computer were also investigated with the States Computer Services Department.

Early in 1983 the Board engineers completed the preparatory work for the specification of its next generation of exchanges. The technology had become clearer over the previous 12 months and digital main exchange systems were now entering production. Maintaining the existing Strowger exchanges was now becoming increasingly difficult since the Board's traditional suppliers had ceased production and thus spare parts were, by necessity, being sourced from cannibalized systems elsewhere. Indeed, a 600 line extension was completed at South exchange using equipment recovered from Guernsey as it completed its exchange conversion to electronic. Philips engineers added a further 512 PRX lines at West exchange. At the same time both West and Central PRX exchanges had updated administration interfaces installed, replacing the paper tape and teleprinter system with a more reliable and faster cassette tape drive, VDU and dot matrix hard copy printer. While these exchanges continued to provide almost fault-free service, the crossbar exchanges were beginning to show signs of mechanical wear. Over 600 relay contacts were replaced at North and East exchanges during the year by way of preventative maintenance.

A budget estimate of £16M for modernization over the next 10 years was presented to the States in June by the Board president, Constable of St Lawrence, Bill Morvan. New switching equipment was estimated at £10.5M while the balance would be spent on network upgrades, and on increasingly sophisticated subscriber apparatus. He revealed that a further £3M would be invested in the underground cable network continuing the programme to reduce overhead distribution which had begun in the mid-1960s with the conversion of South exchange to automatic. This investment had, he asserted, proved its worth in fault reduction and overall network reliability as well as being more aesthetically pleasing than unsightly poles. He also suggested that the Board would initiate a drive to increase provision of telephones in hotel bedrooms, where he claimed that there was only 4% penetration. The budget was approved after a short debate.

Meanwhile the Director of Telecommunications, Leslie May, was appointed as States Treasurer, taking up the position in September, and the vacant post advertised both in Jersey and the UK. In the event the Board's Controller of Planning, Tom Ayton, was appointed in November. Further price increases were approved in the States budget raising calls from 3.5p to 3.8p from the beginning of 1984. With the budget approved the Board finalized work on its exchange specification for tender to European and American suppliers. Invitations to tender were sent in early 1984 to Standard Telephones and Cables, L M Ericsson of Sweden, Philips Telecommunicatie of the Netherlands, CIT-Alcatel of France, GEC Telecommunications and AT&T of the US.

Further computerization of the business administration systems was undertaken introducing forecasting and management software called Wizard. Consultations with the States Computer Services Department resulted in approval for the Board to purchase a Digital Electronics Corporation (DEC) PDP 11/34 mini computer, which would support the ledger systems. It was expected to be commissioned by May 1984.

In early 1984 the Board contemplated relaxing its monopoly over the supply of subscriber apparatus. The market was now being supplied with low cost attractively designed telephones, mainly from the Far East and although it was technically illegal for subscribers to connect these to the network, it was difficult to monitor and control this practice. In the UK, BT had liberalized the supply in 1981 on the introduction of the new British Telecommunications Act. The supply chain had been further improved by the establishment in 1982 of the British Approvals Board for Telecommunications (BABT), which was charged with the duty to test and approve Customer Premises Equipment (CPE) taking over the task previously done by Post Office Telecommunications. Approved telephones were marked with a green circle logo and those not passing the process were marked with a red triangle. Theoretically red triangle telephones

Image by the author

This image shows a Jersey Telecoms branded instrument that was introduced shortly after the liberalization of the Customer Premises Equipment market in 1984. Prior to this date only Jersey Telecoms was permitted to sell or lease phones.

should not be connected to the public telephone network, although quite how this could be policed was never clear. Simplification of the telephone wiring system had been adopted earlier by BT when it introduced the plug and socket terminal[134] (PST), the now familiar telephone connection system. The Board had adopted this method of installation at the end of the previous year and this opened the way for freeing up the market. In order to remain competitive following liberalization the Board had expanded its sales outlet in the Minden Place offices to enable the display of a range of subscriber apparatus.

The proposal was debated in the States on 28 February where the plans were outlined including, bizarrely, the circulation of a list of approved telephone types for States' members' information, the currency of which must have been short lived. It was also stated that other CPE such as Telex terminals, answering machines, call loggers, modems and facsimiles would be individually approved on application to the Board. No date was given during the debate for the relaxation of the monopoly although it was later announced that it would be from 1 July.

The island's off-island links were further secured when in when the Bailiff, Sir Frank Ereaut opened the France-Jersey microwave link at the La Chasse, St Ouen radio station. This £200,000 120 channel system supplied by the French manufacturer Société Anonyme de Télécommunications, Radioélectrique et Téléphonique (SAT) was brought online on 28 May. This new digital system was connected back to Central exchange, St Helier using four 30 channel Pulse Code Modulation (PCM) systems. These PCM systems were part of an initial order of seven systems ordered for the upgrade the island transmission network in preparation for digital switching. The first of these systems had been installed the previous year between local exchanges. On the French side the microwave link transited through Barneville, Lithaire and St George before joining the main French network and on to Paris and the international exchange at Rheims. Half the circuits would be used for normal traffic the remainder for emergencies. After the commissioning of this new link, the pre-war Fliquet to St Lo cable would be

---

134 POEEJ Vol 83 Part 4 - New British Telecom Plug, Socket, Cordage and Line Jack Units

abandoned.

The Board received four responses to its tender for digital exchanges in July. Ericsson declined to bid. STC's bid was again provided by FACE Standard spa of Italy offering the ITT System 12 digital switch. Philips was now in a joint venture with AT&T (Philips en AT&T Telecommunicatie BV) and it offered the European version of the AT&T No. 5 Electronic Switching System (5ESS). GEC offered the BT System X exchange. CIT-Alcatel offered the French E10 digital switch. The Board's engineers would spend the next couple of months evaluating the responses.

The 5ESS was probably not fully compliant with the tender since at this stage of its development it used an analogue switching matrix for the subscriber concentration stage. This was a semiconductor based switch that had been developed by Bell Labs for satellite systems and was known as the gated diode matrix[135] (GDX) concentrator. The ITT and French products were very much oriented towards continental telephone switching and would require considerable adaptation for a BT-like network as operated by the Board. This effectively left only the GEC offer on the table. The Board had previously experienced the problems of adapting continental products to a UK-like environment, although Philips had delivered on time and on budget there were still some reservations over the protracted delays experienced with Pentaconta. Thus on 12 September the Board announced that GEC had been selected as the winning tender, renewing the link with UK suppliers broken during the 1970s. The new President of the Board, Deputy Jack Roche informed the GEC Commercial Manager, Brian Page of the result that morning.

With the supplier settled, preparatory work could begin for the installation of equipment. During the year the construction group replaced Central exchange MDF connection blocks with higher density types to make way for 10,000 digital exchange lines. The manual switchboard was also reduced from 36 to 14 operator positions consequential to the lower traffic demand now that more international calls could be dialled directly. This freed up

---

[135] 100 Years of Telephone Switching (Pt. 2) Electronics, computers and telephone switching (1960-1985) - Robert J. Chapuis, Amos E. Joel IOC Press

space in the manual switch room for additional offices. Nevertheless, the Central PRX switch was further extended by 2,048 lines and 500 Pentaconta lines at North exchange to cope with continued demand.

The finance sector demand for data connections was also rising. The Board introduced the BT X25 based Packet Switchstream (PSS) system installing a network multiplexer (NetMux) in Central exchange. This provided a 4-wire switched data-packet network that was connected to the BT system. Subscribers were connected via dedicated private circuits. The PSS service was used by banks to connect to the Clearing Houses Automated Payment System (CHAPS) computer system. In the Telex world the introduction of new CCITT[136] standards together with liberalization of the teleprinter equipment supply, it became necessary to augment the Telex concentrator with the addition of voice frequency (VF) interfaces to connect the new ranges of equipment on offer.

With the rising demand for telecommunications services the Board requested increased staffing levels. At the end of 1984 it employed 248 full time equivalent staff but would require more as the business expanded. The changeover to new technologies while maintaining existing services drained resources. Already the Jersey finance industry was complaining of the long delays between service requests and installation. The new president, Deputy Robin Rumbold, defended the Board's position claiming that delays were '... no worse than in the UK ...' and that the demand for new sophisticated equipment was 'unprecedented'. However, the finance sector had a point; the continued monopoly was not best suited to swift changes in consumer demand. Statistics later reported in the States[137] showed that 71% of orders were completed within 3 weeks while the remaining 21% could take up to 14 weeks. This lack of consumer focus had been recognized in the UK where the Conservative government had sought to stimulate the market by allowing the Cable and Wireless subsidiary Mercury Communications to start offering competitive services from 1981.

---

[136] Comité Consultatif International Téléphonique et Télégraphique since 1993 called the International Telecommunications Union (ITU-T) was formed in 1956 to advise on standardization in telecommunications.
[137] States debate as reported in the JEP 29 July 1985

The formal signing of the contract to supply System X was made between the Board president and the Managing Director of the GEC Telecommunications Limited switching group, Tony Snoad, on 29 March 1985. The contract would initially be for a replacement trunk exchange unit and a 10,000 line subscriber switch. This would be the first combined trunk/local System X switch installed anywhere. Further contracts would be signed later for the supply of additional Remote Concentrator Units[138] (RCU) to extend digital services to the minor exchanges prior to their wholesale replacement.

Alterations to the existing trunk exchange equipment room on the second floor at Minden Place had started in January with the construction of partition walls and the installation of a raised floor in preparation for the arrival of the first phase of System X in the autumn. A small extension was built out over the Strowger exchange roof to house air conditioning units for the new exchange. The second stand-by generator set installed in anticipation of the additional demand from the new equipment was completed.

A second attempt by Senator Dick Shenton to part-privatize the telephone system was lodged *au Greffe* in April and debated in July. This time the debate was less polarized that the previous occasion and the idea was effectively kicked into the long grass when the States decided to refer the whole matter of the future of all its trading departments to the Policy Advisory Committee for review.

The System X equipment began arriving in August and by November it had been installed ready for equipping and testing. The System X exchange was built in the new BT standard Telecommunications Equipment Practice 1E (TEP 1E)[139] racking practice. This structure was an attempt by BT to standardize equipment cabinets across all types of installation and was of lighter construction than previous systems. The cabinets were available in two heights: 2.2m and 2.6m and thus could be installed in standard office buildings as well as in existing exchanges which were designed for 10' 6½"

---

138 Post Office Electrical Engineers Journal Vol 81 Part 4 1985 The Function and Use of Remote Concentrator Units
139 Post Office Electrical Engineers Journal Volume 72 Part 3 page 160

(3.2m) racks. There were other features of the new equipment that facilitated rapid deployment. The rack units used printed circuit backplanes and inter-module and inter-rack cabling was pluggable, using a factory-wired standard plug and socket system. External connections were minimized and where possible connected via plugs and sockets, thus wiring subscriber lines and trunks to the MDF was much quicker than with traditional systems. The GEC contractors began the installation of the system software and the commissioning testing could begin after the data load tapes had been created from local call routing data.

Demand for new lines required that a small extension be added to Central exchange using equipment recovered from the MNDX. The trailer unit was in poor repair and was scrapped. Additional 30 channel PCM systems were ordered from STC and GEC for further links between Central and the minor exchanges in preparation for the deployment of RCUs to provide Integrated Switched Digital Network (ISDN) services when required by subscribers. System X was now in its System Enhancement Programme 2 (SEP2) phase that would enable direct interconnection to PCM systems for both switching and voice. It would also shortly be one of the first systems to incorporate the CCITT SS7[140] common signalling architecture and thus would be able to connect directly to the UK digital trunk network without additional trunk signalling hardware. The Board was already planning with both BT and the Guernsey authorities a new fibre-optic cable designated the CI No 7 from Dartmouth to L'Ancresse and then on to Jersey over a microwave link. This would enable direct connection into the BT trunk network. Meanwhile, the trunk system would use digital interfaces for the analogue signalling systems AC9 Signalling and AC11 Signalling over the existing cables.

The commissioning work on the System X continued through to the spring of 1986 when the first trunk circuits were cut over from the old switch. This change-over was phased over four weeks as there was no easy way to complete a 'big-bang' solution, this also gave the engineers sufficient time to ensure that there were no serious problems. With the trunk switch changed over, the old electromechanical switching unit was recovered and

---

140 International Telecommunications Union Recommendation Q.700 CCITT Signalling System No. 7

sold as scrap. This cleared space for the installation of the System X local switching unit. The new equipment included the latest Rack Mounted Power Supply (RMPS)[141] system. This was a new diversion from the traditional telephone exchange power system which had formerly required a separate battery room for the open case cells where damaging acid fumes could be kept separate from switching equipment. The new RMPS used gel-filled sealed lead-acid accumulators and rectifier units and each suite of racks was supplied with its own power unit(s) rather than sharing a central battery. All RMPS were linked together to prevent catastrophic failure. This would eventually enable the recovery of the central batteries in the main and minor exchanges to create additional space for the changeover to digital switching. The System X local switching unit was commissioned in November and 2000 additional lines brought into service using the numbering ranges 58XXX and 59XXX. This required some reconfiguration of the East Pentaconta exchange routing plan and routing changes at North Pentaconta and the two PRX exchanges. Routing of traffic from South and Central Strowger exchanges was accomplished by 'tromboning' traffic through East exchange. There was little effect on transmission losses using this indirect routing as the exchanges were now linked over 30 channel PCM junctions which were effectively 'lossless' digital transmission paths. RMPS racks were installed at both South and East exchanges to enable the recovery of the old batteries. At South a standby generator was installed in the vacated battery room, while at East the battery room was refurbished to house an RCU.

The year ended with yet more cable failures. This time two of the three working cables were cut, fortunately the newest and largest, CI No 6, was unaffected thus the disruption was not as bad as it could have been, nevertheless, over 1,800 circuits were lost. Bad weather delayed repair for more than a week. This again sparked a debate on off-island connection security. The Board had investigated and dismissed satellite links as too expensive, estimating that such links were ten times more costly than equivalent cable. However, the Board proposed to open negotiations with BT, Guernsey and the French authorities in order to examine the possibility

---

141 Post Office Electrical Engineering Journal Vol 81 Part 4 January – March 1985

of larger capacity via microwave links.

It was also announced that the Board would no longer have responsibility for on-island radio interference investigations. The Department of Trade and Industry (DTI) would take back the service which had been subcontracted to the Telecommunications Board since the takeover of the GPO monopoly. Consequent to the privatization of BT the DTI had assumed responsibility for management of radio spectrum in August 1984 and now was rationalizing its operations. Answering a question in the States, the Board president noted that the DTI had the authority to rescind its agreement under the Wireless Telegraphy Act, which had been extended to Jersey by Order in Council in 1952.

As a consequence of the introduction of competition into the CPE market, the Board itself expanded its sales shop in Minden Place in order to display an enhanced range of telephones and other equipment. Around the same time it also entered the computer market signing a deal with US personal computer supplier Zenith, and it started supplying and maintaining equipment for local businesses. Because of its track record in the telecommunications business, the Board was easily able to leverage this image of competence into the computer world, rapidly gaining market share as a result.

Early in 1986 the States had approved a budget for the Board to explore the introduction of cellular radio services in Jersey. The Board started negotiations with both the UK suppliers, Cellnet and Vodafone, with a view to providing a licensed service in the island. The system deployed in the UK was based on the European analogue mobile standard Total Access Communications System (TACS) which operated in the 900MHz band. Wireless coverage footprints using this analogue system were considerably better than the later versions of cellular service. Propagation tests showed that at least two and possibly three transmitter sites would be required to get all-island coverage. Negotiations were concluded with Cellnet[142], a joint venture company between BT and Securicor which had won one of the first

---

142 POEEJ Vol 81 Part 2 1985 The Cellnet Cellular Radio Network Part 1 – General System Overview,

licenses issued by the UK government in 1984. Confirmation of the arrangement and budgetary approval for the estimated £1M necessary for its introduction was presented in the States in June 1986. The contract for the network was signed with Cellnet and three transmitter sites were installed around the island at La Chasse, St Ouen and La Collette, St Helier (mounted on the Jersey Electricity power station chimney) came into service by year end while the site at Gorey castle followed later in 1987. Transmission tests were under way before the year end and the system was officially launched on 2 March 1987.

The cellular system was the first service launched by the Board that required the user to purchase rather than rent equipment. Early handsets cost started at around £850 and the monthly rental for the service was set at £25 with a £60 connection fee. Call costs too were high; local and UK calls were set at 25 pence per minute (ppm) at standard rate and 9ppm at cheap rate. Calls to the Vodafone network and to Europe were 80ppm at all times. Calls to other international destinations ranged from 110ppm to 140ppm. Calls from Jersey fixed lines were set at 10ppm at all times. The switch for the local network was part of the Cellnet network at Bristol. Calls thus had to transit via Bristol to Jersey or any other destination. Call connection time was lengthy, around 17-20 seconds. Only 12 simultaneous connections could be made at any one time, this included on-island mobile to mobile calls. Despite the high entry costs around 150 subscriptions had been made by the end of the year. Radio communications were booming as also by year end around 1,600 units were registered on the pager system. Statistics published showed that the Board returned an operating profit of £4.9M during 1986[143] which represented an enviable return on capital employed of 24.2%.

A further 9,000 lines of System X were installed during the latter part of 1986 and early 1987. After testing, the changeover of Strowger lines in the 3XXXX number range was completed in four tranches of 2,000 between March and July. Again to simplify routing from existing Strowger exchanges calls to all numbers in the range 3XXXX were tromboned through the

---

[143] States of Jersey Telecommunication Board annual report 1987.

System X exchange during the changeover. This enabled the staged migration of lines from the old equipment. Installation of RCUs was commenced at South and East exchanges and brought into service in January and April 1988 respectively. All 5,400 lines at South in the 4XXXX number range were transferred to the RCU during the spring. The recovery of the redundant electromechanical equipment was completed by mid year, creating valuable additional accommodation. The RCU at East exchange was initially used as expansion adding numbers in the range 5XXXX.

In Central exchange the manual board accommodation was part refurbished with the introduction of a raised computer floor and suspended ceiling with air conditioning. The subscriber line test desk was then relocated to here from its position adjacent to the Strowger unit and at the same time updated with modern automated test equipment. New desks provided to replace the assistance operator positions were completed during the latter part of the year enabling the transfer of the directory enquiries services to the computerized Directory Assistance System (DAS)[144] the following year.

A new text message paging system was brought into service after in-house field trials. By the end of the year 1,974 pagers were in service of which 121 were of the new type. An average of 1,000 calls per day was being delivered to the paging system.

At the end of 1988 a new cellular mast site was opened at Five Oaks which provided an additional 24 circuits and the number of channels at La Chasse was also doubled to cater for the growing network which had increased to over 300 users. The new mast at Five Oaks had been erected in the spring in preparation for the new microwave link to France. This new 57m (185ft) mast replaced the earlier mast erected by the Department of Electronics for its radio services but incorporated some of its antennae.

More controversially, during the summer of 1988 new adult Premium Rate Services (PRS) appeared. These new services had become available with the introduction of BT's new non-geographic numbering which had begun in

---

144 The BT Directory Assistance System (DAS) came into service in November 1984

November 1985 with the opening of its Freephone and Linkline (0800 and 0345) non-geographic services[145]. This had now been extended to include other ranges on which third party providers offered adult content services, so-called sex lines. However, there was outrage by some members of the public, particularly local religious groups, that these services should be accessible on the publicly owned telephone system. The former president of the Telecommunications Board, Robin Rumbold, had previously set up the Board to be a self appointed guardian of public morality when he had announced in 1987 that such services would not be available in Jersey. From the summer of 1987 the Board had therefore actively blocked calls to all 0898 numbers. In November it tentatively opened access to the 0898 1212XX, 0898 1003XX and 0898 5005XX number ranges following complaints from UK companies that used these numbers for order lines and were inaccessible by Jersey customers. However, the market had changed and other network operators had entered the fray, notably Racal-Vodafone, which had introduced certain more controversial services using the prefix 0836. Although the Board blocked the number ranges associated with these services, other operators began to mix these numbers among other services such as catalogue order lines. This meant that numbers had to be individually blocked which was a far more intensive engineering matter. At this time Oftel had not yet taken control of the UK national number plan and there was every incentive for operators to use numbers for commercial gain. This led to a rather cavalier attitude to their use which caused further controversies later.

By November 1988 GPT contractors (the newly formed group following the amalgamation of GEC Telecommunications and Plessey Telecommunications) had installed more System X equipment and thus began the changeover of the remaining subscribers on Central exchange, with the numbers in the ranges 20XXX to 27XXX. Again this was completed in 4 tranches of 2000 lines and recovery of the old exchange equipment was completed by mid 1989. By the end of 1989 around 31,000 subscribers

---

145 Non-geographic means numbers that are not directly associated with a physical location as is the case with ordinary telephone numbers connected to fixed line telephone exchanges. Such non-geographic numbers can be mapped to one or more geographic numbers which may or may not be in the same country or numbering plan.

were working on System X at exchanges across the island, more than half of which were on Central exchange. A second System X unit was installed in the space vacated by the recovery of the old equipment in Central and this would be used for new line growth. With the closure of the Strowger exchanges the existing exchange maintenance and construction departments were amalgamated into a single group thus creating overall economies and releasing some staff to other operations.

The laying of the CI No 7 submarine cable was completed in January 1990 and opened for traffic in February. This cable terminated on Guernsey and was accessed via the 1,920 channel microwave link at La Chasse. The 50 year old CI No 4 cable to Dartmouth failed in February and was abandoned; it was in any case due for closure at the end of March.

Following a series of confrontations over pay and conditions between the Board and the Electrical, Electronic and Plumbers Trade Union (EEPTU) the two sides entered into a new agreement in May 1989. This was based on industrial practices adopted by the Japanese company Toshiba at its UK plants in which the employees were given more access to company financial data and pay and conditions were settled using a joint committee. It was hoped that his new arrangement would bring an end to the long running disputes which the previous year had resulted in a lowering of service standards with new line installation time extending to 14 weeks. The EEPTU eventually accepted a substantial 18% salary increase. Despite this the Board returned profits of over £6M for the year 1988, only marginally up on the previous year. However, price increases were implemented from July 1989 with call charges rising 12.5% from 3.9p to 4.4p per unit, quarterly line rental was increased from £11.25 to £13 for business lines and from £5.50 to £6.50 for residential lines, while new line installation charges were increased £5 to £45.

The new microwave link from the Telephone Engineering Centre site at Five Oaks to Barneville in France was commissioned by Thompson-Alcatel in July 1989 with 960 circuits shared with Guernsey providing an alternative link through France to the UK. The completion of the recovery of the old exchange equipment allowed the former exchange space at Central and

South to be redeveloped into offices. This enabled the Board to relocate its staff centrally from rented offices in various parts of St Helier. The Board also introduced card operated payphones in the spring at the harbour and airport, although retaining some coin operated boxes at the same sites. GPT Saphire supplied the call office telephones which were operated using special pre-paid cards instead of coins. This had the advantage of lowering maintenance costs as call boxes no longer required cash boxes to be cleared daily and also reduced the potential risk of theft and vandalism. Call boxes were increasingly popular among visitors and itinerant workers as international calls could now be dialled direct. The number of call boxes installed in both public and private premises had increased dramatically over the previous few years with over 200 public kiosks and more than 1,500 rented by subscribers. The new cards themselves generated interest and soon became another collectable item. This secondary market was exploited by the Board as it issued cards with new designs on an increasingly regular basis. Over 150,000 cards were sold during 1990 alone.

By1989 the number of subscribers connected via overhead poles was reduced to approximately 8% of the total. This was near the limit of the economic level of underground deployment as the remaining overhead subscribers were now in increasingly rural situations. This had been a result of a long drive to underground distribution that had lasted 20 years during which hundreds of miles of underground duct and polythene cables had been laid. The Board also provided a fibre-optic network for the States of Jersey Computer Services department, permitting interconnection between various States sites. There was now a noticeable decline in demand for Telex services following many years of sustained growth and during 1989 55 fewer subscribers were in service on its switch. This was as a result of changes in technology and declining costs of facsimile equipment, now around £450, allowing businesses to migrate to a more efficient and convenient messaging system and this marked the beginning of the end for this service.

During 1990 further expansion of the System X RCUs was undertaken to provide for growth and the changeover from the old electromechanical

Image by the author

An example of the Jersey Telecoms pre-paid phone cards, this one inaccurately illustrating 100 years of telecommunications in Jersey. Telecommunications actually started with the telegraph in 1859 but a permanent telephone service was started by the National Telephone Company in 1895.

equipment at East and North exchanges. A new voice mail service was introduced early in the year accessed from numbers in the range 29XXXX and this enabled subscribers to collect messages when absent from their phone which could now be diverted on no reply using the System X Star Services[146] facilities. This augmented the pager system that had been upgraded the previous year to facilitate the display of the number of the caller. At the same time the Board introduced new recorded announcement services using the numbering range 0077XXX, at this stage the international access code was still 010, the EU standard 00 had not yet been adopted in the UK. The new computer based manual switchboard suite was completed and Vanderhof automatic line test equipment commissioned for use both by faultsmen and the test clerks.

The capacity of the inter-island microwave link was increased to 3,840 circuits to cater for increasing traffic and to provide more capacity tfor Guernsey over the French 1920 channel link from Five Oaks. The microwave link at La Chasse was quadrupled in capacity to 480 circuits providing direct connectivity to the French international switch in Paris. The previous year 64 kb/s digital private circuits had also been introduced and the increased demand from the finance industry also required extra capacity on off-island digital circuits. Mobile telephony continued to grow with 938 subscriptions by year end. Prices for mobile handsets had reduced from around £2000 when first introduced to less than £800, making the technology more attractive to business, if not the man in the street, although handsets were still fairly bulky. Meanwhile, CPE sales were booming and the Board requested additional budget to improve its sales outlet in Minden Place to cater for the demand. Although the market for CPE had been partly liberalized earlier, the Board still robustly protected its monopoly by limiting the market to simple devices, maintaining its right to be the sole provider of private switchboards, keyswitch systems and other switched apparatus, and pay phones.

In the summer of 1990 Tom Ayton announced his intention to retire. The

---

146 The advanced software functionality of System X allowed user-controlled facilities to be added after exchange design. The BT propitiatory Star Services product was continually upgraded from its launch at Cheltenham in 1984.

vacant post for Director of Telecommunications was advertised both locally and in the UK. The successful applicant was, however, again appointed from within the organization with the promotion of 33 year old Robert (Bob) Lawrence to the post, he having previously been the Board's Controller of Installation. The year ended on a sour note with a protracted dispute with the EEPTU resulting in an overtime ban and general confrontation between the parties.

In the New Year the unit call charge was increased to 4.7p and it was also announced that call box charges would be increased to 10p as a consequence of the introduction of the new 5p coin. The new minimum call would rise to 10p, however, the 5p unit was retained on pre-paid card phones. The process of modifying some 1,300 public and private coin telephones would be completed by April, over 150 public phone boxes had already been changed during 1990. System X installation continued with demand for new lines still growing. In order to service this demand new 6 digit numbers were introduced initially using the 60XXXX range on the Central exchange unit. This was ahead of the announcement in the spring that there would be more general numbering changes. Following Oftel's move to take over the management of UK national numbering from BT, it was announced that the STD code for Jersey would change from 0534 to 01534. This move was necessitated by increasing demands for numbers as new operators entered the liberalized UK telecommunications market. Further announcements also heralded the adoption of the Europe wide emergency code 112 which would run in parallel with the existing 999 service. The Board also announced a changeover of all numbers within Jersey from five to six digit as System X was rolled out to the minor exchanges. The existing five digit numbering plan was near exhaustion and this necessitated the move as well as making the task of migrating to the new switching technology somewhat simpler. Both North and West RCUs were converted to six digit in July 1991.

The Board had been experimenting for some time with new style aluminium public telephone kiosks. The old wooden kiosks which had been constructed locally were now at the end of their useful life and replacements were now necessary, preferably of a low maintenance design. However, the Board's proposal to use a readily available brown anodized aluminium box fell foul of Jersey Heritage who wanted to preserve the unique Jersey design. The Jersey Heritage chairman, Alistair Lazelle, wrote to the Island Development Committee (IDC) expressing its view that the Board's proposal was too 'high-tech' and inappropriate. It pressed the IDC to consult with the Board to find a better design. On this occasion the IDC agreed and entered into discussions with the Board to find a middle road.

From the 1 June Bob Lawrence formally took over as Director of Telecommunications. This signalled an immediate shift in the business ethic towards a more customer focused philosophy. From the beginning of August the Telephone sales shop would stay open during the lunch hour for the convenience of the public. By fortunate coincidence, BT reduced its charges for UK and some international calls at the same time. This was quickly reflected by the Board, as it underlined the new business strategy which purported value. Further good news arrived with the proposal from the Policy and Resources Committee to give more independence to the States' trading committees, enabling them to manage budgets on a more commercial basis. All these things came happily together with technology changes that would change the face of public telecommunications.

In September the Board introduced basic Integrated Subscriber Digital Network (ISDN 2)[147] which had been introduced by BT earlier in the year. It would for the first time that digital services were offered directly to the subscriber. Data and telephony could now co-exist on the same line although the quarterly rental was more expensive at £51 per quarter, compared with £13 for a normal line, and a substantial £275 installation cost, although the Board claimed that this was cheaper than BT. ISDN significantly increased speed for services such as Fax and dial-up services. It would also in due course permit other advanced functions. Although launched with much hype and promise, ISDN became the Cinderella of telephone services remaining largely a business service in the UK being priced too high for the residential user.

---

147 POEEJ Vol 86 Part 4 1991 ISDN for the 1990s

At this time the Board also looked towards new technologies, particularly at optical fibre distribution for the local network. An experimental Passive Optical Network (PON) was purchased to provide interconnection services between States' departments. In addition an overlay fibre optic network to support private circuits was also being developed, initially in and around St Helier. With its newly acquired commercial freedom, the Board announced a 16.5% increase in line rental at the year end making it £15 per quarter. Its justification for the hefty increase was investment in the future and further it claimed that it was well below the RPI which at that time was running at 20.6% in Jersey.

At the end of the year the issue of the privatization of the telephone system was again raised by Deputy Len Norman who suggested a partial sale. This idea was quickly rebuffed by the Board President Deputy Robin Rumbold who claimed it would be tantamount to selling off the family silver, as the Board delivered good returns to the States and a sale would only provide a short-term gain for the Treasury. He maintained that introducing competition for competition's sake would not benefit the island. He added that the Board had delivered over £45M to the States over the last 10 years. The Board also announced the start of its final phase of digital switch conversion which would result in all existing analogue switches being phased out by spring 1995. The second main System X switch unit from Central exchange would be relocated to East exchange to provide a more secure network structure. This would be completed by 1993 ready for the changeover of the old East crossbar switch. At the same time it stated its intention to introduce more six digit numbers in order to provide for the demand.

Plans were also unveiled to introduce a new submarine cable. After investigating the alternative options, including a link via France and the channel tunnel, it was decided to lay a new fibre optic cable to Goonhilly Down in Cornwall from St Ouen's Bay. The cable would be jointly owned and operated in conjunction with BT and the States of Guernsey Telecommunications Board. While technically the capacity of the existing fibre optic cable (CI No 7) from Guernsey could have been upgraded to provide additional capacity, it was considered prudent to provide an alternative route into the Channel Islands, particularly in light of the earlier experiences with the older analogue cable systems. Indeed, following another fault in March 1991 it had been decided to withdraw the ageing No 5 cable to Dartmouth. The old analogue cables had continued to suffer regular damage, while the new No 7 cable had been buried into the seabed, and consequently was less susceptible.

The Board was becoming ever more commercially successful and in February 1992 it was able to reduce certain trunk call tariffs and also reduce local calls from 4.7p to 4.5p. Profits on a turnover of £26M were a staggering 25% (£6.5M), clearly illustrating the exceptional return on investment of the States controlled business. Changes to the information services were also put in place together with a two pence per minute price rise. However, the local weather forecast rate was increased to 28.2 pence per minute. The information services were now hosted on the voicemail platform which was a digital computer based system and had consequently been expanded over time. To accommodate the additional announcements the codes had been changed from 18XX to the 'Super Call' 0077XXXX range. However, changes by Oftel to the national numbering plan meant that the code 00 would in future be reserved exclusively for international access in line with the whole of Europe. Therefore the code 06966 was later substituted.

On 1 April 1992 West exchange was cut over from the PRX onto a System X concentrator. All numbers in the range 81000 to 84999 were changed to six digit prefixed by the digit '4'. On the 11 April North exchange was cut over from the Pentaconta switch to a System X concentrator hosted on Central. The numbers in the range 61000 to 63999 were changed to six digit prefixed by the digit '8'. The same month the Board announced controversial plans to introduce local call timing in order to maintain the current level of profitability. This would be technically possible when all electromechanical exchanges had been phased out. This statement coincided with the publication of the Telecommunications annual report which again showed increases in profitability. Needless to say this proposal did not sit well with either business or the general public, particularly the pressure group Age Concern, given the revelation of the latest profit levels. The following month the proposal was played down in the face of the general opposition.

The same month Bob Lawrence unveiled the new £1.5M computer based Administrative and Billing System (ABS). This new system developed by UK consultancy, Logica, was designed to integrate several disparate computer based systems in order to improve the business efficiency. The new ABS would for the first time enable the production of itemized billing for all lines connected to the System X exchanges, although a charge of £1.50 per line per quarter, or a one-off fee of £10.00 for a retrospective request, would be levied. Additionally efficiencies could be made for both the Board and business by the consolidation of business line accounts onto a single bill. At the same time it was revealed that Logica and the Board had entered into an agreement to promote the system on the world market.

At the same time the Board was becoming more business focused and increased sales from its retail shop reflected this. Reduced prices for fax and answering machines had spurred take-up with subscribers, answering machines were now affordable having reduced dramatically in price over the previous few years and rapidly growing in popularity even among residential users. Small businesses were taking fax machines as prices dropped. Bizarrely, the popularity of answering machines necessitated a change to the Telecommunications (Jersey) Law 1972 to proscribe the use of offensive language on announcements.

The telephone network itself continued to grow with a year on year growth of 4% subscriber lines to 53,000, each of which attracted an average revenue of £53.00 per quarter, despite the prevailing economic recession. Similarly the mobile network was growing with over 2000 handsets in service which was approximately 2.3% of the island's population. Upgrade work on the Cellnet network carried out in August increased the channel capacity by 50% to meet demand. The range of Star Services, an additional feature to the System X exchanges that enabled subscribers to add facilities such as last number, call transfer and call barring, were promoted mid-year coinciding with the new telephone directory which had a full page of Star Services instructions. Although having been technically available for nearly seven years, the Board had not actively advertised this feature until sufficient subscribers had been transferred to System X and were able to take advantage of them. Similarly, when first introduced, many subscribers would not have been able to use Star Services with dial telephones, but liberalization of the CPE market had rapidly increased the growth of tone dialling telephones. Not all Star Services were free and some needed activation prior to use with a subscription added to the telephone account. The Board also launched an £85,000 programme to convert 78 public call boxes to card phones and introducing automatic anti-vandal alarms at the same time. The Telex switch was also updated to a fully digital version. Earlier modifications to the Telex network had enabled the introduction of low voltage (CCITT V24) terminal machines and personal computer Telex emulation cards. This latest upgrade required the recovery and replacement of the last remaining legacy machines.

At the year end the Board announced details of the new £15M fibre-optic cable project recently approved by the States. This was to augment the existing northbound fibre-optic link from Guernsey which was backhauled onto Jersey via the inter-island microwave link. The Goonhilly project would provide a second cable between the islands to replace the existing analogue cable and to improve the digital bandwith currently provided over the microwave link between the islands. The new 260km UK section would be at the time of its commissioning the longest unrepeatered fibre-optic submarine cable in the world. Surveying for the route would begin in 1993 followed by laying in 1994 with a service ready date early in 1995. The Board, however, attracted criticism for not following Guernsey and BT in reducing the cost of Sunday telephone calls for the run-up to Christmas. In its defence it claimed that this was a marketing tactic employed by BT to 'wrong-foot' its UK competitor, Mercury Communications. When pressed on Guernsey's decision to follow suit, it further claimed that it had no scope in the current financial climate to make such reductions. This was in the light of increased profits over the year to more than £8M. Finally, to round off the year, Bob Lawrence announced the introduction of the Japanese concept of

Total Quality Management. This initiative encouraged all employees to take individual responsibility for the provision of service in order to improve the subscriber's experience, whether in the provision of new lines, sales or fault repair. An initial team of 12 volunteers would try to ensure all departments worked in concert to improve results.

In the spring of 1993 Jersey Telecoms exhibited its new videophone range. Videophone technology had been available for some years but the introduction of ISDN had improved the quality and utility of the product. However, videophones were expensive and ISDN would mainly be of interest only to the corporate market, where it firmly remained. The technology was cumbersome and never inspired interest from beyond business. Videophone technology would only filter down to ordinary users in the Internet age. The cost of ISDN was always a hindrance to the deployment of this and many other advanced services which probably contributed to its narrow market, largely as an interface product for private branch exchange systems. In recognition of this a small reduction in charges was made in May 1994, lowering the cost of a 1 hour video call to the UK from £24.00 to £18.00.

The success of the mobile network in Jersey caused the Board to reconsider its joint venture with Cellnet. The analogue network had paved the way for mobile telephony but now new technology was on the horizon. At this point over 2,300 mobile handsets were in operation on the island and the Board estimated that this would grow to more than 7,000 by the turn of the century. It therefore decided in June 1993 to invest £2M in a GSM[148] network over which it would have sole control. Negotiations with potential suppliers were already under way. In the autumn Jersey Telecoms hosted the GSM Association meeting at the Hotel de France where 50 representatives from 23 countries discussed business over a two day conference. The Board used this opportunity to announce the signing of a contact with Alcatel Network Systems UK, a subsidiary of the French

---

148 **Groupe Spécial Mobile** later **Global System for Mobile** the system developed by European manufacturers and promoted by the GSM Association to supersede the analogue system. This system operates on the 900MHz and later 1800MHz frequencies in Europe allocated by agreement between EU spectrum authorities. GSM became the dominant mobile system throughout the world.

telecommunications group.

In July the Board signed a tripartite agreement for the new submarine cable system with BT and the Guernsey Telecommunications Board. The supply contract was awarded to STC Cables Limited. The Board president, Robin Rombould and his Guernsey counterpart, Barry Lovell were signatories alongside the BT representative. Survey work on the new route began in August with the arrival of the BT Marine[149] cable ship *CS Iris* which would survey the UK link route and in September the BT Marine *CS Discovery* would survey the inter-island passage.

In the autumn Oftel announced changes to the UK numbering plan that specifically affected the so-called sex lines that had been barred to Jersey consumers since the introduction in 1988. The Board acting as a moral policeman had decided to bar access to these numbers which had caused some inconvenience to subscribers when non-sex services, such as catalogue order lines which had numbers within the same ranges. Oftel had now allocated the access code 0891 to these lines in place of 0898 which would, nevertheless, continue to provide these explicit services until existing numbers had been migrated to the new range. However, the Board decided to continue its ban despite opposition from the general public and in particular the Jersey Right's Association, which asserted that as self-appointed guardian of the public morals it was itself unaccountable to the public which it claimed to protect.

At the 1993 year end the issue of local call timing was again revived. The Board defended its proposals which would increase peak charges for a local call to 3.84 pence per minute asserting that to recover the equivalent receipts would require increasing line rental by 200%. New tariffs for trunk calls were announced at the same time reducing UK peak calls from 18ppm to 15ppm (which it claimed was 1.5ppm below the BT rate – although the BT rate included VAT at 17.5%). Needless to say these proposals were not popular with either business or residential subscribers, particularly with Age Concern, whose cause was taken up by Senator John Rothwell. The Jersey

---

149 BT Marine Limited was formed as a subsidiary of BT plc following its privatization. It is now part of Global Marine Systems Limited, formerly Cable and Wireless Marine Limited which acquired the BT subsidiary in 1995.

Small Business Association accused the States of double standards as it was at the same time trying to implement anti-inflationary measures. When the matter was debated in the States, Senator Dick Shenton described the increases as stealth taxes. However, the Board president stood firm and refused to amend the proposals while conceding that some services such as the Samaritans would be given Freephone access. However, despite the bullish performance in the States, two day later he announced that the Board had dropped the local call timing from its proposed tariff changes. He refused to admit that this was a climbdown, instead claiming that the 'complex matter of call charging required more analysis'. However, this débâcle lost Rumbold his presidency and he was replaced in the annual States committee elections by Senator Nigel Querée and although he had asserted in the previous week's debate that local call timing was a 'non-starter' he nevertheless conceded that tariff review was a top priority for the new Board. During that debate Senator Shenton had called for wealth taxes to replace the shortfall as a result of the withdrawal of the proposal. This clearly illustrates that at that time the States treated Jersey Telecoms not just as a communications provider but a convenient source of public revenue.

Early in the New Year of 1994 the Board did indeed revisit the issue of tariffs. The new States committee made a hasty review of all charges and in February new proposals were unveiled. Despite its earlier U-turn, the Board did not abandon the principle of local call timing but instead produced a watered-down version that would go some way to generate the same revenues as the previous proposals. The minimum charge for a local call would be raised from 4.8p to 7p for a minimum period of 30 minutes and then charged *pro-rata* per second thereafter. The 7p minimum initial charge was also reflected across its trunk call tariffs with only the time allowed varying according to destination and time of day. At the same time line rental was raised from £2.50 per month to £4.00 per month, while business customers would face an increase to £8.00 per month. To counteract these hefty increases low user tariffs were also introduced at £2.50 for residential and £5.00 for business users. The existing low tariff for old aged pensioners would remain unchanged. It also introduced an innovative new product by way of its packaged local call rental tariff which for £15.00 per month would allow unlimited local calls. The Board claimed that despite these increases the average bill would actually fall by £0.25 per month based on the dubious claim that the increases in local calls and rental would be off-set by reductions in long-distance call charges. The Board had found it necessary to reduce the differential between its long-distance charges (especially for calls to the UK) since businesses were able to bypass the Jersey Telecoms charges by either getting their UK offices to call back at lower cost or with the use of Freephone (0800 or 0500) bypass call selection provided by Mercury Communications and other UK operators. This was clearly impacting on trunk call revenues.

The proposed £15.00 tariff received much criticism as it was quickly noted that it would require a lot of local calls to generate the £11.00 differential from the regular tariff (almost 160 per month). Overall, the public was not convinced by the claim of overall lower charges (which may have actually been true for business users with high long-distance call content) but unlikely for the residential user. The Board also pointed out that its new charges were less than BT charged in the UK (although unlike BT it was not subject to aggressive competition) but this argument was weakened when the same week BT withdrew its peak rate for local and national calls. It is interesting to note that Jersey Telecoms chose to compare its charges against the large UK incumbent rather than a more relevant comparison with, say, Guernsey. However, in an effort to quell the public's dismay at the increases, it was decided to include free itemized billing for all customers connected to System X exchanges if they so requested. Despite the public opposition to the increases, the proposal was nevertheless approved overwhelmingly by the States at its March sitting. The house was convinced by the Board president's argument that the increases were necessary to meet the demands of the Finance and Economics Committee, causing Senator Dick Shenton to reiterate his assertion that this increase was tax by stealth.

Following the approval a 'hotline' was set up to assist subscribers with billing and tariff queries. This was in parallel with the new system developed to describe the new tariff structure which allocated numbers rather than names to individual products. This, it was claimed, would simplify the structure and to avoid confusion.

At the end of March 1994, the BT Marine Cable Ship *CS Discovery* arrived off Greve de Lecq to begin laying the Jersey to Guernsey section of the new fibre optic submarine cable containing 12 strands which would carry six 155Mb/s SDH systems. The fibre would follow a similar course as its predecessors and be landed at Saints Bay on the south-west of Guernsey. The shore end of the cable was landed and brought ashore up the slipway and jointed to the on-island cables back to Central exchange, St Helier. Now that these new cables required no specialized submarine transmission equipment, it was no longer necessary to take the cables to the repeater station. The northern section to Goonhilly, Cornwall would be laid from Le Braye slipway in St Ouen's Bay and cover 237km to land at Kennack Sands near Kuggar, Cornwall and was completed in mid-November 1994. Large portions of the cable were ploughed in to the seabed to avoid the damage from dragged anchors and trawling that the old analogue cables often experienced, indeed, shortly after the completion of the new cable in January 1995 the two remaining old cables were again severed by dragged anchors.

In July further trunk tariff reductions of 10% to 13% were announced. Following the lead from BT the peak call rate was withdrawn leaving just two different rates set by time of day and the lower rate all weekend. At the same time the Board started to block access to the Freephone bypass numbers. Naturally, this was an unpopular and a blatantly protectionist move by the monopoly operator and was also in conflict with the rules set out by Oftel, published the previous month, which now had control of numbering in the Channel Islands by virtue of the extension of the relevant parts of the Communications Act 1984. Following complaints from businesses, the Board entered in to negotiations with Mercury Communications to come to an arrangement for charging for access to its Freephone numbers, again in conflict with the Oftel rules. In order to legitimize its position the Board should have applied for permission from

Oftel to add an Access Deficit Charge[150], but as the monopoly provider in Jersey, this was unlikely to have been granted. The Mercury Communications press officer, Ed Staples, expressed concern that Jersey Telecoms had unilaterally blocked access. Prior to the cut-off, Mercury's rates were almost half those charged by Jersey Telecoms. Nigel Querée, the Board president, misrepresented the relationship between Mercury Communications by stating that 'free calls have to be paid by subscribers'. This ignored the access refund provided by the Freephone service provider to the originating network operator, which was passed on to the Board by BT and it sought a larger slice of the cake to protect its own revenues which it had recently substantially increased. This latest incident did nothing to allay the public perception that the Board's charges were too high. The Board had to fight a rearguard action to try to justify its stance over the following months.

In August 1994 the biggest single line fault incident since the 1960s and the modernization of the network from overhead to underground occurred when an arsonist set fire to an underground cable jointer's tent at Georgetown, resulting in the meltdown of a main local trunk cable disconnecting 1,600 subscribers. Technicians had to work around the clock for 4 days in order to restore all the connections, and further disruption was subsequently caused during the replacement of partly damaged cable sections over the next few days. It was estimated to have cost the Board in excess of £100,000.00 to rectify, including lost revenue and compensation payments.

Also in August the roll-out of the new GSM system was under way and new wireless transmitter sites were being deployed in order to ensure adequate coverage of the island. Although GSM uses similar frequencies to the existing Cellnet system, GSM could handle more calls and thus the area covered by each transmitter would be smaller in order to allow for more traffic. Planning applications were placed for sites at Trinity and St Peter in

---

150 An Access Deficit Charge (ADC) is sometimes added to interconnections on networks to assist the dominant incumbent operator to fulfil its Universal Service Obligation. This is normally only the case where the telephone network is new or underdeveloped, as in the third world.

addition to those already deployed at Five Oaks, Gorey, Fort Regent, St Ouen, St John and St Brelade. While some of these sites used existing lattice mast structures, in some cases telephone poles with monopole aerials were used. It also intended to negotiate with the BBC to use its mast at Les Platons. Meanwhile the main Mobile Switching Centre (MSC) was to be installed at its Five Oaks facility in the existing microwave terminal building. The GSM system was tested over the following few months and opened to the public at the beginning of December. On opening roaming agreements were only in place with UK operators but it was hoped to add French, Swiss and Irish networks by Christmas. Oftel allocated the access code '0979' to Jersey Telecoms GSM system.

In September a new telecommunications market emerged with the establishment of a company called Interactive Telephony Limited. This company, headed up by local entrepreneur Nick Ogden, started to exploit the developing Internet business and set up access services for local users. Access was through normal telephone numbers using a dial-up modem. At this time modems where a new product and were relatively expensive and thus initial take-up of services was relative slow.

In the spring of 1995 following the reorganization of telephone numbering in the UK by Oftel as previously notified, all national codes were changed by the addition of the digit '1' after the initial '0' thus the Jersey code changed from '0534' to '01534'. At the same time a number of the larger cities had completely new area codes allocated. This change was necessary as the liberalization of the telecommunications market in the UK had greatly increased the demands on the numbering system. New operators had entered the fray and more and more numbers were required to meet their needs. While there had been no impact in Jersey as the States still had a monopoly over the telephone system, membership of the UK numbering plan meant compliance. This was nothing more than a nuisance and consequential expense for local telephone users, especially business which needed to update stationery and reprogramming PABX's and Fax machines. A new international dialling access code was also introduced, changing from

'010'[151] to '00', which would now be the standard across the EU. This coincided with the final stage in digital exchange conversion when on the 8 April the Central PRX satellite exchange was cut-over to the System X exchange. This affected subscribers with numbers in the ranges 38XXX, 39XXX, 58XXX and 59XXX while the subscribers in the range 7XXXX were completed on 18 April. At the same time these numbers were changed to six figure with the addition of the digit '8' in front of the existing number.

---

[151] '010' had originally been introduced in the 1963 with the opening of international calls between London and Paris. When the service was extended outside the London area all international traffic still passed through London. At that time the London STD area code was '01' and thus the old electro-mechanical exchanges directed calls to London where the final '0' of the access code routed the call to the international exchange. Later international exchanges were installed in many UK locations, but the access code remained the same.

Chapter 20 - Towards Competition 1995 – 2002

Call costs were continuing to fall as a consequence of competition in the UK. BT was the main supplier of interconnection for Jersey Telecoms and it was constantly under pressure to reduce charges as competitors entered the market. International calls were reduced in June following lower rates to UK numbers. Overall, the Board claimed, call charges had dropped 35% over the last year, notwithstanding the substantial increase in local call charges. ISDN charges were also reduced. At the same time an independent survey suggested that Jersey Telecoms' charges were among the lowest in Europe. This was probably as a consequence of price following to BT's tariffs and the absence of VAT on local bills. In September a new GSM rental tariff was introduced. Branded Lo-Talk it was set at just £10 per month. Since its launch the GSM network had attracted 1,200 subscribers and with this lower tariff and with reduced handset prices through subsidies, this had increased to 2,500 by the year end.

Jersey Telecoms hosted the GSM conference in November at which its representative was Philip Ainsworth, the Board's Financial Director, who was appointed chairman of the Business and Accounting Rapporteur Group. The GSM association was an important building block in the deployment of the system worldwide and a major player in the standardization of services and technical issues. The accounting group was a key component in setting up roaming and international charging agreements that enabled the ubiquitous use of the mobile phone.

At the year end the Board also announced the introduction of additional Star Services onto all analogue telephone lines including the calling number display facility. Most of these services would now be available free, although there would remain a charge for the caller display option and some others.

In February 1996 Jersey Telecoms introduced its Synchronous Digital Hierarchy[152] (SDH) 'sealed ring' over the submarine cable to the UK and the microwave link via France in conjunction with BT. This effectively meant that a single link failure would not interrupt the telephone service or private circuits going to the UK. The sealed ring meant that the digital traffic would be automatically and seamlessly rerouted over the remaining path. This was a huge improvement in service since prior to this technology circuits would have been rerouted manually at each end of the cable. The same month the Board took another step in its tariff rebalancing strategy with an increase of £1 per month on line rental. Falling wholesale call charges due to competition on international services had permitted stable or falling prices for off island calls, although the minimum call tariff remained unchanged at 7p. In order to maintain revenue streams the Board had instead increased the unavoidable fixed line rental costs in a strategy that was being increasingly adopted across Europe as competition reduced call revenues. The overall effect of this change was mostly neutral for business and those that made many calls, but reflected an increase for low users. Later in the year the Board introduced per-second billing for its GSM network, giving further relief to consumers by removing the minimum call charge. This led it to claim that its network was among the cheapest of the 15 EU countries. At the end of August there were over 5,100 users connected to the GSM network.

The telecommunications market continued to evolve. In June the Board further relaxed its monopoly by permitting competition in the supply of customer premises equipment to private switchboard supply and maintenance. This followed intensive lobbying by the Jersey Electricity

---

[152] Synchronous Digital Hierarchy (SDH) is defined in ITU Recommendation G.803: Architecture of transport networks based on the synchronous digital digital systems.

Company which had recently announced its intention to lay a telecommunications fibre optic submarine cable alongside its proposed power cable connection to the French electricity supply grid. At the same time the Board upgraded its retail outlet in Minden Place, St Helier to give it a more customer focused appearance. This move, it hoped, would enable the Board to counteract its easing of the supply monopoly by presenting a better image to the buying public.

In September the long awaited proposal to incorporate States' trading committees was ready for presentation to the States assembly. Pierre Horsfal, when president of the Finance and Economics Committee, had consulted on the proposal to incorporate the Telecommunications Board, the Postal Committee and the Harbours and Airport Committee trading operations. The proposition to be presented to the States had, however, dropped the Harbours and Airport from the paper, at least for the time being. The Telecommunications Board, while welcoming the increased commercial flexibility of incorporation, stopped short of embracing full privatization (which had been first proposed in 1982) claiming that the States' best interest would be served if the business remained wholly owned by the government. A publicly owned utility could operate with more freedom and would be able to make strategic investment decisions unencumbered by the present requirement for States approval of budgets and the consequent bidding against more politically sensitive social project budgets.

When the matter was formally lodged in the States sitting of 23 September 1996 it was broadly welcomed by the presidents of the Postal Committee, Senator Frank Walker and Senator Nigel Querée of the Telecommunications Board. They both welcomed the ability to act more competitively and to have commercial freedom to diversify the businesses. Querée also welcomed the prospect of a separate regulator for telecommunications as he claimed that the Board had always felt uncomfortable acting in both roles as it did under the present law. (Notwithstanding this unease, in December the Telecommunications (Jersey) Law 1972 was amended to give the Board authority to act on behalf of the

States in the issuance of licences). The propositions for both incorporation proposals were passed on a standing vote.

At the end of the year the submarine cable between the islands and onward from Guernsey was upgraded using Alcatel supplied equipment to complete the SDH 'self healing' ring technology that would allow circuits to automatically and seamlessly switch over between the Guernsey cable and the direct submarine cable to Jersey which had been upgraded earlier. This would improve communications and available bandwidth for both telephony and for business data circuits as no perceptible delays would be incurred in the case of a single cable failure. A further System X unit was installed as a platform for testing and as a backup facility in case of major failure of an operational switch. A second Operator Services Subsystem was also installed as a backup unit. At the same time a major project to replace the Central exchange MDF was also well under way. This was intended to reduce the footprint of the existing frame and to increase the efficiency of connections for new work. The work was carried out by the supplier and local staff.

In January 1997 Jersey Electricity Telecom (JET) launched its private exchange supply and maintenance business, marketing L M Ericsson equipment. Mike Liston of the JEC continued to lobby for yet more relaxation of the monopoly, but the Board maintained its stance and insisted that it would await an independent regulator to make such decisions in the future. Around the same time another local IT supplier, CDP Sigma, also entered the private switching equipment supply market. Like many other similar companies it was already supplying Fax machines and modems to business.

More tariff changes were announced the following month. International calls continued to fall, particularly to specific destinations such as most of Europe and Australia and New Zealand, however, this was offset by a further increase of £1 per month on residential line rental while business lines were increased just £0.25 to £8.75 per month.

In April, Jersey Telecoms reacted to the increased competition in the supply market with the launch of its Centrex[153] service, which enabled it to offer an alternative to PABXs or key systems for small businesses.

In June a new Internet Service Provider (ISP) entered the market. Interactive Communications opened offices in Waterloo Street, St Helier and started marketing its services under the brand name *LocalDial*. The new company headed up by Nathan Wright offered access from both analogue and ISDN lines and utilized a private circuit to an Internet backbone supplier in London.

In July Jersey Telecoms was criticized because of problems on its GSM network. It initially blamed the Island Games for an unusual surge in mobile usage. Its network was said to be designed to manage 9,300 subscribers although at this point only 8,500 were active locally. It announced in April that the addition of new subscribers would be frozen until July while a software upgrade was implemented and until a £3M upgrade to the network was completed towards the end of the period. However, in mid-May it removed the freeze claiming that the recent software upgrade would be sufficient. When the island games opened, the number of inbound roaming mobiles surged to 1,800, which was twice the norm, thus the network congestion rate increased dramatically. Later it was claimed that at the peak of the games, 3,000 call attempts per hour had been reached. The upgrade was carried out by Alcatel over the weekend of 19 July and further problems were experienced for a short time following the return to service on the Sunday morning. The system crashed again on 21 July which caused Jersey Telecoms to offer half a month's rental rebate to its subscribers 'as a reasonable offer in view of the recent issues'. This was on top of a month's rental rebate to its subscribers given earlier in February following protracted network problems. The Board defended its position saying that the network

---

153 Centrex is a service supplied by telephone companies that uses a subdivision of the main exchange to emulate the features of a PABX. It is economic only where the monthly cost of exchange lines required to service the extensions is less than the monthly rental of a PABX plus any main exchange lines required. Centrex was originally conceived in the US by the New York Telephone Co in the 1960s and was a feature added to System X exchanges at the request of BT after it had experimented with the product using AT&T No. 5ESS switches.

was growing rapidly and that further work would be carried out soon. The town area would receive a new transponder that would improve capacity and that in September the system would be expanded to cater for 25,000 subscribers which would ensure growth of the network over the next year. Already the mobile penetration rate in Jersey was nearly 11% and with the rapid growth of GSM elsewhere, it could be expected to reach that number within a year or so. The Board was now aware of the fragility of the present system and thus it also planned to install a second switch to minimise the repetition of recent problems.

The Jersey Telecoms monopoly was coming under increasing criticism. The travel industry complained that it was cheaper for local business to get their offices on the mainland to call back than to make calls to the UK. It was claimed that local business was paying between 27% and 100% more for calls to the UK than the other way round. A lack of choice was identified as the problem. The JEC also jumped into the fray as it wanted to see more deregulation of the market. The Board promised further call cost reductions in the autumn, but in the meantime it continued blocking Freephone bypass numbers (0800 or 0500) that businesses could otherwise use to get reduced call costs. The Board defended its position claiming that it was illegal to make such calls unless a contractual agreement was signed between the supplier and the Board (however, note that this was in contrary to the Oftel numbering conventions). It also claimed that the quality of the circuit was likely to be poorer than its own connections which, while likely true, was not of prime concern to users. The contractual arrangements referred to here meant that the Board added a premium to the 0800 call origination that effectively severely reduced the available margin for the supplier, thus maintaining artificially high prices for the service compared to the same service in the UK.

However, in August the Board announced cuts to UK calls and some international destinations as a result of new interconnect agreements with Cable and Wireless Communications and France Telecom. At the same time it announced that per-second billing would be introduced on fixed line calls and an upgraded accounting system would produce semi-itemized billing from October. Nevertheless, it maintained a minimum call charge regime which would ensure that all calls would still be at least 7p.

In September Nigel Querée resigned as president of the Board claiming other commitments prevented him from fulfilling his role. Telecommunications Board member Paul Routier stepped up to act as president and was later confirmed in the position by the States. By now the process of incorporating the Telecommunication Board was under way. Nevertheless at this point the Board was also acting as regulator and the commercial pressure to open the market was causing some conflict. Routier continued to assert that an independent regulator would be desirable but that would probably be delayed until the incorporation process was completed.

In November the Board announced that it was to enter the ISP business as a joint venture (or 'virtual' company as Bob Lawrence put it) with the Guernsey Telecommunications Board. This move would put the Board into direct competition with the existing local ISPs *SuperNet* and *LocalDial*. Despite this broadening of its portfolio, the Board came under increasing attack from local businesses. In a report the Jersey Institute of Directors (IOD) criticized the States for not having a credible IT policy and claimed its continuing monopoly in telecommunications was stifling innovation. It also wanted a faster approach to freeing up the market so that telecommunications costs to local businesses could be reduced. This report did not fall on deaf ears as Senator Frank Walker affirmed that the States needed an IT policy and that it also needed to rethink its strategy on telecommunications. However, the Board rebutted this report by claiming that it was always looking for better and cheaper ways of providing services. Bob Lawrence pointed out that costs for private circuits were falling and saving business thousand of pounds per year (although this only applied to large businesses). Routier stated that before any major changes in competition could be considered the 1972 Telecommunications Law would need to be reviewed. The Board, he claimed, was already working with commercial operators in order to provide better services.

The new GSM network continued to be fault-prone and another outage at the end of November allowed the Board to announce that it was taking delivery of a second switch that would help alleviate these problems. The additional switch would be located on a different site and would enable both pre-testing of software and hardware upgrades as well as a fail-over in the case of problems with the on-line switch. The new switch would be located at North exchange in the space vacated by the recovery of the old crossbar local switch.

In January 1998 Routier accepted that the IOD report had some merit but 'was disappointed' that these had not first been directed to the Board. Nevertheless, it was recognized that a more proactive approach to business was needed and Bob Lawrence promised that it would engage in dialogue with users. The same month details of the proposed joint venture with the Guernsey Telecommunications Council were revealed. The official name would be Cinergy Communications but it would trade simply as *Cinergy*. It was hoped that the new services would be opened by the end of March. Dial-up access to ISP services through an 0845 number would become available and this form of access would also be offered to the existing ISPs for users to access local services from the UK. Calls would be charged at 2ppm peak and 1ppm at other times compared with the BT Lo-call[154] standard access rate of 5ppm from the UK. Existing ISP's did not react with much enthusiasm to this new offer. In an answer to a question in the States, Routier assured that *Cinergy* would be operated at 'arms-length' from the telephone business and would be housed in separate premises. The set-up and operational costs would be shared by both the Jersey and Guernsey Boards and this was formalised in a signed agreement on 21 January 1998.

The new venture was officially launched on 8 April at Channel House, Green Street, St Helier. The initial offer from *Cinergy* was a 5 hour per month package at £8.50 or 15 hours for £14.75 per month. This monthly allowance was upped to 10 hours and 99 hours respectively in August when

---

[154] The BT Lo-call rate was a marketing brand for its 0345 so-called local call rate services first introduced in 1996. Oftel had permitted BT to offer its standard local call rate (then 5p for a 3 minute call) on a per-minute basis for certain call types. BT also operated a 1ppm and 2ppm charging rate for ISP access. Mercury Communications offered a similar service using the code 0645. They were both later changed to 0845 by Ofcom which also introduced 0844 for ISP access.

at the same time it introduced its off-island access option using an 0845 number. Concurrent with the initial launch the Board announced a special tariff offer to local schools. A standard analogue dial-up connection would be £395 per year and ISDN at £730 per year both with unlimited access. The Board again claimed that this was cheaper than the comparable offer from BT, although it made no mention of the cheaper offers available from the other local ISPs. The same month the Jersey Telecoms website was launched.

In reaction to the announcement, *SuperNet* introduced price reductions on its monthly subscription and then bizarrely announced the sale of the company and its 2,500 subscribers to UK ISP *PSINet*. The following month the other local ISP, *LocalDial*, entered a merger with *Guernsey.net*, expanding the overall customer base to 1,800 across the islands.

Also in April the Board completed its SDH self-healing ring configuration between the island and the UK in cooperation with BT. The island digital network was connected via the CI No 7 and CI No 8 submarine cables to the BT network which it claimed to be the first in the world to complete this configuration between two independent providers, although it had earlier completed a similar SDH interconnect between Jersey and Guernsey. Two cable faults during the year proved the worth of the system as no perceptible outage of services was experienced.

The Board again came under attack again for blocking call bypass though 0800 number access. On this occasion it defended its position by suggesting that calls made over these services were of lower quality and that it was simply protecting its own network (from what?). At the same time the annual report revealed that it had posted record profits of £9.3M. Its GSM mobile network had grown to 12,000 users and there were 1,665 payphones in use, all contributing to the bottom line. Despite these record profits the Board again increased residential line rentals above inflation from £6.50/month to £7.50/month and business rental was increased by just 25p, continuing the strategy of rebalancing the tariff structure to a single line rental product. Calls at weekends to the UK as well as some trunk call rates were reduced, although this had little effect on profits since lower rates tend to lead to increased traffic. Full call itemization would be available from July 1998 at an additional charge of £1.50 per month.

In May 1998 Newtel Holdings, the owner of the cable television provider Jersey Cable, announced that it had entered into partnership with the JEC to form a new telecommunications company, *Newtel Limited*. The new company wished to obtain a licence from the Telecommunications Board to provide certain telecommunications services. The JEC had already proposed a fibre optic submarine cable to be laid alongside its electricity connection to the French grid and this new venture wanted to exploit this.

Also in May the States debated the proposed incorporation of the Telecommunications Board to turn Jersey Telecoms into a limited liability company. The proposition was passed by a narrow margin. The debate outlined the proposal to incorporate the business and for the States to retain 100% of the share capital. A new independent regulator would be formed to oversee the telecommunications sector. No details of the new laws required to implement these far reaching changes were presented at this debate. Senator Dick Shenton, who had previously attempted to privatize Jersey Telecoms, insisted that only a full sale would serve the best interest of the States and to island businesses.

In June the Board started testing advanced broadband services after a £250,000 investment in Asynchronous Digital Subscriber Line (ADSL) equipment. Board employees were connected to the Internet using the new ADSL equipment to evaluate the and test the functionality of the new system. The ADSL system was connected to the Internet through the Board's recently installed Asynchronous Transfer Mode (ATM) packet data switch. Initial speeds were set at 250 kb/s.

In September the Board took a decision to outsource it directory enquiries service to the BT call centre at Ryde on the Isle of Wight. This was clearly an economic decision designed to reduce overheads on its network in readiness for incorporation. The basic decision was brought about because the local access equipment was considered to be not Millennium Bug[155]

---
[155] The Millennium Bug was a problem that could affect some older computer programs when the century chaged from 1999 to 2000, thus confusing two digit dates. (which

compliant and also because Oftel had allowed BT to increase the charges it levied for access to the database which Jersey Telecoms' own enquiries operators used as it also included all Jersey numbers as well as UK and international information. The resulting charges to the Board would have been in the order of £1M annually. At the same time the cost of making a directory enquiries call would rise by 350% most of which would accrue to BT. Overall, the Board would save £250,000 in operator salary costs as a consequence of the change. The transfer was completed in the spring of 1999.

Despite its monopoly, competition in the off-island call market was coming and reduced prices were promised. Two new simple voice reseller companies entered the local market towards the end of the year. Eurofone Limited and Long Distance International Limited (LDI) began offering cheaper calls using Freephone access numbers. The competition was initially dismissed by the Board as a second-class service. The Board could take this aloof approach since it charged a premium to these suppliers to let them access their own Freephone numbers.

In the spring of 1999 Newtel Limited placed a formal application for a licence before the Telecommunications Board. The new company headed by Peter Funk, the chief executive of Jersey Cable had substantial local backing from both the JEC and the Guiton Group, owners of the Jersey Evening Post and its local IT company, Matrix. Agreement had been reached to utilize a substantial part of the new fibre-optic submarine cable planned to run alongside the new electric cable connection to France. It was announced that Newtel planned to develop a broadband network and to offer Internet access and other services to local business.

In April the Board introduced a new standard public telephone kiosk. The structure in anodized yellow aluminium had been designed with the cooperation of the Environment and Planning Department, the *Centre Ville* town planning group and the disabled support organization the Jersey Access Group. The design was based on the earlier lantern roofed model but

---

were commonly entered in computer databases which "assumed" the '19' part of the year.

Image by the author

An example of the final design of the new aluminium telephone kiosk manufactured by Prosider Prodotti Siderurgici SPA , Italy. Some kiosks of this design also housed small GSM transponders. In the photo the microwave warning sign can just be made out on the roof peak.

was wider to permit wheelchair access. The first prototype kiosk had been installed the previous year in Ruette Haguais (a small road connecting King Street and Broad Street) for evaluation and comment before embarking on island-wide deployment. Also in April the Board announced the closure of its partnership with the Cellnet analogue mobile network which would cease service at the end of June. The growth in GSM and the dwindling sales and traffic on the old network made its continued operation uneconomic. However, the GSM switch was doubled in capacity to prepare for growth as traffic increased 80% over the previous year. At the same time the GSM network coverage was enhanced to improve reception in some 'black spots', with the introduction of additional masts at Rozel, St Peter's Valley, Gorey, La Moye, Rue des Pres, Beaumont and at the Airport. Meanwhile, profits continued to rise. An increase of 7.9% boosted overall profit to £10.4M for 1998, which represented a healthy return on capital invested of 16.3%. This increase in revenue was driven mostly from the new mobile network. Fixed call revenue dropped by £2M over the previous period and revenue from other fixed services was also declining as tariffs were readjusted.

There were also numbering changes introduced. All remaining Jersey five digit numbers were replaced with six digit numbers by adding the prefix digit '7' to existing 2XXXX and 3XXXX numbers. At the same time Oftel introduced standardization of Freephone numbers to begin 0800[156] and premium rate 090XX numbers. This realignment also meant that the Jersey GSM code was changed from 0979 to 07797 as Oftel realigned all mobile and other personal numbers behind the 07 prefix. The two GSM codes would run in parallel until April 2001. These changes would impact upon local business as stationery and advertising required updating and many fax machines, private telephone systems and programmable telephones would need to be changed. All this was part of the Oftel 'Big Number' programme

---

156 Until Oftel standardized 0800 as Freephone, operators had used a number of access codes. Before Oftel took control of the UK National Number ing Plan BT had continued as the numbering manager, a role inherited from the GPO. It reserved 0800 for its owh Freefone service and allocated 05XX to its rival Mercury Communications in 1982. Unable to use the BT 0800 number range, Mercury allocated 0500 as its own Freephone number. A declining number of these legacy connections continue today.

which would again reallocate national dialling plan numbers, particularly of main cities throughout the UK. This numbering change was phased in over two years to ensure minimum disruption.

In May 1999 the Board announced the launch of JT Highway, based on the BT Highway[157] product. This would offer users the option of up to 128kb/s data connection to the internet or access to 64kb/s and a simultaneous telephone call. Although the uptake for small business was good, it failed to impress the home market as it was more costly than an analogue line and the marginal increase in download speed was not significant. Trials continued with the ADSL equipment which promised to bring even faster download speeds. A new Frame Relay switch solution was also deployed in November to manage private circuits more efficiently. A new agreement with BT permitted the introduction of a new Capital Connect product that terminated private circuits in London for onward connection at a lower cost than previously available.

The incorporation process was now progressing well. Consultants had been employed to report on the Jersey telecommunications market and also to examine the proposed legislation. A similar process was under way in Guernsey as both islands States' struggled with the transformation of the telecommunications market in line with developments elsewhere. This proved difficult for the Board as it continued in its dual role of operator and regulator. The JEC's new fibre optic connection to France from Archirondel was nearing completion and the company applied to the Board for a licence to operate as a telecommunications provider. In response to a question in the States August sitting from Jean Le Maistre, a States' appointed director of the JEC, on why the licences requested by both the JEC and Newtel were so delayed, the Board president, Paul Routier, responded with the diplomatically cast answer that these matters had been referred to the Policy and Resources Committee for evaluation. This was a useful delaying tactic. The following month the Board announced a £77M five year investment plan to upgrade its network for frame relay data packet

---

[157] BT Highway introduced in late 1997 was a product based on ISDN2 and offered flexible access to two telephone lines or one line with access to a data connection via an RJ45 connector.

switching (which opened in December 1999) and broadband services.

In October 1999 the Board announced the dissolution of *Cinergy*, the jointly operated ISP set up with its Guernsey equivalent, after just 20 months of operation. Answering questions in the States, Telecommunications Board vice-president, Deputy Phil Rondel, claimed that the market dynamic had changed and with 'free-Internet[158]' access it made more sense to absorb the operation back into its core business. The Board had written off the not inconsiderable set-up costs which Rondel tacitly alluded to when he compared the cost against the proposed £77M new investment.

The new Internet provision would be marketed as *JT Freesurf*, would be a plain vanilla offer with no added value components, such as web space and additional email accounts, unlike the Cinergy traditional ISP model. It was launched officially at the end of January 2000 concurrently with a new web portal, Jersey Insight which it had developed in conjunction with Jersey Post. Indeed, the Board was able to piggy-back on the recently launched Jersey Department of Postal Administration free email service, *JerseyMail*, which clearly enabled one States department to subsidize another from public funds. The new email service was free and, surprisingly, included a POP[159] interface at no extra charge unlike most of the other 'free' email offers available at the time.

In February 2000 it was confirmed in the States by Policy and Resources committee president, Pierre Horsfall, that the necessary legislation to deregulate the telecommunications sector was now in the process of preparation. Jersey Telecoms would be incorporated as a limited liability company with 100% of the shares remaining in States ownership. The new liberalized market would be overseen by an independent regulator. Horsfall went on to suggest that from now on the telecommunications market

---

158 Free Internet is somewhat misleading. In the UK a number of ISPs had transferred from a subscription model to a pay-as-you-go model when BT introduced its 0845 (later 0844) shared revenue options. This enabled users to access ISP services without a monthly charge, although of course there was still a payment through the phone bill.

159 Post Office Protocol (POP) is an application-layer Internet standard protocol used by local e-mail clients to retrieve e-mail from a remote server over a TCP/IP connection.

should be gradually opened up to competition.

With the pressure now on to move to a liberalized market, the Board started to re-evaluate its position. In March it issued a Class Licence at no charge that would permit companies to enter the e-commerce business without specific permission from the Board. The same month it finished its tariff realignment process by making increases to residential lines and business lines to bring them to the same level. This effectively eliminated specific residential and business line types. With pressure from other suppliers of international calls the Board now had little room for manoeuvre in terms of upward call price increases. In general, with the exception of local calls, all call revenues were falling and only the increase in revenues from rental and the GSM business, which now had over 25,000 subscriptions, maintained the continually growing profit which reached £11.7M for the year 1999. Also in March Newtel entered the business Internet connection business offering services in conjunction with UUNET, a WorldCom company. It also extended its GlobalCall offer to include reduced call costs to UK numbers.

In May the Jersey Telecoms GSM system added Short Message Service to its network. This service had been available on other networks for some time and local callers had found ways of sending and receiving them without charge by 'hacking' into other providers. The new service would enable legal messages and would be charged at a flat rate of 7p per message or 15p for inbound roaming users.

In June the final report on the liberalization of the telecommunications and postal sectors was placed *Au Greffe* for debate in the States. The final recommendations included the establishment of an independent Jersey Competition Regulatory Authority (JCRA) which would oversee the application of the new liberalized laws to these sectors and in time would also oversee the application of general competition law. Under the proposals to be put before the States neither Postal Administration nor the Telecommunications Board would be permitted to make major business decisions without the approval of the Industries Committee to ensure future regulation should not be compromised. The following month questions were raised in the States by Deputy Phil Rondel with regard to a new office lease for the Board. It proposed new management offices at Grenville Street at an annual rental in excess of £400,000 per year with an additional £200,000 for parking spaces. The new offices enabled the senior management to move out of disparate existing exchange buildings into a single more prestigious accommodation. The staff migration would be phased over six months.

In September 2000 the new Pay-as-you-go mobile product was launched under the name of GSM Freedom after a delay caused by difficulties with the charging software. This offered a SIM pack for £35 which included £5 of credit. Local calls were charged at 15ppm and UK calls at 50ppm. Initially no international calling was available. The same month the new JT Rapid ADSL residential broadband product was launched at a monthly rental of £50. The initial offer was for 256 kb/s download at a 40:1 contention ratio.

In October the inter island fibre optic cables CI No 7 and CI No 8 were upgraded with new terminal equipment enabling a doubling of capacity. This followed on from the opening in July of a microwave SDH circuit to France which would provide a resilient backup service for Internet connection capacity for business and ISPs. The same month the proposed law to establish the JCRA was lodged *Au Greffe*. This was at a time when the JEC was blaming the States for delaying the licensing of its new fibre optic cable to France. It claimed that the Telecommunications Board was abusing its exclusive right in not addressing the issue. On its part the Board pointed the finger to the Industries Committee which was now overseeing postal and telecommunications strategy.

In November, Newtel Solutions the company partly owned by the JEC opened up in the refurbished Jersey Cable offices at No 1 Colomberie, St Helier. Newtel was now positioning itself ready for the proposed liberalization of the market. Throwing down the gauntlet it announced that it would offer half price calls over the Christmas period.

At the late November sitting of the States the JCRA law[160] was debated

---

160 Competition Regulatory Authority (Jersey) Law 2001

and apart from an amendment proposed by Deputy Mike Vibert that the appointment of the JCRA chairman be approved by the States, the proposition was passed by 37 votes to 12. The chairman's appointment was in line with the existing Financial Services Commission procedure which was felt appropriate for a nominally independent organization. The creation of the JCRA would pave the way for the sectoral legislation necessary for the liberalization of both the telecommunications and postal markets. Much work would be required to separate the existing States departments responsible for these sectors from the government machine in order to enable them to act as commercial entities. The question of the pension schemes, which were linked to the civil service scheme, was of prime concern, since much of the existing funds for these two departments had been States supplied. Thus it would be necessary for this to be recoverable after liberalization.

In the New Year of 2001 a new controversy was raised over the siting of GSM masts. The Board wanted to enhance the coverage of its network but public concerns over the safety of the microwave radio emissions had led to a secretive approach whereby the site locations were not made public. In a question in the States it was noted that the information was available from the Planning department but the Board, while denying that there was any danger from its masts, decided on grounds of security to keep a low profile.

January also saw the launch of the JT Rapid Business products, providing high speed internet for businesses. This was an ADSL based product offering a better contention ratio (20:1) and the option for higher bandwidth. The connection fee for the JT Rapid product was £115 was set at £350 with a £80 or £115 monthly rental depending on download speed, the highest available was 512 kb/s.

Also in January internal fixed network number portability was introduced. This enabled subscribers to move to a new home with their existing telephone number even if the new address was on a different exchange. Prior to this move local geographic numbers were linked directly to the exchange locality and a number change was necessary for anyone moving to a different exchange area. While this move permitted analogue and ISDN lines to be moves, technical restraints prevented Centrex system subscribers from using this service.

In February the discussions on how the telecommunications market would be liberalized were hotting up. At a conference organized to discuss local commerce, Bob Lawrence cautioned that the local infrastructure could be compromised if the States got its strategy wrong. He advocated that the States should retain 100% share ownership in order that it could maintain a cash flow into the Treasury. A sale would generate a windfall, estimates of the sale value of between £200M and £300M were suggested, but this would only be a one-off boost to the States coffers. Also in February the Board announced that it would begin phasing out phone cards after the release of its latest design to celebrate the Year of the Jersey Cow. The rise in popularity of mobile phones was blamed for the 50% drop in the sale of cards since 1998. A Board spokesman hinted that a new smart card may be introduced as a replacement.

With the impending liberalization of the market, Newtel announced a tie-up with UK telecommunications and cable television company, Telewest. This agreement would boost expertise and investment in the telecommunications sector and enhance Newtel's bid to become a competitive telecommunications operator on the island, so claimed the Newtel chairman, Peter Funk. However, this new partnership did not prevent Newtel from announcing staff redundancies the following month, blaming the delay in market liberalization.

In May the first signs of liberalization began to show with the appointment of the new JCRA chairman, Dr Patrick McNutt, who promised a gradual approach to market liberalization. McNutt had previously been chairman of the Irish Competition Authority in Dublin and had published a number of papers on economic and competition law and policy. However, the same month a delay in the drafting of the competition law was announced because of prioritization in the legal drafting department. Nevertheless, the telecommunications law was on track and would soon be debated in the States.

June saw the unveiling of the Board's re-branding exercise in preparation for the market liberalization. The existing Jersey Telecoms yellow and blue logo which had been introduced in the early '90s was replaced with a new green and blue stylized version, proclaiming simply Jersey Telecom, the removal of the 's' was supposed to create a divorce from the previous States committee structure. The £100,000 re-branding was designed to create a new corporate image, together with a new staff training programme for the Board's 440 employees that was meant to increase efficiency and service quality ahead of impending competition. The strategy had been some time in the making as the 2001 telephone directory had not included either the new or the previous logo. The residential ADSL product monthly rental was cut 25% to £40 in order to stimulate demand as initial uptake had been slow.

The following month the Telecommunications (Jersey) Law was finally debated in the States. After a delay in presentation, the law had an easy ride and was passed on a standing vote. Because of the delays the original incorporation and liberalization schedule had been disrupted, and so it was unlikely that the target date of 1 January 2002 would now be met. At the same time as the passing of the Law, which would now require Royal Assent to come onto the statute book, the Board announced a new round of tariff changes, reducing some international calls by an average of 48%., although this claim was challenged by Newtel which asserted that it remained the overall cheapest provider.

Following the debate Patrick McNutt outlined the likely principles of future licensing by the JCRA. A consultation process would be undertaken to get views on the shape of the licensing and regulation regime and the conditions that would be included in the licenses. He said that a simple transplant of UK or European practice may not necessarily be appropriate for a small jurisdiction but it was nevertheless important that monopolies did not abuse their positions.

In preparation for the new competitive market Jersey Telecom launched its first fixed line bundled products. This included re-branding the products from its previous numbered products scheme into more user friendly names, for example the JT2400 standard line was now renamed *Coreline*.

With this basic offer a number of packages that included various offers of bundled minutes for local calls and Internet access were launched. During the year a new Network Operations Centre was commissioned in Central exchange. This enabled centralized control and monitoring of the entire island network, a move that would simplify network management and, it was hoped, improve overall quality. Its GSM network was enhanced to provide General Packet Radio Services (GPRS) which would enable faster data access for mobile users. By the year end over 71,000 mobile SIMs had been registered on the network, however, this did not equate to actual users since with the prior introduction of a pay-as-you-go offer, there was no certainty that all SIMs remained operational.

Towards the end of the year a row erupted over the Board's stance towards the new submarine telecommunications cable laid as part of the joint Jersey Electricity Company and Guernsey Electicity Generating Board Channel Islands Electric Grid (CIEG) project. The Board had distanced itself from the matter by again passing the issue to the Industries Committee which was acting as watchdog over the incorporation Jersey Telecom and in the role of the JCRA prior to the introduction of the new telecommunications law, due to be enacted in 2002. The Board president, Paul Routier, also expressed concern that the cable also required permissions from the French authorities before entering service. He also cautioned against competition for competitions sake and maintained that the Board provided a high quality network designed to serve the islands needs. In response to this the chairman of Newtel, Peter Funk, pointed to the delays experienced in obtaining licenses for its own business. He denied that the delays in France were legal but were in fact technical, the French market being fully liberalized. He asserted that the Board was merely employing delay tactics in order to maintain its monopoly as long as possible.

The new Telecommunications (Jersey) Law 2002 was registered at the Royal Court on 16 January 2002 and immediately thereafter Newtel started talks with the JCRA in order to obtain a telecommunications licence. It also simultaneously began the process in Guernsey when it approached the

Office of Utility Regulation (OUR) to obtain a telecommunications licence under Guernsey law.

In April the process of incorporation of Jersey Telecom began in earnest with the appointment of John Henwood OBE as its first board chairman. Henwood was the recently retired head of Channel Television and was also the chair of the local Institute of Directors. Until the legislation came fully into force Jersey Telecom would continue to be run as a very profitable States trading committee. Until then, Henwood would act as a shadow chairman., Profits of £14.9M (16% on capital) for the year 2001 were returned. This not inconsiderable increase did not prevent it from continuing its policy of rebalancing tariffs with line rental increasing by 3.1%.

On 1 July 2002 the Telecommunications (Jersey) Law 2002 was brought partly into force. This gradual introduction was necessary to enable the licensing process to take place, ensuring that when the law was brought fully into force there would be no 'legal gap' created. A consultation on licence conditions and charges was also launched.

Jersey Telecom again reduced the price of its Rapid broadband service from £40 to £30 in July. This price drop, it claimed, brought it into line with BT, against which it benchmarked all its pricing prior to competition. Whether this was an economic adjustment against local network costs was open to debate. At the same time it introduced a wholesale version of the product priced at retail minus 40%. This allowed *LocalDial*, one of the island's rival ISPs, to enter the broadband market. The row between Newtel and Jersey Telecom again erupted when it complained again about its call bypass Freephone numbers being blocked. The Board defended its position relying on the 1972 law and the absolute monopoly that this conferred.

The interim licence issued by the JCRA to Jersey Telecom Limited came into force on 1 November so that it would enable the new company to operate legally after the law came fully into force on 1 January 2003. Meanwhile, delays in the introduction of the law was blamed on squabbling between States committees over land and facility access rights. One of the main bones of contention was the mast at the Board's Five Oaks site which was shared with the Airport's Department of Electronics. The total value of assets transferred to the new company would amount to £58M which would be represented by shares in the order of £20M held by the Treasury on behalf of the States.

In the meantime in Guernsey, Jersey Telecom had been awarded a licence by the OUR to operate as a rival to the now Cable and Wireless owned Guernsey Telecom, the former Guernsey States' department. The new licence was awarded to Wave Telecom, its Guernsey registered limited liability subsidiary company. The first managing director of this new venture was Tim Ringsdore, who had previously been with the Channel Television subsidiary, the Creative Channel. The new company installed one of the world's first so-called soft-switches[161], supplied by Marconi Telecommunications[162] to support its Guernsey operation, however it would rely heavily on its parent network for support of the services it intended to offer and on the wholesale provisions of the Guernsey incumbent. This was a model that would become the norm for new entrants in the islands as competition developed. Newtel Limited was also awarded a licence by the OUR.

The Telecommunications (Jersey) Law 2002 appointed days act was debated in the States on the 21 November. The same day the new full board of Jersey Telecom Limited was named as John Henwood (chairman), Bob Lawrence (executive director), Philip Ainsworth (finance director) and three non-executive directors: John Boothman, former MD of Deutche Bank International, Jersey Advocate David Le Quesne and Dr Nigel Horne a professional engineer with much experience in the telecommunications and computer industries. The States approved the 1 January as the day that the new company would begin operations and the Telecommunications (Jersey) Law 1972 would be repealed at midnight 31 December 2002. Deputy Gerald Voisin of the Industries Committee stated that it had approved a temporary loan of £10M to cover its contributions to the pension fund and this would be repaid through the normal shareholder arrangements. He also promised

---

[161] A SoftSwitch is a telephone switching device for use with IP based packet switched networks.

[162] Marconi Communications plc was the latest iteration of Jersey's traditional equipment supplier originally GEC Telecommunications. Marconi was saved from administration by L M Ericsson in 2005.

that the States would not interfere in the running of the new company and that the telecommunications industry regulation would now become the responsibility of the independent JCRA.

There was some concern expressed by the outgoing Telecommunications Board president, Paul Routier, over the level of licence fees charged by the JCRA. However this was defended by both the JCRA executive director, Charles Latham, and Deputy Gerald Voisin who both asserted that this was a result of the licensing process necessary to make the legal changes. Future costs would likely be lower as other licensees would also be contributing. Newtel would be granted its license on 15 January 2003 and on 1 January three smaller operators would also be licensed. However, licensing would continue to be a bone of contention for the former States department in this new era of competition.

Following a public consultation Ofcom, the UK regulator with responsibility for telephone numbering on Jersey, brought in a new range of numbers for directory inquiry services on 17 December 2002. The former number 192 was now replaced with a range of competitive services using numbers begin the format 118XXX. Jersey Telecom had already subcontracted it services to BT and its new enquiries number would be 118534 at a call fee of 25p.

Following on from the States approval of incorporation of Jersey Telecom the debate begun on whether the company should be sold off, as the States of Guernsey had done earlier in the year. This was initially sparked by a speech in the States from Senator Frank Walker president of the Finance and Economics Committee. He proposed that a realization of the value which he estimated at £200M (against fixed assets transferred of £58M) could bolster the States strategic reserve fund. However, the prospect of a sale did not sit well in a number of quarters, including that of the new chairman of Jersey Telecom, John Henwood, who described the proposal as 'barmy'. In his view the trading conditions were not right for a sale. This view was supported by the drop in value of Cable and Wireless on the UK stock market following a review of its credit rating. Cable and Wireless was seen as the logical first choice following its purchase of Guernsey Telecom.

The Jersey Telecom unions were also opposed to the sale as it was seen as a threat to employment and pensions. The debate on the sale of the new company would continue for many years following its incorporation at midnight on 31 December when the new competitive era of telecommunications began. Jersey Telecom Limited now took over the operation of the island's telephone networks, it being a subsidiary company of the Jersey Telecom Group Limited, which also included a number of other subsidiaries into which various assets were distributed from the original Telecommunications Board.

Telephone Managers

Incumbent	Company	Title	From	To	Status
John Durham	South Western and Wales	Superintendant	June 1888	May 1891	Moved on closure
T A Bates	NTCo	Superintendant	March 1895	1899	Promoted
John Lemon	NTCo	Superintendant	1899	August 1900	Transferred
Howard Eady	NTCo	District Manager	August 1900	February 1913	Transferred to GPO
Howard Eady	GPO	District Manager	February 1913	May 1915	Promoted
A G Mackie	GPO	Telephone Manager	May 1915	June 1919	Transferred
A O Forrest	GPO	Telephone Manager	June 1919	March 1923	Transferred
Alfred R Bennett	States Telephone Department	Engineer Manager	April 1923	August 1924	Retired
Kingsley	States Telephone Department	Engineer Manager (acting)	August 1924	December 1925	Stood down
John H Stanhope	States Telephone Department	Engineer Manager	January 1926	June 1940	Resigned to evacuate
Percy K Luxon	States Telephone Department	Engineer Manager	June 1940†	July 1952	Died
Stanley G Syvret	States Telephone Department	Engineer Manager	July 1952	January 1960	Retired
Harry W Coppock	States Telephone Department	Engineer Manager*	January 1960	December 1974	Retired
Leslie May	States Telecommunications Board	Director of Telecoms	January 1975	December 1983	Resigned to become States Treasurer
T F (Tom) Ayton	States Telecommunications Board	Director of Telecoms	January 1984	May 1991	Retired
R P (Bob) Lawrence	States Telecommunications Board	Director of Telecoms	June 1991	December 2002	Transferred to Limited Company

Table 30

† Luxon only officially appointed to post on 22 June 1945 having acted throughout the occupation

* title changed from Engineer Manager to Director on 1 January 1973 when monopoly assumed

Telephone Statistics

Year	Lines	Stations	Local Calls	UK Trunk Calls	Inf services	Telegram	Waiting
1932	3,287	3,793	2,646,631	11,484			
1933	3,486	4,033	2,763,930	16,478			
1934	3,701	4,347	3,181,953	21,173			
1935	3,948	4,707	3,244,778	25,600			
1936	4,188	5,015	3,549,744	31,786			
1937	4,338	5,232	3,807,445	44,787			
1938	4,530	5,505	3,877,375	53,937			
1939	4,682	5,746	3,993,346	57,237			
1940	4,744	5,877	4,046,542	28,042			
1941	3,790	4,869	2,849,946	0			
1942	3,459	4,711	1,728,835	0			
1943	3,263	4,505	1,751,143	0			
1944	3,374	4,645	1,743,589	0			
1945	3,481	4,267	1,544,338	0			
1946	4,101	4,972	1,900,985	0			
1947	5,357	6,658	4,160,505	158,333			
1948	6,052	7,926	5,283,235	230,788			
1949	6,598	9,046	5,881,148	263,794			
1950	7,166	9,790	5,873,888	274,030			
1951	7,662	10,529	6,243,398	344,257			
1952	7,997	11,043	6,506,351	390,969			
1953	8,410	11,542	6,988,395	486,987			
1954	8,956	12,231	8,648,526	585,315			
1955	9,544	13,159	10,246,980	685,509			
1956	10,178	14,647	11,300,446	735,089			
1957	10,690	15,874	10,938,567	834,822			
1958	10,982	16,199	10,542,299	808,165		29,853	353
1959	11,130	16,644	8,103,780	735,519	-	32,382	578
1960	11,711	17,889	9,065,339	813,007	45,405**	30,633	603
1961	12,325	19,098	9,955,395	480,018*	53,064	31,822	557
1962	12,844	20,040	10,424,398	532,475	60,044	24,239	653
1963	13,338	21,157	11,110,055	552,830	63,089	19,962	625
1964	13,860	22,281	11,760,320	611,212	91,104	19,548	677
1965	14,593	23,584	12,780,665	697,475	292,333	19,216	772
1966	15,614	25,164	13,374,485	845,338	354,908	18,494	487
1967	17,273	27,629	14,000,000†	1,031,827	443,864	18,104	463
1968	17,807	27,318	15,068,000†	1,295,118	450,442	19,314	371
1969	18,913	29,359	16,216,000†	1,606,975	402,036	19,270	417
1970	19,704	30,657	16,543,735	2,250,000	520,000†		
1971	20,501	32,638	16,857,047	2,500,000	696,974		
1972	21,624	34,555			750,000†		
1973	22,947	37,103			825,000†		
1974	23,763	38,940			950,000†		
1975	24,445	39,930			1,000,000†		
1976	25,637	42,438	24,220,000	3,443,000	1,100,000†		
1977	27,004	44,894	26,548,000	4,677,000	1,250,000		
1978	28,656	47,678	31,765,000	5,233,000	1,300,000		
1979	30,183	50,282	33,000,000	5,304,000			
1980	31,497	52,658	33,000,000	5,454,000			
1981	32,605	55,382					
1982	33,736	58,335					
1983	35,120	61,447					
1984	36,529	64,291					
1985	38,047						
1986	39,835						
1987	41,582						
1988	43,885						
1989	46,300e						
1990	50,900e						
1991	53,000						
1992	54,485						
1993	56,534						
1994	58,069						

Year	Lines	Stations	Local Calls	UK Trunk Calls	Inf services	Telegram	Waiting
1995	59,308						
1996	64,471						
1997	65,500						
1998	68,700						
1999	70,300						
2000	74,100						
2001	73,900						
2002	74,300						
2003	73,200						
2004	72,100						
2005	72,000						
2006	71,500						
2007	58,700						
2008	58,321						
2009	57,708						

Table 31

* After 1960 records of incoming trunk calls no longer kept

** Prior to 1960 no record was kept of such enquiries by the operators

† Estimated as data unavailable

Chapter 21 - The Telegraph the Telephone, the Railway Companies and Way Leaves

The Western Railway

A way leave is an easement or right-of-way granted by a land or property owner for access by another person or entity for the purpose of conducting legal trade or passage. The history of the way leave goes back into at least Saxon times and was used extensively by the railway and telegraph companies throughout Britain for the construction of their networks. Indeed, until the Electric Telegraph Act of 1863, private way leaves were the only means of construction for the telegraph companies. The Act did not, of course, extend to Jersey and different arrangements were made locally for the digging of streets. Nevertheless, the use of private way leaves by the telegraph and telephone companies was extensive.

When the Jersey and Guernsey Telegraph Company started building its new telegraph line they were unable to use the earlier Henley underground cable as this had been deliberately damaged. It is also probable that it had fallen into general disrepair through being unused for some time thus causing a general degradation of the insulation.

Consequently, it negotiated a way leave with the Jersey Railway and Tramways Company Limited to erect poles along the length of the line as far as Beaumont where the route to Plémont would be taken through St Peter's.

The Jersey Railway company itself was only established in 1870 and the railway board probably thought this an opportune request to enable signalling to be carried out at little or no cost to themselves. In fact, both the Jersey Railway and the J&GTCo shared a number of board members. In order to secure the way leave, the telegraph company agreed to provide a telegraph for the railway free of charge. This required the erection of poles all the way to St Aubin. A Breguet ABC[163] telegraph instrument with 6 Leclanche Gravity cells was installed at both the St Helier terminus and the one at St Aubin. No formal contract appears to have been formed, but rather a gentleman's agreement between the company boards.

This arrangement continued into the Post Office era and there seems to have been no problem with the casual arrangement until, in 1883, when the Jersey Railway made a request to the local Post Office superintendent for an upgrade from the telegraph instruments to a telephone line. The matter was brought to the attention of the PO southern Area Manager who discovered that there was no formal relationship in place and that the local PO engineers had been continuing the practice without question since the takeover of the system from the old company some 10 years earlier. An investigation was mounted and the PO realized that an upgrade to a telephone circuit would require a second wire to be erected along the whole length of the line and so instead offered to replace the existing system with more modern Wheatstone ABC instruments at cost. At first the railway refused this offer but, at the insistence of the PO, negotiations to put into place a proper way leave agreement were initiated. A formal agreement was signed on the 2 October 1884 and as part of that agreement new instruments were provided free of charge. The agreement granted an exclusive way leave to the Postmaster General 'in perpetuity'. The railway also provided free transport for PO maintenance engineers and inspectors.

In 1893 the railway company requested that an intermediate instrument be installed at the Millbrook station. The PO considered this offer and

---

163 http://www.telegraphsofeurope.net/page145.html last accessed 18/05/2017

Image by permission of Societe Jersaise

The above photo shows the relationship of the GPO and NTCo telephone poles along the Western Railway at Bel Royal station. The poles heavily loaded with 8 way arms belong to the NTCo while the GPO poles are only carrying a few telegraph circuits. The photo was taken shortly after the closure of the railway in 1936.

initially sent details to the PMG office in London. In August 1894 the PMG offered to install an instrument for the sum of £11.4.0d with a further annual maintenance fee of £2.0.0d. The railway rejected this offer and after some negotiation they finally settled on an installation charge of £10.0.0d with free maintenance as per the original agreement.

This new way leave agreement was thought to have settled matters but in June 1896 the local PO superintendent noted that an A to B private telephone line had been erected by an ex-PO employee on the railway. The PMG challenged the railway company on the agreed exclusive way leave contract, but the railway's advocate responded that as the agreement had not been registered in the Royal Court of Jersey, it had no validity. Therefore, the railway was free to permit anyone a way leave. After some considerable investigation, the solicitor acting for the Post Office advised the PMG that this was indeed the case, and that Jersey law required any agreement to do with land which extended over a period greater than nine years required registration in the Royal Court. This was a shock to the PMG and an investigation was started on the viability of removing all PO lines from the railway property. The investigating engineer, however, concluded that there was no easy solution as so many way leaves would be required in order to move all the wires that the time and expense was disproportionate.

The PMG, therefore decided to reopen negotiations with the railway company to secure a new agreement. The railway was now reluctant to grant an exclusive privilege to the PO since there was already an existing line and the NTCo had already started erecting further poles and circuits during 1896/97 to connect their new exchanges at St Helier, Millbrook and St Aubin. The negotiations continued into 1898, the railway rejected the PO request for a permanent way leave claiming that they did not hold absolute secure tenure over some parts of the line. On 13 May 1898 the railway offered a 14 year contract which would take the agreement until the end of the NTCo licence term, 31 December 1911, after which the PMG claimed the right to purchase all NTCo lines. The PMG was reluctant to accept but finally, the need to maintain their telegraph circuits and those circuits provided by the PO for the Admiralty and the War Office forced them to accept. On the 1

March 1899 the agreement was duly signed, but not until after the way leave agreement with the NTCo had been concluded for the same term on 1 January of that year.

As part of the agreement, the PO agreed to continue the relationship with the railway for provision of private circuits on PO poles. The railway extracted an agreement that the PO would supply, free of charge, 5 telephone stations along the railway at St Helier, Millbrook, St Aubin, Don Bridge and Corbiere. The railway company also wanted an increase in the yearly fee paid for each pole, the PO successfully fended off this request but, as a result, the agreed free transport arrangement was scrapped and the PO employees were forced to pay for each journey made on the railway for maintenance except if the repair was on a railway instrument. The new agreement was duly presented to the Royal Court.

Later, the PO negotiated a fixed yearly fee for two passes for their maintenance engineers in the sum of £20.0.0d. This arrangement continued until January 1911 when the railway included a clause removing liability for the PO employees using these passes. The PMG instructed the PO solicitor to send a tract of the Cambrian judgement (a case that required railways to have a duty of care for passengers) and the condition was withdrawn.

The new agreement lapsed on 31 December 1912, the year after the acquisition by the PMG of the National Telephone Company. The poles owned by the NTCo had now become the property of the PO and this meant a substantial increase in both pole numbers and miles of wire supported on them. The PO had some 184 poles carrying 50 miles and 700 yards of wire along the length of the line from St Helier to Corbiere (although this excluded those circuits used exclusively by the railway); the NTCo poles numbered some 158 poles with 158 miles and 1160 yards of wire along the same route. It should be noted that the NTCo poles were on the seaward side of the track and so, by necessity, were taller than the PO poles in order that the distribution wires could span the PO wires without interference.

The railway now felt that they had the upper hand and stuck out for a better agreement than the previous ones. The started out with a new demand for the way leaves. At this point the PMG again instructed the

Engineer in Chief Southern Region to investigate the possibility of removing the poles from the railway property and reconstructing along another route. The local superintendent engineer set out the costs of relocating the poles in absolute terms, and these are shown in Table 32 below.

	Poles	Dressings*	Totals
Existing PO Poles	£215.0.0d	£525.0.0d	£740.0.0d
Existing NTCo Poles	£370.0.0d	£785.0.0d	£1155.0.0d
New Construction	£1020.0.0d	£1570.0.0d	£1960.0.0d

Table 32

* Dressings include insulators, wires, crossarms, spindles, steps and screws etc.

The existing routes construction pole types were as in Table 33 below:

Post Office Route				National Telephone Co Route	
Number	Stout	Medium	Light	Number	Medium
1	26'			32	20'
8		28'		24	24'
27		30'		16	26'
19		26'		2	28'
36		34'		44	30'
93			24'	7	32'
				1	34'
				17	36'
				8	40'
				4	45'
				2	50'
184				167	

Table 33

In addition there were 70 PO pole stays and 95 NTCo stays. The average

age of the poles along the routes was said to be 18 years, this would indicate that the poles were new at the time of original construction in 1895 and that many of the PO poles had been replaced since the original route was constructed by the J&GTCo.

(Note: The disparity between the 158 poles mentioned above and the 167 poles in the above table is because the NTCo poles from Millbrook to Bel Royal were along the public highway).

The routes were constructed using the following wire types and gauges:

Wire weight	Material	Length
100 lb/mile	Copper	27 ¾ miles
150 lb/mile	Copper	4 miles
200 lb/mile	Galvanized Iron	10 ¾ miles
400 lb/mile	Galvanized Iron	117/8 miles
40 lb/mile	Bronze (Cadmium-Copper)	54 miles
		213 miles

Table 34

The material costs only are shown and no labour charges are taken into account. This is an interesting observation and illustrated the operation of the nationalized industry, that is, labour is a necessary overhead which cannot be removed. The local engineer also noted in his report that there was no certainty about obtaining all the necessary way leaves for the move. The actual difference in cost of providing a new route was marginal, just £65.0.0d, but the effort and uncertainty required in making the move could not be accurately measured. In this light, the PMG decided to continue the negotiations with the railway company.

The new agreement for a period of 20 years was not finally concluded until 13 March 1915 and was backdated to 1 January 1913. The PO eventually negotiated a fixed fee of 1/- per pole, stay or yard of underground duct, except for those fixtures used only by the railway. The yearly pass was withdrawn and the railway reverted to charging per journey.

This may not have been a good move by the railway, since in the first 14 months they invoiced the PO for only £9.10.6d, somewhat less than the previous fixed £20.0.0d per annum. The PO also managed to remove the clause that covered the provision of free circuits and made an annual charge of £12.0.0d per circuit per annum. A separate agreement was made for the provision of call offices along the railway. Either party could discontinue the agreement at 6 months notice. The final agreement registered in the Royal Court was drawn up by the Engineer in Chief, Mr Eldridge and signed by A T Kinsey for the PO and Henry Edward Le Vavasseur dit Durell for the railway on 1 January 1916.

The circuits provided by the PO to the railway are shown in the table below:

Corbiere	Don Bridge	St Aubin	Millbrook	St Helier
Free of Charge*	Rented	Free of Charge	Rented	Free of Charge
	25 yard spur	40 yard spur	10 yard spur	
2M 550 yards	1M 840 yards	1M 1160 yards	1M 760 yards	

Table 35

* This circuit was presumably negotiated when the Government signal station at La Moye was connected.

It was noted that the 2 wire circuits were provided over 100lb per mile copper wire.

At the time of the takeover of the NTCo by the Postmaster General the circuits carried on PO poles along the railway line are shown in Table 36 below:

Corbiere	La Moye	St Aubin	Beaumont	Millbrook	Piquet House †	Elizabeth Castle*	St Helier
1350 yards	3M 40 yards	1320 yards	1600 yards	1720 yards	780 yards	120 yards	
3 PO ccts	3 PO ccts	5 PO ccts	8 PO ccts	9 PO ccts	9 PO ccts	9 PO ccts	
	2 Admiralty		6 WD ccts	6 WD ccts	22 WD ccts	20 WD ccts	

Table 36

Notes: * A Railway section designation
 † West of West Park
 WD = War Department

The total mileage of the PO circuits was 21 miles and 160 yards, the Admiralty circuits 6 miles and 80 yards and the War Department 22 miles and 760 yards.

The NTCo distribution along the railway line at the time of the PO takeover is shown in Table 37 below:

Location	Distance (Yards)	Circuits	Exchange
Corbiere			
Distribution Point	270	4	La Moye
Old Corbiere Station	720	6	La Moye
Distribution Point	770	10	La Moye
Distribution Point	630	6	La Moye
Don Bridge	1540	4	La Moye
Distribution Point	80	6	St Aubin
Pont Marquet	760	12	St Aubin
Distribution Point	160	18	St Aubin
Distribution Point	1200	30	St Aubin
Portlette	350	45	St Aubin
Distribution Point	300	50	St Aubin
Distribution Point	140	55	St Aubin
St Aubin	100	80	St Aubin
St Aubin's Derrick	630	50	St Aubin
La Haule	280	34	St Aubin
Distribution Point	140	30	St Aubin
Distribution Point	70	25	St Aubin
Distribution Point	210	24	St Aubin
Distribution Point	280	14	St Aubin
Beaumont	840	16	Millbrook
Bel Royal	630	45	Millbrook
Millbrook	140	95	Millbrook
Distribution Point	140	90	Millbrook
Distribution Point	70	84	Millbrook
Distribution Point	70	78	Millbrook
Distribution Point	70	74	Millbrook
Distribution Point	70	70	Millbrook
Distribution Point	140	68	Millbrook
First Tower	70	62	Millbrook
Distribution Point	280	35	Millbrook
Distribution Point	140	32	Millbrook
Distribution Point	140	28	Millbrook
Picquet House	770	30	St Helier

Table 37

This agreement continued in place until the PO sold the telephone system to the States of Jersey in 1923. At this point 129 poles along the railway were transferred to the States Telephone Department, leaving 47 under the ownership of the Post Office. The PO way leave agreement for these remaining poles ran until 1933. These would have been for the poles carrying the telegraph circuit along the railway, branching to Plémont Cable Station at Beaumont, and to St Aubin, Millbrook, St Peter's, St Ouen's, St Mary's and St Johns post offices. The PO maintained the overhead telegraph routes to the outlying post offices beyond the handover of the telephone system to the States. Millbrook and St John's were connected to the railway route at pole No 38 (located near Millbrook railway station) and shared the use of some of the State's poles and underground cables en-route. The PO telegraph poles were equipped with 4-way arms, while the NTCo poles used 8-way arms for wire support and distribution. This is evidenced from contemporary photographs of the railway lines.

The Jersey Eastern Railway

The problems that the PMG experienced with the Jersey Railways and Tramways Company Limited did not seem to exist with the Jersey Eastern Railway Company Limited (JER). This probably stems from the fact that the JER, unlike its sister company, did not have external agreements with any of the telegraph companies and instead provided its own signalling systems between Gorey and St Helier. A telephone circuit was installed for the line in September 1880 by the India Rubber Gutta Percha and Telegraph Co (Ltd) of London and Bristol. The Submarine Telegraph Company (STCo) also had an office at Gorey pier but this line had been installed in 1864, some eight years before the railway was in existence and used its own pole route from Belvedere Hill through Longueville and across Gorey common. The railway initially did not extend as far as Gorey Pier but terminated at Gorey Village.

There is no doubt that the NTCo agreed a way leave with the JER sometime during 1895, since they opened the La Rocque and Gorey exchanges in July of that year, connecting them via junction circuits to St Helier using the railway as a pole route. The PO also agreed a way leave in formal document signed in, 1896, by Thomas Blampied of the JER and C B Bullswell of the PO. A new agreement was signed on 4 March 1914, subsequent to the PO acquisition of the NTCo. This agreement was signed by the then PMG the Rt Hon Charles Edward Henry Hobhouse MP. At this time there were 128 poles along the railway line which carried both telephone and telegraph circuits.

The PO may have entered into a way leave agreement around 1896 in order to provide a new circuit to the Fliquet cable hut, although there is a possibility that Fliquet had been served from overhead wires from an earlier time. This cable had previously been provided by the underground Henley cable installed by the STCo in 1860 and may have failed, or progressively deteriorated, after some thirty odd years of use. Certainly, the Jersey to Rennes telegraph circuit via Fliquet was provided over poles along the railway, as this circuit was mentioned during the handover of the PO telephone system to the States of Jersey Telephone Department in 1923. The circuit ran from pole No 145 at Gorey pier along a route of 67 poles to the cable hut at Fliquet. This route was also handed over to the States in 1924 when the then chief engineer of the States Telephone Department, A R Bennett, offered to reconstruct the route, free of charge, during the expansion programme of the telephone network, incorporating the PO circuit into the telephone route. The deal was struck and the States purchased some 25 poles which were in good order for £82.10.0d while offering £20.0.0d for the remaining 42 as scrap. The route was reconstructed because of its 'wavering path and many road crossings'. The deed of sale was signed by the Rt Hon Vernon Hartshorn MP the then PMG and by the Deputy Greffier of the States John Edward Le Huquet on 18 August 1924. The PO agreed to the States maintaining the single wire circuit but also included a clause to permit access by PO engineering staff at any time. There was also a provision for a further circuit to be installed, if requested by the PO, along the pole tops.

Other Wayleaves

At the time of the handover of the island telephone system to the States on 16 March 1923, the PO also noted a number of other private wayleave agreements. These were mainly for poles and stays in and around St Helier and had been largely inherited from the NTCo in 1912.

Table 38 below lists these wayleaves:

Quantity	Type	Owner	Site	Cost per year
1	Pole	FG De Faye	David Place	£1.0.0d
1	Standard	Grandin's Iron Works	Commercial Buildings	10/-
1	Pole	H Hescott, 9 Kensington Pl	Lewis Street DP	5/-
1	Pole	H Le Brun, La Cordine, St Ouen	Beaumont	2/-
7	Poles	GH Le Rossignol	Les Niemes Farm	10/-
1	Pole	Dr AP Nicolle, Upper Kings Cliff	Goodlands Cottage	1/-
2	Poles	H Tarr, Belozanne	Mont Couchon	2/-
1	Standard	A Amy	Waterloo Lane	£1.0.0d
1	Pole	AW Grandin, 13 Burrard Street	Vine Lane	10/-
1	Bracket	FF Le Conteur, 67 Lee Rd, Blackheath	Rouge Boullion	2/6
2	Poles	JD Bauche	Beaconsfield	£1.0.0d
1	Pole	JT De La Perrelle, 18 Colomberie	Dorset Street	£1.0.0d
1	Pole	J Balliene	La Chasse	1/-
1	Pole	Geo Colman Ltd, Covent Garden	Commercial Street	1/-
1	Pole	AE Freaut	Belville Street	1/-
1	Pole	States Education Department	St Paul's School	5/-
4	Poles	RR Marett	La Haule Manor	4/-
1	Poles	RR Marett	La Haule Manor	1/-
6	Poles	RR Marett	La Haule Manor	6/-
4	Poles	MBF Bryant, Millbrook	The Firs	6/-
3	Poles	Allain & Co	Yard	30/-
1	Standard	BJ Blampied, 70 Esplanade	A yard in Kensington Plce	10/-
	Stays	Luce Cologne Ltd	42 King Street	£2.0.0d
1	Pole	JP Britton, 10 Bath Street	Minden Lane	£1.0.0d
1	Pole	D Langlois	102 St Saviours Road	1/-
3	Poles	W Norman, St Lawrence	Grassdale	1/-
1	Pole	A Turpin, 26 Vauxhall	Tunnell Street	5/-
1	Pole	M De La Perrelle	5 Seaton Place	5/-
1	Pole	JG Gallie	Colomberie	£1.0.0d
1	Pole	A Le C Blampied, Passmoor, Vallee des Vaux	12 Parade	£1.0.0d
1	Pole	Sir Jesse Boot	6 King Street	£3.0.0d
1	Pole	States Piers & Harbours Ctte	Harbour	£2.5.0d
1	Pole	J Morris	Gasworks	1/-

Table 38

Standards were normally built with substantial cast iron or steel and mounted on suitable rooftops and usually carried very high densities of open wires.

The States Telephone Department continued the practice of obtaining way leaves from private property owners and offered 'peppercorn' rentals of for each pole or stay where possible.

Chapter 22 - The Post Office Telegraphs and Cables 1945 - 1972

Although during WWII the occupation of the islands meant that the cables to the UK mainland were not used, they remained in place and were apparently not interfered with in any great way. Indeed, the War Office had planned before D-Day to utilize the existing Channel Island cables to support the invasion. Plans were laid to lift and divert these cables to landing places in Brittany and Normandy before the liberation of the islands. In the event, these plans were never carried through[164].

Following liberation, one of the the immediate concerns was the restoration of the telephone and telegraph service. On 12 May 1945 a party of GPO staff arrived on Guernsey to re-establish the wireless link to Chaldon[165]. A mobile 6 channel radio system was installed at Fort George and was working by 14 May. This was followed on the 16 May with the opening of a VF telegraph circuit to London and by 10am telegraph traffic was passing. The remaining channels were commissioned soon after giving a further telegraph circuit and four telephone circuits.

The paragutta cables from Compass Cove, Dartmouth to Guernsey were picked up and tested, the cable ships running the gauntlet of extensive minefields. They had suffered no serious deterioration from disuse during the war and end to end testing and restoration of the first cable was

---
164 Signal Venture – Brigadier L H Harris CBE TD Msc MIEE, Gale and Polden 1951
165 CI Communications and the War, Post Office Electrical Engineering Journal Vol 38, p102

completed on 25 May. The inter-island cables had been used by the German occupying forces and were in good order. However, it was decided at this time to abandon the two older telegraph cables, as new technological developments meant that all services could be provided over the newer telephone cables.

To expedite the rapid return to normal communications, simple ex-military 1 + 3 (that is to say one physically connected service plus 3 carrier services) and 1 + 4 carrier systems and quantities of stores for repair of the underground plant were landed at Guernsey and Jersey on 20 May using an American landing craft. The cables from Fort Doyle had been diverted by the Germans to a military communications centre at Oberlands Road in Guernsey, and this site was used temporarily while the existing PCQT[166] cable which had been diverted by the Germans was reinstated and repaired. The 1 + 3 systems were established at the cable ends at Oberlands Road and the 1 + 4 system was located at the St Peter Port telephone exchange and connected over the existing cables to Saints Bay cable hut. In Jersey the cables had not been moved, although the Germans had planned to relocate the repeater equipment into a special bunker near the existing repeater station at Trinity Gardens, this had not been completed before the liberation.

Prior to belligerencies 12 channel systems had existed on both the Guernsey to Jersey submarine links, but these appear to have been removed, possibly because of the lack of spares, by the occupying forces who had established their own systems on the cables. The 1 + 4 systems were opened between the islands and by 21 June full communications were available on both cables to Dartmouth.

Portable 6 channel MCVF telegraph systems, of a type extensively used by the military, were installed at London, St Peter Port and the HPO in Broad Street, St Helier, Jersey for the re-establishment of the telegraph service and to provide the necessary services for interconnection between the islands' airports. The systems were connected to the audio channel of of the

---

166 Paper Core Quad Trunk (PCQT) a type of cable used for telephone junction and transmission circuits.

carrier systems. The system was later moved from the HPO to the Trinity Gardens repeater station when this was returned to service. Later the radio equipment was moved to Guernsey after the cables between the islands were returned to service. Guernsey, being closer to the UK, was a better location for the wireless equipment.

In May 1945 Luxon, the States of Jersey Telephone Engineer Manager, noted in correspondence with the GPO that there were only two circuits available to Guernsey and four from Guernsey to the UK. The Jersey to Guernsey circuits were initially on the physical audio frequency circuits of the cables connected to the radio links. He requested 8 UK circuits by 1946. However, by June 1945 a total of 8 multiplexed circuits existed between the islands, the telegraph being operated over the audio sections of the cables. These 8 circuits would also have supported inter-island traffic and so the number of circuits to the UK would have been less. Work in the autumn of 1946 doubled the capacity of the UK cables, perhaps by adding multiplex equipment on the second cable, at least, it was reported in the JEP on 22 November that the number of circuits available to the UK was now 8.

The continental service was not restored until 26 May 1946. Calls were charged for minimum 3 minutes at 12/6d to France (Zone 1) and 9/- Zone 2. The Fliquet cable providing services to Rennes was not reopened until April 1949 when calls to local French numbers were charged at 6/- for 3 minutes. Local French-speaking operators were paid a supplement of 10/- per week. Other examples were Vatican City 20/- and Sicily 22/6d. It is assumed that the Siemens multiplex equipment installed in 1939 was still in working order having been used throughout the war by the occupying forces.

In addition, there were now 6 wireless telephony circuits between Fort George in Guernsey and Chaldon Down in Dorset, making a total of 17 trunk circuits shared between the islands.

Later, 12 channel multiplexing equipment was reinstated on both the Dartmouth to L'ancress and Saints Bay to Plémont sections bringing the total circuit capacity up to 24. Not all of the circuits were available for telephony as some were required to supply bearer channels for inter-island airport telegraphy, news services and the GPO telegram service. After the installation of the multiplex equipment, the telegraph operating on the audio circuits was removed and operated over a 12 channel VF telegraph system on a telephony channel. In addition, there was a complex signalling arrangement to enable the existing trunk switchboards to operate over the multiplexed channels which required an additional trunk circuit using 12 channel VF telegraphy circuits to handle the supervisory signalling[167].

During 1951, the Post Office reviewed the trunk circuit needs of the island. Meetings were held with the States Telephone Department with a view to determining the traffic requirements and the possibility of suitable accommodation for the new equipment. The UK trunk network was moving to semi-automatic working and the GPO were introducing new Sleeve Control[168] switchboards at the new Auto-Manual centres which served as the trunk exchange network using a mix of manual and automatic switch trunk connections. The new switchboards were different in concept from the old Central Battery (CB10) type boards currently used as trunk positions by the States. In addition, the new auto-manual service employed the AC1[169] signalling system which required different battery voltages from those supplied by the States 40V CB10 system. The sleeve control switchboards required a 50V supply and the associated AC1 signalling relay sets required 6V and 140V battery supplies for the amplifier valves. The existing trunk switching consisted of an island suite in the Central switchroom of seven modified CB10 'B' positions with three dummy positions. These positions used bridge circuit signalling and required a VF telegraph circuit to repeat the supervisory signals over the carrier trunk channel circuits in addition to a speech channel which added further cost to the trunk network.

The amount of traffic from Jersey had increased dramatically after the cessation of rationing in 1951 and 300 calls per day were being originated in Jersey during the busy potato season. This was generating annual revenue of between £50,000 and £60,000 from both islands at the current call charge rates. These had now risen to 1/6 (7.5p) daytime and 1/- (5p)

---

[167] CI Communications, Post Office Electrical Engineering Journal Vol 42, p141
[168] Telephony: A Detailed Exposition of the Telephone Systems of the British Post Office : Volume II – Atkinson J
[169] ibid

evenings for a 3 minute call. The share of the revenue paid to the States for operating the service had been reduced from 20% to 13.5% after the war.

A meeting was held on 7 September 1951 between the Post Office and the States Telephone Committee to discuss the proposed new trunk exchange. The GPO stated that they would require sufficient accommodation to provide a switchboard suite with an eventual design capacity of 40 positions. At this meeting the States' proposed that additional accommodation could be provided at the recently constructed Northern exchange building at Herupé, St John. This building had been built during 1949/50 at a cost of about £9,000 but installation of the new CB10 exchange had not yet begun.

This proposal was considered by the GPO as there was certainly sufficient land available to build a second switch room above an extended equipment room to the south side of the building. The GPO sent two representatives to Jersey to assess the proposal. Mr G H Farnes and Mr V Roberts visited the island on 20/21 September 1951 to make a report on the feasibility of the offer. The report concluded that in order to provide sufficient growth capacity for the 25 year planning period, it would be necessary to spend between £12,000 and £15,000 on civil works for the construction of ducts between the repeater station at Trinity Gardens, St Helier and Herupé and in addition, initial cabling costs were estimated at a further £22,600 for the 2.5 mile cable haul. Although there was a suggestion from Luxon, the States Telephone Manager, that half of the duct construction costs may be borne by the States, the report concluded that this option was very expensive.

Further negotiations continued following the report. The GPO asked the States if accommodation could be found nearer to the existing States Central exchange and Trinity Gardens repeater station. Luxon considered that this would not be possible as suitable building land in St Helier was at a premium. The GPO themselves investigated this route and questions were asked as to whether there were funds available for building accommodation in Jersey. Some consideration was given to accommodation in the newly leased Lyric Hall in Cattle Street, which the States had recently acquired to provide a relief exchange to Central. Lyric hall measured 57' 6" by 24' with a smaller annex of 14' by 9' to the rear. However, the engineering group decided that this was too small for the planned final growth and there was also insufficient height on the ground floor for the AC1 signalling equipment racks, which stood 10'6½". Finally, at a meeting on 17 June 1952, the States offered to provide accommodation by building a second storey on top of its Telephone House administration building on the corner of Minden Place and Cattle Street. The GPO looked at the proposed extension and agreed that it would be sufficient provided that no floor space was lost by the inclusion of an internal stairwell. Instead an external metal staircase was proposed which would allow operator access. A smaller extension on the first floor of a building opposite the rear of Telephone House provided accommodation for the operator rest rooms. This proposal was considered to be the best option and the GPO agreed to go ahead. Final negotiations were delayed by the untimely death of Luxon, but the agreement was concluded on 11 October 1952 with the acting Telephone Manager, Stanley G Syvret.

The GPO 25 year circuit growth predictions as of 1951 for the Jersey trunk exchange is detailed in Table 39 below.

Year \ Route	BS-JE	TS-JE	GY-JE	JE-Rennes
1957	28	37	12	3
1962	34	45	17	4
1967	40	55	23	5
1974	48	70	28	6

Table 39

Notes: Bristol = BS; Taunton = TS; Jersey = JE and Guernsey = GY

The GPO planned to upgrade the cables between Dartmouth and the Channel Islands by the introduction of submarine repeaters into the existing co-axial cable. This would enable the capacity to be increased to carry two supergroups. A supergroup consists of 5 x 12 channel groups, thus 60 circuits, bringing the total capacity up to 120 circuits per cable. At this point the radio circuits would be recovered. The radio circuits were, in some weather conditions, liable to fading, and at all times their quality was less than that expected for normal inland trunk calls. In any case the frequencies had now been reallocated for the use of the new BBC television service. This upgrade would suffice for the period up to 1962 according to forecasts.

During October 1952 the GPO upgraded both cables between Dartmouth and Guernsey. Each cable was picked up by the *CS Alert* at equidistant intervals and three submarine repeaters were inserted. It was reported in the JEP that the first repeater was sunk at a depth of 384 feet near the Hurd Deep off Alderney, the remaining two at around 215 feet depth between there and Dartmouth. Each repeater installation required interruption to the trunk service as the cable had to be taken out of service each time it was cut. The change of configuration also meant that new power-feeding terminal equipment had to be installed at Guernsey and Dartmouth to power the submarine repeaters, and additionally new multiplex equipment of the GPO 51 Type Practice[170] was installed at each station. This new equipment practice was the GPO's first attempt in developing a modular installation method. All equipment was mounted on standard pitch 19" rack type construction using a pressed steel structure.

The Jersey to Guernsey section was not upgraded with repeaters as it was only some 22 nautical miles, but new line amplifiers were installed at the Saints Bay and Plémont cable huts and new terminal equipment was installed at the Trinity Gardens repeater station.

The construction of the new building was completed during the latter part of 1953 and ready for installation of equipment by January 1954. The trunk switchboard equipment was installed by Messrs GEC of Coventry in the second floor extension. The AC1 signalling equipment was accommodated in an equipment room adjacent to the switchboard. A power and battery room was constructed on the ground floor of the building on Cattle Street. This was accessed through the garage doors and contained the 50V signalling battery and the 140V and 6V batteries necessary to operate the AC1 signalling equipment. Maintenance support was provided by Telephone

---

[170] POEEJ Vol 51 p 197 Oct 1958 An Improved Form of Mechanical Construction For Transmission Equipment- 51-Type Construction

Department engineers under a contractual arrangement with the GPO. The new exchange was officially opened on 17 June 1954 by His Excellency the Lt Governor Admiral Sir Gresham Nicholson. An inaugural call using a new ivory coloured GPO Telephone No 332 was made to the PMG the Rt Hon Earl de la Warr. At the time of opening 51 bothway trunk circuits were provided from Jersey to the UK, these being provided by the GPO automanual trunk exchanges enabling most inland calls to be dialled directly by the originating operator. Trunk routes were terminated at both Bristol and Salisbury automanual centres, as well as direct circuits to Guernsey.

On 28 December 1954, the PMG modified the telephone licences of both the Jersey and Guernsey administrations in order that they could manage their own circuits between the islands, independently of the GPO, although the GPO maintained control of the cables and equipment. This was celebrated in Guernsey by offering free inter-island trunk calls over the Christmas and New Year period. This generous offer was not reciprocated by the Jersey Telephone committee.

At the 1956 the annual trades' exhibition held in November at Springfield Stadium, St Helier, the GPO demonstrated its public inland Telex system to local companies in an effort to popularize the service. Although the GPO private inland Telex service had been available since before the Second World War, much disruption had been caused to the service during hostilities and now a new commercial realism was being pursued by the GPO. It had decided to popularize the service in an effort to increase its telegraph department revenues in the face of falling usage of the inland telegram service as a result of the rise in telephone trunk traffic.

After WWII the GPO public Telex service had been established as a separate entity from its inland telegraph service[171]. A limited international service via a manual switchboard was established in London in 1947. On 15th November 1954 a new (manually switched) public inland Telex Service opened. This enabled a group of (local) subscribers to share a pool of telegraph junctions to the nearest Telex manual switchboard. By 1955 more

---

171 Telex: A Detailed Exposition of the Telex System of the British Post Office, R W Barton, Pitman 1968

than 2,000 UK subscribers were connected to the system and later that year 6 Telex exchanges with manual boards were established at Birmingham, Bristol, Glasgow, Leeds, Manchester, and London. In February 1956 the GPO announced its intention to automate the Telex service[172]. As a consequence a whole new raft of Telex switching centres would be established, including one at Bournemouth, which would also serve the Channel Islands. The Bournemouth automatic switch did not open until 1959. Telex subscribers in Jersey were connected using Multi Channel Voice Frequency (MCVF) GPO type AC6[173] circuits over voice channels on the Dartmouth-Guernsey-Jersey submarine cable. The MCVF system could support 18 and later 24 Telex circuits over a single telephone voice channel, thus was economic in deployment. The AC6 equipment was installed at the GPO repeater station at Trinity Gardens, St Helier and thereafter connected to subscribers over States Telephone Department provided private circuits using the standard GPO +80V/-80V DC signalling system.

In February 1957 in answer to a question in the House of Parliament the assistant PMG Kenneth Thompson stated that various new European cables, including one to Jersey, would be laid following the completion of the first Transatlantic submarine telephone cable (designated TAT1). This cable had been proposed as early as January 1956 and the landing site for a cable from Tuckton Bridge, Bournemouth was selected at Le Dicq slipway.

Civil works for the backhaul cable from Le Dicq to Trinity Gardens repeater station was shared with the States Telephone Department, which was also upgrading its network facilities in the St Clement area during the beginning of the year. Work on the submarine section did not begin until the spring of 1958 when surveys of the proposed submarine route were conducted. The GPO cable ship *CS Alert* arrived off Elizabeth Castle on 26 June 1958 but work was delayed by bad weather. Meanwhile, GPO cable engineer E J French (Area Engineer, Bournemouth) carried out tests using a Mole Drainer[174] on the beach at Le Dicq. The Mole Drainer was a tool

---

172 The New Telex Service, Post Office Electrical Engineering Journal Vol 48, p17
173 Telex: A Detailed Exposition of the Telex System of the British Post Office, R W Barton, Pitman 1968
174 Bournemouth – Jersey Submarine Cable, Post Office Electrical Engineering Journal Vol 51 p243

Image by permission of Lloyd Le Gresley

Cable jointers preparing the cable at Plemont Bay shore end of the Plemont – Saint's Bay cable ready for jointing in 1964. The damaged shore-end replacement section was laid by the *CS Poolster* which also repaired the Fliquet – St Lo cable at the same time.

normally used for burying telephone cable in soft ground, particularly along ditches in rural areas. This was the first time it had been used on sand. Also attending the trial were W E Mockridge, also of Bournemouth area and R R Maynard form the GPO Regional HQ at Bristol. The operation was watched by a large crowd of holidaymakers.

The shore end of the cable was landed on Sunday 1 July as soon as the weather conditions were suitable. The cable was brought ashore with the assistance of the States' tug, *Dutchess of Normandy,* aided by the lighter *La Mauve.* The cable was drawn up the beach by a tractor and finally into the jointing chamber at the top of the slipway. The operation proved to be an entertaining diversion for holidaymakers, some of whom assisted in the hauling of the cable. The sea earth cable was laid the following day and the whole buried into the sand while reinstatement of the slipway was completed.

The *CS Alert* set off to lay the sea section which was completed in two sections, the second being laid from Tuckton Bridge and joined mid channel. The new cable was 134 nautical miles in length and was fitted with 10 repeaters at 13 nautical mile intervals. The complete submarine system including terminal equipment was supplied by Messrs Standard Telephones and Cables. The terminal equipment installed at either end of the system was of the latest transistorized equipment mounted in GPO 56 Type transmission equipment practice, an updated version of the earlier 51 Type with minor changes including a different unit to rackside connector type. The new system, designated the CI No 5 by the GPO, comprised 2 supergroups totalling 120 channels. This was the first submarine telephone cable connected directly from the UK to Jersey. Previously all submarine cables had been routed via Guernsey, including all earlier telegraph cables with the exception of the diverted Borkum cable which initially only terminated at Jersey. Although the cable was completed in the autumn, it was not until the following year that the channel equipment was brought into service.

In 1961 Channel Television (CTV) was awarded the franchise for Independent Television in the Channel Islands. Consequently the

Independent Television Authority (ITA) began construction of a new transmitter at Frémont Point to cover the islands. The GPO still retained an absolute monopoly over wireless telegraphy and thus became responsible for the delivery of the programme content to ITA. An off-air receiving and relay station was required on Alderney since the Band III signal could not be reliably received on Jersey. The construction of the off-air receiver was completed by the ITA but the delivery to the transmitter was the responsibility of the GPO. A microwave circuit was constructed from Alderney to Frémont Point for this purpose. Because of the distance over open sea, the circuit was fitted with diversity switching using a pair of receiving aerials, together with a redundant circuit for backup in times of failure of the working link. Echo suppressing equipment was also fitted at the Jersey end to reduce signal degradation, although the whole circuit was subject to the vagaries of the weather at the Alderney end. On the broadcast side, although nominally the responsibility of the GPO radio interference duty, little could be done about co-channel interference from adjacent French stations in adverse weather conditions.

A separate service was provided from the CTV studios in Rouge Boullion, St Helier to Frémont Point switched via the Trinity Gardens repeater station. This switch point also facilitated a test point for both the monitoring and repair of the circuit. An off-air monitor point was also included to check the quality and integrity of the signal delivered to the transmitter. The coaxial cables were laid from the studio and to the transmitter using ducts laid in a shared operation with the States Telephone Department. Again, redundancy was provided for security. The contemporary technology for these services still relied in part on thermionic valve amplifiers and thus the normal ageing process of the valves could cause problems in service. The transmitter was completed in March 1962, CTV's studios were completed some time later. The Alderney relay station was commissioned during August after a delay caused by the Alderney States.

Meanwhile, growth was continuing in trunk calls. The GPO started planning a further cable for Jersey during 1964. Surveys were carried out and a new route from Tuckton Bridge, Bournemouth to Greve d'Azette was planned. The contract was awarded to the Submarine Telephone Company (a subsidiary of Standard Telephone and Cables) in the sum of £500,000 in October 1964.

Before this new cable could be installed, however, additional space would be needed at the Trinity Gardens station. A two storey extension was commissioned to be built on the eastern side of the existing building which would comprise a new power room on the ground floor with an equipment room and office and staff amenities accommodation on the first floor. This work was completed in 1967. Meanwhile, the number of Telex circuits was growing as demand from Jersey's burgeoning finance industry increased. New 24 channel MCVF equipment was installed in the existing station accommodation which connected local Telex subscribers to the Bournemouth switching centre over channels on the CI No 3 cable via Dartmouth. This new equipment was provided in the latest GPO 62 Type equipment practice which was the first approved GPO usage of slide-in printed circuit board construction.

In early October 1966 the Dutch cable ship *CS Poolster* arrived but cable laying was delayed because of bad weather. When the weather improved on 17 October the shore end was prepared for laying but the ship ran aground on the beach near Le Dicq slipway while laying the shore end. The shore end was in fact successfully laid and the ship was pulled clear on the next favourable tide on the 23 October. During this manoeuvre the ship fouled the sea-earth cable and this had to be repaired. A new sea-earth was laid and taken one mile out to sea while the shore end of the cable was taken out two miles ready for laying the remaining repeatered sections to Bournemouth. Although this cabling work was completed by the end of the year, the repeatered sub-sea section was delayed until December 1967 when the *CS Monarch* picked up the shore ends and laid the remaining 137 nautical miles of the sea section. The new CI No 6 cable comprising 4 by 8 supergroups (1920 circuits) came into service during February 1968. The power feed and terminal equipment was provided by Messrs Standard Telephones and Cables Limited and 4 groups of 480 circuits of channel equipment by Messrs Pye TMC Limited of St Mary Cray, Kent. All new

Image by permission of Lloyd Le Gresley

Dutch Cable Ship Poolster ran aground at Greve D'Azette during the laying of the shore end of the CI No 6 cable in October 1967.

equipment was in the GPO 62 Type practice finished in Light Straw and located on the first floor of the new repeater station extension. The power supply equipment, batteries and standby generators associated with the new system were located on the ground floor.

During 1968 talks between the States of Jersey and the Post Office began on the proposed divestment of the GPO's responsibility for the provision of trunk services to the offshore islands. This had resulted from the Wilson government's review, under the leadership of Postmaster-General Anthony Wedgewood-Benn, of the operation of the Post Office. This would ultimately result in the Post Office Act of 1969 which changed its status to that of a public corporation. This process had been begun by the committee under Lord Bridgman in 1932 but that report had been largely ignored by successive governments. The object of the review was to increase the efficiency and profitability of this dinosaur of a government department and would result in the separation of post and telecommunications into separate operations. The principle of the takeover of posts and trunk telecommunications was agreed by the States in July 1968. The postal operations were taken over by the States on 1 October 1969, but the negotiations on the telecommunication operations took somewhat longer.

During 1968 planning was undertaken to increase the inter-island capacity. A cable was ruled out on grounds of cost and instead it was decided to install a 960 channel 6.4 GHz diversity switched microwave link between Monument Gardens repeater station in Guernsey and a new building to be sited at the Jersey Airport telecommunications department site at La Chasse, St Ouen. New 30 metre lattice steel masts were erected at each site and the multiplex equipment installed. The new capacity enabled sharing of the CI No 5 cable capacity with Guernsey. The fully redundant terminal equipment came into service in the spring of 1970. Meanwhile planning of further off-island capacity for Guernsey (which could also be shared with Jersey over the new microwave link) was undertaken. The CI No 6 1,380 circuit cable came into service between Bournemouth and Guernsey in March 1972.

On the 1 October 1969 Post Office Telecommunications became a public corporation and thus no longer enjoyed the protection of the British Government. It was not realized for some time that this change of status removed the legal indemnity of government surety over its activities conferred by previous legislation and thus the operations of Post Office Telecommunications was now effectively uninsured in Jersey. Staff had been operating in ignorance of this fact and had unknowingly been breaking the law by driving the GPO vans, provided for Telex maintenance and radio interference duties, uninsured. In addition all the buildings and staff were similarly uninsured. This matter was hurriedly addressed in the spring of 1970 during which time the driving of vans was belatedly suspended.

The original timescale set by the States for the takeover was February 1970 and a budget of £750,000 was voted towards the acquisition of the GPO telecommunications assets on the island. In preparation for the takeover the States Telephone Department began training staff in the event that the existing staff did not transfer after the takeover. This programme of placement of engineers began in 1969. Negotiations dragged on for some time and it was not until 1971 that the law draughtsmen were instructed to produce the new enabling law. Meanwhile, after the takeover of the postal operation, GPO Telecommunication services was subcontracted by the Postal Committee to continue maintenance of the teleprinters on the telegraph system and, for a while, some other postal equipment. This was carried out by GPO staff located at the Trinity Gardens repeater station.

The draft Telecommunications (Jersey) Law was debated by the States in June 1972. After some discussion the report was adopted and sent for its third reading without amendment and approved at the next sitting. The law next appeared before the States in November after it had been approved by Her Majesty in Council during August. At this sitting the Telecommunications (Channel Islands Consequential Provisions) Order 1972, which dealt with the transfer of property and equipment, was debated. The new Act would come into force on 1 January 1973 when the GPO operations and assets on Jersey would be incorporated in the the newly formed States Telecommunications Board. The new law would vest the monopoly for all island telecommunications with the States of Jersey. The total assets in the new

Board amounted to £1.8M and included all the GPO transmission equipment on the island, shares in the sub-sea cables to Guernsey, France and the UK, including a share in those passing via Guernsey, the GPO properties and associated underground duct and cables located at Plémont, Fliquet, Trinity Gardens and leases on floor space at the ITA transmitter station at Frémont Point and the associated microwave equipment and at the Channel Television studios, Rouge Boullion. The GPO presence on the island ceased at midnight 31 December 1972 and all 8 staff employed at that time transferred to the States Telecommunications Board on favourable terms. The Telecommunications Board continued the maintenance of the Postal Department's telegraph services and also assumed responsibility for the monitoring of radio interference under contract to the UK Ministry of Posts and Telecommunications.

Tables 40 and 41 below summarize the faults experienced on the telegraph company cables taken under Post Office control. The STCo cable repair ship *CS Lady Carmichael* was renamed *CS Alert* after acquisition by the Post Office.

Jersey & Guernsey Telegraph Co Submarine Cable to UK 1870 – 1884 Cable Outages

Date out	Date repaired	Days	Reason	Repair Ship
27 February 1874	19 April 1874	51	Jersey - Guernsey	International
31 May 1876	26 June 1876	27	Submarine fault	?
14 June 1877	21 July 1877	37	Off Guernsey	International
27 July 1877		1	Straightening cable	International
14 October 1877		1	Shore end Dartmouth	Not required
20 February 1878	7 April 1878	46	3 places GY-Dartmouth	International
19 November 1878	16 December	27	Off Dartmouth	?
2 February 1881	23 March	49	Compass Cove, Dartmouth	?
19 October 1881	27 October	8	Dartmouth bridge	Not required
4 August 1882	8 August	4	Plémont shore end	Not required

Table 40

GPO Telegraph Submarine Cable to UK 1884 – 1940 (GPO UK – CI No 1)

Date out	Date repaired	Days	Reason	Repair Ship
11 February 1884	21 March	39	New 3-core cable laid JY - GY - DM	Monarch
7 January 1894	24 January	17	North of Guernsey	Monarch
30 March 1896	14 April	15	Off Plémont	Monarch
23 April 1897	19 May	26	Off Plémont	Alert
29 November 1898	13 December	14	North of Guernsey	Alert
13 April 1901	?	?	Repairs	?
10 February 1902	15 February	5	Guernsey-Dartmouth	?
8 June 1902	11 June	3	Jersey-Guernsey	Alert / Monarch
18 December 1902	8 January 1903	21	Guernsey-Dartmouth	Monarch
31 December 1903	17 January 1904	18	Guernsey-Dartmouth	Dacia
13 February 1904	24 February	11	Guernsey-Dartmouth (12ml off Gy)	Dacia
9 November 1904	14 November	5	Jersey-Guernsey (2ml off Jy)	Alert
27 February 1905	12 March	13	Guernsey-Dartmouth (Hurds Deep)	Alert
10 March 1909	13 March	3	Jersey-Guernsey	Alert
13 February 1910	23 February	9	Guernsey-UK	?
9 April 1910	24 April	12	Jersey-Guernsey	?
? January 1912	25 April	~ 90	Intermittent fault Jersey - Guernsey	Monarch
27 January 1913	4 February	8	Guernsey – UK in Greater Deep	Monarch
30 May 1913				?
9 June 1915				?
4 October 1923				?
7 December 1929	16 January 1930	50	Guernsey – UK off Casquets	Monarch
18 December 1929	10 January 1930	23	8 Miles off Plémont	Monarch
18 May 1931	4 June	17	Guernsey – UK section	Monarch
29 May 1933	31 May 1933	2	Repairs	?
20 August 1932	31 August 1932	10		Monarch
6 October 1932	7 October	1	Cable dragged river Dart	
? February 1933				
9 October 1933			General repairs	Monarch
17 September 1935	27 September	10	42 Miles off Dartmouth	Monarch
26 February 1937			Dart Estuary	Alert
? October 1938	3 November		Jersey shore end	Monarch
27 January 1939	? February		Jersey shore end & Dartmouth	

Table 41

Chapter 23 - Civil Defence Communications

The Civil Defence committee was established under the Civil Defence (Jersey) Law 1952. This law was a direct reaction to the cold war and in line with civil defence activities elsewhere, particularly the UK. The Island Defence Committee (which was largely responsible for the maintenance of sea wall defences and other related matters) was given the responsibility of administering the law. It would appear that little work was carried out to implement any warning, and in particular, nuclear war preparations. It had been assumed that the small size and lack of strategic advantage of the island would largely prevent it from direct attack. However, five power operated warning sirens had been installed in various strategic locations around the island before the start of WWII.

In 1959 the general tensions brought about by the Cold War, it was decided to review the civil defence strategy. J H Cabot was appointed as a part time organizer to run the civil defence network. A connection to the BBC transmitters at Les Platons was arranged to enable local emergency communications over the airwaves (this facility was also extended to the Guernsey civil emergency services). A draft civil emergency procedure was developed and it was intended to make this available to the general public at any time when the international situation deteriorated to any appreciable extent.

In February 1961 J E F de Faye took over the part time post and immediately initiated a review of the whole civil defence strategy. This resulted in the recommendation to request a senior UK civil defence officer to review the local set up and to advise on the formulation of plans for a protracted emergency situation. Consequently, the Bailiff approved the scheme and J B Howard of the Home Office Civil Defence Department compiled a report which was presented to the Defence Committee. Among the recommendations was the appointment of a full-time official to oversee the emergency strategy. Meanwhile, a review of the distribution of sirens had been implemented and a number of new equipments, bringing the total to 17, were installed to ensure better island coverage and these were all connected back to a central control centre.

Therefore in January 1963 Lt Colonel (retired) B R Pearson was appointed as Civil Defence Coordinator with a view to developing a public safety policy and to establish an effective communications system to coordinate States departments in the event of an emergency. Pearson set up the civil defence headquarters in a disused German bunker located near the telephone repeater station near Trinity Gardens, St Helier. This bunker had been constructed by the occupying forces in 1944 with a view to transferring the entire submarine transmission system equipment from the nearby GPO building. However, the D Day invasion and the subsequent disconnection of the submarine cables to France rendered the move unnecessary and the bunker remained unequipped. Therefore, the building was ideal for use as a command centre, all the rooms being designed with high ceilings to accommodate the transmission equipment racking.

Pearson set about the task of repositioning the warning sirens control to the bunker. These sirens were controlled from the command bunker and connected over States Telephone Department private circuits. The siren sites were also equipped with mains power and batteries for backup. Radiation detectors were procured and located at the bunker. A new ventilation system was designed and installed as was a standby generator in case of mains power outage and all was completed by December 1963. Recruitment of emergency officers was undertaken and these were mostly drawn from the local civil service and States trading departments on a part-time basis. Initial training was given and this was recommended to be a maximum of 48 hours per month which should be during normal working hours. The public communications system was extended to the recently opened ITV transmitter at Frémont and following this it was estimated that 95% of the population could be reached over the airwaves (presumably under the assumption that the emergency occurred after working hours and that the power station was still operational). All this work was carried out in time for the local organization to join a UK/NATO exercise in November 1964.

Part of the Home Office recommendations was to ensure a continuing telephone service through an emergency. This meant that all key personnel could be reached at all times. The present arrangement of the islands telephones was not up to the job. Each exchange relied on batteries for back-up in the case of mains failure, thus in a prolonged outage the maximum time a service could be guaranteed would be dependent on the life of the exchange batteries. No standby generator facility was available ordinarily and only a limited number of portable devices suitable for connection to telephone exchanges were available on the island, none of them belonging to the Telephone Department.

Of the existing exchanges in 1963 only Central exchange had an emergency facility that could be enacted in special circumstances. This facility could be introduced such that a key on each subscriber line circuit rack could be configured so that its operation would disconnect all lines not considered to be essential in an emergency. Operation of these keys would, therefore, reduce the exchange traffic to that of the essential lines, thus prolonging the potential battery life. This feature had not, however, been configured. The remaining exchanges had no such feature and although manually operated exchanges could easily be arranged to instruct the operators to only service designated emergency subscribers, the remaining lines would still have a connection that would, even unanswered, tend to drain the exchange batteries.

Consequently, in order to meet the Home Office recommendations a solution was needed. The Telephone Department came up with a proposal

Image by the author

The Civil Defence bunker at Robin Hood was originally constructed by the German occupying forces in 1944/5 with the intention of relocating all submarine transmission equipment from the nearby repeater station in Trinity Gardens (50 metres to the right of the bunker). However, it was overtaken by events and the transfer never occured. The bunker was ideal for the purpose of civil defence as it was constructed as an air-tight building with ample accomodation for long occupancy.

that would enable all island-wide essential subscriber lines to be connected to a central switchboard. This would become known as System E. In 1966 a new switchboard was constructed using three CB10 switchboards that had been recovered from the Lyric Hall relief exchange after its closure in 1959. These had been stored with a view to using them for future extensions on the remaining manual exchanges. However, it was decided to utilise them as a special switchboard that would be located at Central exchange, beneath the Main Distribution Frame of the newly built trunk exchange building adjacent to the existing Central exchange. This room was designated as the 'protected area', and although it had no special air conditioning or radiation proofing it was considered to be a relatively safe area as its walls were somewhat thicker that a normal room.

A special single-sided distribution frame was constructed and banks of remnant relays[175] installed that would switch lines considered essential on the exchange from its normal connection to the System E emergency switchboard. Similar banks of these relays were installed at the remaining manual exchanges. Lines designated as emergency (such as police, honorary police, medical personnel, firemen, politicians, etc.) would have their telephone line wired via these special relays and when the system was activated the line would be diverted over telephone junction circuits from outlying exchanges to the central switchboard. All relays could be controlled from the switchboard. The switchboard would be operated by suitably trained civil defence volunteers. The switchboard was connected over circuits to the command bunker. Once the switchover to System E was complete, in theory, the normal exchanges would be closed down to save their batteries. While this system was obviously capable of providing the required service, the lack of substantial shelter and the reliance on unfiltered air for the operating staff would surely have led to a short-lived solution.

The command bunker was fully equipped with emergency telephones connecting it directly to other important departments such as police, fire

---

175 Remnant relay is the term given to a telephone relay that requires no permanent holding current. The relay is operated or released by a polarizing electrical current and the relay remains in its latest state until the operation is reversed.

and ambulance. There was also a fully equipped sound studio and control room latterly operated by the BBC for broadcasting emergency instructions. Pearson conducted his last CD exercise in October 1979 and was replaced by Lt-Col Bill Clayden. Later more stand-alone wireless communication facilities were introduced. The Honorary police were connected to an all-island network in May 1982, which enabled a more convenient means of communication under emergency conditions. The System E switchboard was quietly retired at the beginning of the 1980s.

In 1985 a refit of the command bunker was undertaken starting with a refit of the broadcast studio facilities. Meteorological equipment was also installed that would give regular updates on current weather conditions. This was a fax-like machine connected to the UK Met Office. Further emergency tests were carried out during 1985 but by now the Cold War was beginning to thaw. In 1988 three automatic radiation monitoring stations were constructed in former German bunkers at Ville du Bas, St Ouen, Trinity and near Victoria Tower, Gorey. The data collected from these stations was connected by wireless telemetry to the States Met Office at Jersey Airport and the data was also passed to the States Analyst.

By the end of the 1980's, however, the end of the cold war signalled the beginning of the end for civil defence in the UK. The main nuclear threat now came from the nearby French power station at Flamanville near Cherbourg. By 1995 most of the island's warning sirens had been removed and the control network had been closed down.

In 1994 the island-wide Terrestrial Trunked Radio (TETRA) system was launched that connected all the emergency services over an integrated voice and data wireless network. This ensured full communication between all emergency services and is a standard adopted by many jurisdictions, including the UK, for management of civil emergency and major incidents.

Image by the author

A general view of the operations room asused by the Civil Defence. This was originally intended as the main equipment room for the German repeater station.

Chapter 24 - Cable Radio and Television Relay Services

Local relay services

Relaying of radio broadcasts began soon after the start of commercial broadcasting by the British Broadcasting Company. In the early days, many radio receivers were still of the crystal set type, broadcasting was in its infancy and few people owned any radio receiver as the technology was still developing. During 1922 the British Broadcasting Company was set up by a consortium of leading British radio manufacturers. The BBC then took over the Marconi 2LO London station and started regular broadcasts from Marconi House in the Strand on 14th November 1922 on a wavelength of 369 metres (813 KHz). Interest in broadcasting started with electrical retailers who quickly adopted the new technology.

Although valve radio sets were available they were fickle and required a reliable power supply, which in the early 1920's was not yet readily available to the majority of households. Those with valve sets needed to send their battery accumulators for regular charging to the local garage or electrical shop. Those with crystal sets found that they were inconvenient and difficult to use, requiring frequent tuning to maintain the station. Soon, electrical suppliers around St Helier started offering radio relay services, maintaining a good quality receiver in their shop and distributing the audio output via a suitable amplifier around the houses in the immediate area for a modest weekly subscription, one such supplier was Hoddis and Le Maitre, David Place. The supplier provided a loudspeaker housed in a wooden box for each subscriber. There were many such shops in different areas of St Helier providing this service to local housing as the popularity of radio grew and the BBC provided a reliable and good quality service.

Such providers continued to offer these services up until the Second World War, when the German occupation forces banned the use of radios by the general population. Following the cessation of hostilities, the technical advances made during the war brought cheaper and more reliable valve radios, therefore there was little incentive for local shops to revive their pre-war service on the same small scale.

Rediffusion

When broadcasting first began in earnest in the UK in 1922, Rediffusion, originally called the Broadcast Relay Service Ltd, was formed in 1927 and started to relay radio programmes to subscribers by negotiating an agreement to use an existing network of urban power cable standards belonging to city tram operator British Electric Traction (BET), which provided a number of tram services in major towns and cities around the UK. The tram network passed by many homes and it was therefore simple to distribute along these lines. Down streets not passed by the tram lines private way leaves from landlords and home owners were used. It therefore had access to a ready-made market of potential listeners-in who wanted radio entertainment – if only it were affordable and non-technical. In 1947 BET acquired a substantial minority shareholding in the company[176] and eventually assumed a majority in 1967. It provided the service for a small weekly subscription and quickly profited from the public's interest in the new medium.

Rediffusion was almost immediately profitable. The company soon branched out from simply 're-diffusing' radio, into the manufacture of radio sets. From there, the sale and hire of sets in the High Street followed.

As broadcasting opened up in the overseas Dominions of the British Empire, Rediffusion followed on, using the tram wires, or bespoke 'pipe radio' systems, to provide the local broadcasting stations to the cities, as well as the new BBC Empire Service (now BBC World Service Radio).

When the BBC began the first regularly scheduled high-definition television service in the world in the mid-1930s, Rediffusion was again well-placed to provide television sets for sale and rent, plus a 'pipe-TV' service to those not well-placed for broadcasts from Alexandra Palace, or reluctant to

---

[176] Competition report on proposed merger The British Electric Traction Company PLC and Initial PLC http://discovery.nationalarchives.gov.uk/details/r/C1731650 (last accessed 11/06/17)

have such a gauche symbol as a VHF TV aerial on their roofs. When 405 line television was first introduced to the UK in 1936, there were many places where reception was impossible (notably Brighton on the South coast, which was shielded from the London transmitter by the local terrain). Rediffusion developed early cable TV technology to allow TV signals to be carried over twisted pair cable to subscriber's homes. Nowadays, we associate twisted pair with telephony, but it is possible to transmit a TV signal by modulating a short wave carrier at around 3.5 MHz with vestigial sideband (VSB) amplitude modulated video provided the cable pairs are well balanced. When ITV opened in 1956, Rediffusion simply used another twisted pair in the same bundle.

World War II interrupted television, the growth of wired distribution, and much of the peacetime activities of both BET and Rediffusion. For the duration, the company devoted its energy to the war effort using its research facilities for a number of top-secret projects. This meant its expertise in reception and rebroadcasting suddenly became of essential national importance. Even now full details of Rediffusion's activities during the war are held under the protection of the Hundred Years Rule.

Immediately post-War, the world changed dramatically. BET was included, with some justification, in the list of companies the new Labour government planned to nationalize. The tram systems started to disappear too, partially because the necessary nationalization of the electricity companies meant that the old local generators became part of the new Central Electricity Generating Board. The councils who ran the trams had to pay for the electricity, and coupled with the dilapidation of the systems during the war, it was cheaper and easier abandon them and replace them with ordinary bus services. Within a decade of the end of the war, almost all of the tram networks had disappeared. BET, however, avoided nationalization, due largely to Labour losing the election of 1951.

BET which bought into Rediffusion in efforts to avoid nationalization had started to diversify and looked for other opportunities, especially overseas. With the Dominions disappearing, it repeated previous successes by starting Overseas Rediffusion, offering wired television and radio, and later wireless broadcasting stations in the remaining colonies. That expansion brought Rediffusion to Jersey when on 18 June 1949, the operating company Rediffusion (Channel Islands) Limited was registered. Although registered as a CI company, Rediffusion never attempted to enter the Guernsey cable market but did open a retail outlet in the 1960s. Nevertheless, a petition was placed before the Jersey States in the spring for licenses to provide cables across public roads. The JEP published a virulent editorial against the company on 9 August 1949, the eve of the States debate. The JEP asserted that the system was unnecessary as the Telephone Department could provide the necessary cables. It also asserted that there would be no control over what the company broadcast and there would be a loss of revenue from the some 14,400 licences issued on the island. It further sought to protect local radio retailers from competition. It is clear that the JEP had little understanding of the control exercised by the Telegraph Act in its extension to Jersey, nor the nature of the distribution systems for telephony and wireless relay.

Deputy E H Le Brocq of St Saviour brought the proposition before the house, declaring no personal interest in the company. He clarified the terms of the petition stating that it had originally been drafted for Malta and that the only stations rebroadcast would be those of the BBC under the terms of its GPO licence. He recommended the system as it would have little impact on the environment, being installed only in built-up areas and there would be no noise from the loudspeakers as the maximum volume was controlled from the central station. It would provide an affordable means for the average man to have access to the wireless without the normal interference and would also provide local employment. The proposal was seconded by Deputy John Le Marquand who also declared no personal connection with the company.

Deputy Hettich (chairman of the Wireless Retailers Section of the Chamber of Commerce) spoke against the company but Deputy Kruchefski noted that although the debate had been tabled for some time, no radio retailers had petitioned against it. Hettich claimed that the Chamber of Commerce had objected. Kruchefski responded that the speaker had regularly complained against everything that required a GPO licence but there was nothing that could be done about it since the States had adopted the UK law. Personally, he was in favour of the proposition.

Deputy Morrisson (Telephone Committee) was against. He said that he did not always agree with JEP editorials but on this occasion he did. He said that this proposal would merely create yet another monopoly. Deputy Avarne said that originally he was against, but as he now understood that the programmes would not be originated by the company he had changed his mind. He felt it would be a boon to the poor. Deputy Venebales agreed stating that he was in favour of private enterprise. Deputy Le Marquand suggested that crystal sets may well be cheap enough and had served well during the occupation. Senator Collas recalled that 14,000 radio sets had been confiscated by the Germans and that by now these had all been returned or replaced and more. He estimated about 16,000 or 17,000 on the island or one for every three persons. That, he claimed, meant everyone had access already.

Some members complained that the new cables would be unsightly, being thicker than telephone wires. Senator Hind of the Beautés Naturelles Committee said that this was not so. The cables would no thicker than the wires that the Jersey Electricity Company had already on its distribution poles.

At the end of the debate it was pointed out that the licence to operate was in the gift of the Post Master General under the Telegraph Act, the only question was whether the States should permit the crossing of streets. There was no question of the States interfering with wires fixed to private property. Finally, Deputy Le Brocq pointed out that the company could be licensed to operate by the GPO and that in order to cross streets they would need permission from the Main Roads Committee or the Constable of the parish. He therefore recommended that the States adopt the preamble. However, the house voted 26 to 14 against the proposition.

The company was undaunted and the bill was again placed before the States the following year at the sitting on 31 March. The ensuing debate was short and it would seem that much background work had been done to allay the fears of States' members because the preamble of the bill was adopted. A second reading was scheduled for 11 April 1950 which passed 'on the nod'. Finally, after a minor amendment proposed by the Legislation Committee in May which would allow parish Constable discretion to give permission for cable crossings, the amendment to the law[177] was passed at the end of June.

On 27 September the JEP reported that Rediffusion engineers had established a test site at Les Platons and that two caravans with special equipment were on site testing radio reception. At this time the bill was still awaiting Royal assent, but Rediffusion engineers in company red and grey livery vans were already busy erecting private way leave cables. The JEP also reported that Rediffusion engineers were testing television reception in Alderney on the 2 February 1951 and around the middle of April the company began advertising its services.

Rediffusion established its Jersey offices in 1951 at 1 Library Place, St Helier. The first General Manager of the local company was Quentin L Cazalet. It acquired the warehouse formerly used by merchants W Dennis and Sons at 52 Esplanade in 1953 and in 1954 it also had administration offices at Colomberie Close. Initially, the company offered the BBC wireless stations, the Light Programme, the Home Service and the Third Programme, with the addition of Radio Luxembourg, receiving the Long Wave French programme broadcast during the daytime but in the evening relaying the famous Medium Wave 208 service. By September 1951 it had connected over one thousand customers to its relay service.

In the first instance, Rediffusion only relayed radio broadcasts. Their distribution hub used high quality wireless receivers and efficient aerials to minimize interference, such facilities were usually beyond the domestic radio listeners' ability. The Rediffusion cable network was at first almost entirely supported by private way leave, with almost no cabling beneath public roads, only aerial crossings. The radio service was carried over balanced pair quad cables manufactured from PVC, a plastic developed as cable insulation during WW2. Cables were clipped in pairs, using galvanized

---
177 Jersey Law 20/1950 Loi Modifiant La Loi (1914) Sur La Voirie

buckle strap clips, along the fascia boards or soffits of buildings in the built-up areas and across streets suspended from catenaries using galvanized or piano wire bound with uninsulated soft copper wire. The catenaries were tensioned using fence line braces. Often, street crossings were necessary where a way leave was not obtained. These were made either from the house gutter level or by using chimney brackets in order to get sufficient clearance above the highway. As the network developed, the company used its own telegraph poles to span larger distances between buildings at the edges of built-up areas.

The network was in effect a large public address system. The signal was distributed from its offices using large valve amplifiers and boosted, where necessary, with street mounted mains powered valve amplifier units. The amplifiers necessary for such a large network were substantial and thus the network had to be subdivided in order that the amplifiers could be designed to a manageable size. Consider that a network of 1,000 subscribers would need an effective input power of approximately 2kW allowing for line transmission losses, then it can be seen that using valve technology would require several amplifiers to attain that level of output. The subscriber speaker sets, which were made of Bakelite, were fed in parallel from the main distribution network. The network was bridged at each subscribers premises and a balun fitted to the line, the feed was thus spurred off this connection while maintaining the transmission path. The connection was housed in a Bakelite or, later, plastic junction box which was fixed to the subscriber's fascia, soffit or house wall as convenient. The cable was often clipped down the wall using steel and lead cleats which was a quick method of fixing. In the customer premises the cable was terminated in a Bakelite switch box that had each channel marked with a letter and the required service was selected using a rotary switch. The volume control was also connected to the box and, for the sake of reliability, used a stepped incremental rather than continuous method for increasing the speaker volume.

The relay services were at once a success in the main built up areas. Surprisingly, even up to the 1960s, a large number of houses in St Helier

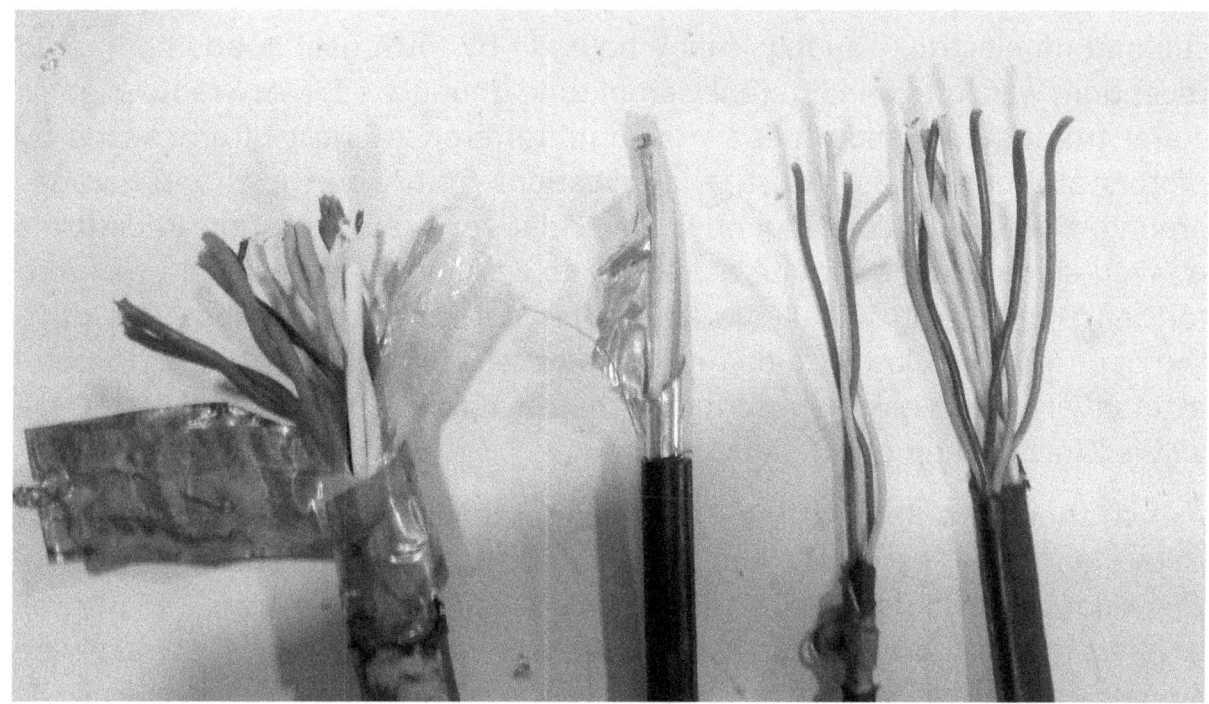

Image by the author

Examples of the PVC cables used by Rediffusion for television and sound distribution. From left to right a trunk cable with six television and three audio pairs, a two pair local television distribution cable, a two pair television connecting cable showing the earthing slip still in place and a six pair flat audio cable. All cables were provided with foil screens except the audio cable which distributed radio services. The centre two cables used balanced pair confuguration while the others were twisted pair.

still had no electricity supply, many being lit by town gas. In addition, radio reception was often indifferent, amplitude modulated (AM) radio signals suffer from 'static' and other switching interference, most often caused by motors and other light industrial applications from workshops and garages around the town. Consequently, the relative quality of the Rediffusion relayed signals made it quite attractive to listeners. The company was also fortunate in gaining way leave access to States social housing. This gave the opportunity to low income groups to access to radio broadcasts through a relatively low weekly subscription thus avoiding the capital outlay for a wireless set of their own.

A further advantage of the Rediffusion network was the introduction of television relay services. Rediffusion had established an experimental television receiving station at Les Platons as early as November 1951.It was able to receive AM broadcasts off-air at its receiving site, first receiving signals from Crystal Palace then from the Wenvoe transmitter when it opened in August 1952 broadcasting on channel 5. This site consisted of a receiving aerial constructed from telegraph poles spliced together end to end to provide a mast of around 80 feet in height. Specialized television receivers were installed in a building close by (also two bungalows were constructed for use of senior management). The rural location well away from any industry or housing complexes enabled comparatively interference-free reception, thus providing a quality often not available to television set users, particularly those in low lying areas such as St Helier. At this time there was much interest in television reception among the radio retail community.

In April 1953 it applied for permission from the GPO to relay BBC TV broadcasts in time for the Coronation of Queen Elizabeth II. The company quickly began to upgrade its network to include TV distribution in addition to wireless. This required the addition of an additional screened cable pair to carry the modulated TV signal and sound carriers, although for practical manufacturing purposes, a two-pair cable was used. Although the TV receiving site was well outside the planned service area of the Wenvoe transmitter, the small number of subscribers and the crowds that assembled outside its Library Place shop were able to view the coronation broadcasts. The licence application was not forthcoming in time for the coronation, nevertheless, the PMG waived the requirement for a licence in order to allow the Jersey viewers to see the spectacle. However, almost immediately after the celebrations on 4 June 1953, the PMG demanded that the relay service was suspended until the license was approved. This placed the management of Rediffusion in an embarrassing position and they offered to refund subscriptions or hold the payments over until the service could be resumed.

Before reviewing the licence, the PMG requested that the States of Jersey should be canvassed for opinion. A debate was proposed but the local radio dealers association again began lobbying the States claiming that Rediffusion's system would damage their businesses. They claimed that the relay service would be a 'monopoly' and, somewhat paradoxically, that it was not in any case necessary as the BBC was proposing a Channel Islands transmitter within a couple of years. The Rediffusion management countered this claim, stating that their system would allow television now and that local radio dealers would be able to buy sets directly from Rediffusion at competitive rates which could be used for either service.

Rediffusion had, by now, also established a manufacturing facility in the Island in a converted warehouse on the Esplanade, St Helier. This was under the name of Television Research (Jersey) Limited (TVR) and was a separate, though affiliated, company from the relay system. The TVR board consisted of Q L Cazalet (also Managing Director of the Rediffusion (Jersey) Limited), W T Scarborough, V Vibert and the Earl of Jersey. The factory manufactured television sets on the island, even making the wooden cabinets locally, in addition to developing and manufacturing products for the parent company.

The States debate on whether Rediffusion's relay license should be extended to television as well as radio was held on 18 January 1954, and in addition the house debated the legal changes necessary for Rediffusion to have shared use of the overhead distribution network of the Jersey Electricity Company Limited (JEC) in order to bring their TV signals from Les Platons to their St Helier distribution hub (for which Rediffusion paid the JEC an annual fee). This was duly approved by the house and the arguments proposed by the radio dealers association were roundly rejected. An amendment to the JEC Law[178] was made and came into effect on 21 December 1954, although in practice Rediffusion and the JEC had installed the cables in ignorance of the legal requirements some time earlier. The PMG issued the extension to their relay licence on 1 April 1954.

The Rediffusion TV service began again on 1 May 1954 heralded by a full page advertisement feature in the JEP. This was followed on the 17 June by the official opening which was an elaborate affair with many important guests including Sir Wavell Wakefield MP, a member of the Rediffusion UK Board, Mr A R A Rendall, chief of designs for the BBC, the Lt Governor Sir Edward Grasset and many local dignitaries including C J Syvret the Telephone Department's Engineer Manager. The main ceremony was at the purpose-built receiving station at Les Platons where a plaque was unveiled and the many guests were able to see the special receivers and watch television broadcasts from Wenvoe. It was stated at the time that Rediffusion may offer alternative services when the French television service was available. The company started distributing the BBC TV channel more than a year before the BBC off-air service was available in the Channel Islands.

Having failed in their attempt to prevent Rediffusion from gaining a TV relay licence, the radio dealers turned their attention to the JEC. The relationship between the JEC and Rediffusion for sharing the JEC distribution network was extended by the JEC offering Rediffusion televisions in its sales room. The radio dealers association again took the view that the JEC was abusing its 'monopoly', a view which was also shared by the JEP editorial and numerous letter writers. A petition was presented to the States requesting that, as a major shareholder of the JEC, it should prevent it damaging the radio dealers by instructing the company to withdraw from the television market. The petition was reviewed and rejected by the Finance Committee and talked-out of the States. The animosity between the dealers and Rediffusion nevertheless continued unabated.

---

178 Jersey Law 19/1954 Electricity (Amendment) (Jersey) Law, 1954

With access to the overhead routes of the JEC, Rediffusion was now able to expand its network rapidly throughout the built-up areas of the island. By mid 1955 it had extended its reach along the west coast to St Aubin and Red Houses using a mix of JEC and its own poles along the railway walk, as well as along the roads from Les Platons to St Helier. In October 1956 the company closed its Library place offices and began trading from a new showroom at 52 Esplanade, the site of its distribution and TV Research Limited. At the beginning of 1957 the cable distribution network having been upgraded to distribute 2 TV channels, the newly opened French TV service was added to the inclusive radio and TV package available to its subscribers. During the year it also switched its radio receivers over to the new BBC VHF service which it was now receiving off-air from the UK.

However, it was not until August 1958 that it was able to offer its subscribers an ITV station, although its parent company had been among the first to start commercial broadcasts from its London franchise in September 1955. Because of the higher frequencies allocated to the ITV network, it was unable to receive reliable off-air signals at Les Platons. Therefore it set up a receiving station on Alderney to receive broadcasts from the new ITA transmitter at Rowridge which relayed Southern Television and began relaying the test transmissions to its subscriber. The Alderney station was connected to the Rediffusion control room at Les Platons by a GPO provided microwave circuit, as at this time the PMG claimed an absolute monopoly on the provision of telecommunications circuits. At the beginning of September it claimed to have over 4000 subscribers watching two channels. This amounted to almost half the number of issued television licences at that time, statistics up to the 1 June 1958 showed that 9,079 licences had been issued by island post offices[179]. A sales drive offered existing television set owners the option to convert to cable at 5/6d per week as an incentive to grow its customer base. It continued to expand the reach of its network, adding St Brelade's Bay in early 1960. Although there was now a substantial viewing audience for ITV the JEP did not provide any reliable listings in its radio and television guide until the opening of Channel Television.

---

179 Jersey Evening Post 1 June 1958

In 1960 Channel Television won the ITV franchise for the Channel Islands. Plans were set in place for the building of studios and transmitter facilities and it was announced that ITV would relay the Westward Television signal locally. Consequently, in April 1961 Rediffusion switched it relay service to Westward in preparation for the change. Channel Television opened on 1 September 1962 and thereafter its Alderney relay link was closed without any redundancies.

After 1962 Rediffusion's market position changed. It was no longer the sole provider of alternative television services and now both television channels were available across most of the island. However, it still retained some advantage in areas that were 'shaded' from the Frémont Point transmitter by the island geography. Helpfully for Rediffusion, those areas where reception was poor coincided with its cable network coverage, notably along the south facing coasts, much of St Helier and Gorey. In these areas it was both able to retain and increase its base, particularly given that the higher frequency channel used by ITV was the weaker of the signals. It also took advantage of the extensive housing developments at Quennevais by pre-wiring new houses along side other services. Nevertheless, it began to reposition its retail arm by offering aerial television options in its Esplanade shop. It also beefed up its advertising image with the introduction of 'Rediffusion Regie' a comic parrot cartoon strip. Its retail ambitions were further bolstered by a move to a town centre location in Beresford Street in October 1967. This coincided with BET's increased ownership of the group and also spawned a new company image replacing the existing red and grey vans with a fresh new yellow and white livery.

Meanwhile its TVR business was still growing and plans were passed to build a new factory and research facility at La Pouquelaye, St Helier. The new buildings were officially opened in October 1968 despite stiff opposition from local residents during the planning stages. The company took the opportunity to relocate and at the same time upgrade its equipment. New distribution amplifiers were installed at the new site and at the same time a programme to replace the network cabling was begun. This was in line with the new BBC 2 colour service on 625 lines which was opened in London in

Image by the author

This picture shows a disused Rediffusion distribution network amplifier installed in Highlands Lane, St Saviour. This example is a transistorised amplifier, it required a mains supply to operate.

July of 1967. The existing cabling was not of sufficient quality to carry the higher bandwidths required for the new service and as there was to be a changeover to the new standard for all transmitters a complete new system was required. New self-supporting multicore cable was installed in some cases along side the existing cabling in order to maintain continuity of service. This cable consisted of 3 video pairs and 4 audio pairs. A new modulation system designated TD.80 was developed to carry the additional channels were sent over the same pairs modulating a higher frequency carrier (around 8 MHz) with a vestigial sideband transmission system using the lower resulting sideband. This approach, known as *tete-beche*, minimized interference between the two carriers and would eventually permit six channels to be transmitted.

Although it experimented with colour since the beginning of public trials, it was unable to receive and thus distribute a reliable signal from the UK at its Les Platons station. Had it retained a station on Alderney, the story may have been different, but effectively it had to wait until the opening of the service on the island before it could offer the additional feature. Nevertheless, in order to attract custom to its wired service, it offered black and white renters a 'colour rebate' or future discount on an upgrade when the service became available from local broadcast transmitters.

In May 1971 the MD of the company Quentin L Cazalet died aged 65. He had been with the company since its inception and had been an active contributor to the Jersey business scene having had interests in many local companies. The UK parent company later appointed one of its directors, Stephen W Wells to the post.

In the event colour TV did not arrive until June 1976 when the ITA opened its service, the BBC service, which shared the same transmitter, opened in time for Christmas that year. Over the past years until this event, the company had concentrated on its retail business where it had the advantages of a large parent company on the wholesale purchase side. During this period technologies were changing and although Video Cassette Recorders (VCR) using the Philips system were then becoming available it was not until 625 line transmissions were available that they could be sold in Jersey. Audio-visual retail, therefore, was largely in the television and stereo audio system market. Following the opening of colour broadcasts the company put into place a programme to upgrade its cable subscribers to 625 line working deploying four conversion teams. More than 4000 rented cable sets needed modifying or replacement, although dual standard sets had been provided for some time there were still 405 line units in place.

In 1977 Redifon computers were provided to the JEC as part of an updated billing system. Redifon Computers Limited was a Crawley based subsidiary of the parent company that had developed from telecommunications equipment manufacturing into computers. It was later renamed Rediffusion Simulation Limited as it took a position in the aircraft simulation and training market. The computer was provided through Rediffusion Reditronics, the new name for TVR since the previous year. TVR and Reditronics had provided both research and development as well manufacturing for many years. Much of its output was exported to other group members as well as outside customers. It also supplied the local market with products such as hotel music systems using multi-track magnetic tape based machines as well as public address systems and specialist equipment for the cable network.

From 1972 the UK parent company had entered the community access television (CATV) market, providing local content based services to several UK towns. In 1976 in Hastings it also tested the world's first optical fibre based television distribution system. Later in 1980 the UK government issued licenses for a number of cable television subscription channels, of which Rediffusion was awarded 5 channels which were grouped under the name of Starview. Initially this service was not available outside the UK mainland because of the unavailability of bandwidth for distribution on submarine cables to Jersey. However, in 1983 the geostationary telecommunications satellites operated by EUTELSAT (owned by a consortium of European telephone operators including BT) made channels available to broadcasters for the distribution of programming. This enabled Rediffusion Jersey to receive the UK cable channels directly via satellite communication. Consequently, the company proposed a three year experiment to introduce additional cable channels subject to a licence from the Telecommunications Board, which since the 1973 law had the exclusive rights to licence telecommunications systems including cable television systems. The company planned a three year experiment by investing £250,000 on two 3.7m satellite receiving dishes and decoding and distribution equipment at its La Pouquelaye site. Its existing network could carry up to 6 channels, meaning an additional 2 channels could be added following the startup of Channel 4 in the previous November. At that time around 6,500 homes were passed by the cable network in St Helier and a further 5,000 elsewhere. A certain amount of additional network equipment would be required, such as local amplifier systems and minor modifications to the main distribution network trunks. The experiment would first enable St Helier residents to take up the service and if successful a more ambitious investment would upgrade the entire network to up to 100 channels. This upgrade was estimated at around £10M. The monthly subscription for the two channels, one of which would mostly show films, would be an additional £8.

However, before the experiment could begin a number of official hurdles had to be negotiated. Planning permission for the dishes was granted early in 1984. In May the Broadcasting Committee gave its approval while reserving its right to vet programmes. Later that month the Telecommunications Board granted a two year licence beginning on 1 December but applied conditions that prevented the transmission 'adult' programmes and also the inclusion of local advertising, following lobbying from Channel Television. The two receiving dishes, one directed at Intelsat and the other at European Communications Satellite, were commissioned in July. Everything went according to plan and the CableVision service on Rediffusion channels 5 and 6 was launched on 1 December, with the company claiming over 400 pre-booked subscribers rising to over 1,000 by the year end. At the official opening ceremony the first face seen on the new channels was that of June Allez, who had also been the first presenter on Channel Television some 22 years earlier. Also at the ceremony were members of the Rediffusion (CI) board, members of the Broadcasting Committee and the Telecommunications Board.

Image by the author

Rediffusion channel selection switches as installed after the CableVision upgrade added extra channels to the network.

The restriction on local advertising in the licence was challenged in February 1985 by Senator Dick Shenton. He observed that the States Policy Advisory Committee had considered the inclusion of the restrictive clause was wrong, provided the advertising content was adequately monitored. The Telecommunications Board, however, refused to delete the clause saying that there was a lack of clarity of how the adverts could be monitored. There was some confusion on whose responsibility it would be and what laws applied. However, Shenton condemned this response as 'high handed' and avowed to bring the matter to the States. The motion was presented in April and after a long debate the proposition which was robustly attacked by the Telecommunications Board members was passed by 22 votes to 21 and the Board was duly required to remove the restriction, replacing it with a clause that referred to the IBA code of advertising practice and the UK Cable Authority, and the discretion of the Broadcasting Committee. Cablevision replaced its Screen Sports channel with the Sky Channel on 1 May 1985 which was launched on UK cable networks prior to the launch of the satellite service. By the end of 1986 the Rediffusion retail division was also offering satellite television to its customers as an alternative to its Cablevision service. An eclectic mix of a dozen English and continental channels were offered at £28.00 per month or outright purchase of equipment at £605.00. While this was not a cheap alternative it was however, a sign of things to come.

By now, however, things were changing at the top of the organization. In this respect Rediffusion was finding the competition in the UK cable TV market increasingly stiff. Following its ability to win just one of the UK governments cable franchise licence's at Guildford, Surrey, Rediffusion's parent company, BET, started restructuring the business and selling-off those parts which it did not deem as central to its activities. Its networks were now lagging the new technologies introduced by rival cable companies and so it decided to dispose of its research and development subsidiaries. Thus Reditronics and its workforce of 106 was acquired by SCK Holdings Limited, a venture company headed by UK MEP Sir Jack Stewart-Clarke, former managing director of Philips' subsidiary Pye of Cambridge.

Rediffusion continued to share the La Pouquelaye site with the new owners. The takeover was not however a success. The loss of Rediffusion contracts, particularly of the pre-recorded taped background music business, as a result of the reorganization of BET's structure, saw the closure of the business by the end of 1987. However, Channel Television purchased the site for £1.8M favouring the move over upgrading its existing studios.

BET's interest in cable television had now virtually ceased. In March 1988 its remaining shares in Rediffusion (Jersey) Ltd cable division were sold to a consortium of investors including Peter Funk, Deputy Derek Maltwood, the Guiton Group and Channel Television. The new company headed by chairman Peter Funk was renamed Jersey Cable Limited and Steve Wells remained as MD. Plans were announced to increase the cable channels on offer to more than 30 with network expansion to cover more than 85% of the island's population. At the time of the takeover the existing network passed over 5,000 homes from Gorey in the east to St Brelade in the west and served 1,450 subscribers. The takeover came as a surprise not least to the Telecommunications Board which was somewhat wrong-footed as the current licence only permitted 6 channels.

BET finally disengagement with the island in July 1988 when it sold its remaining businesses to Channel Television. Channel had already entered the retail and television rental market in June 1982 when it acquired the business of Regent Radio, Beresford Street, St Helier (opposite the Rediffusion retail shop) re-branding the business as Channel Rentals. Channel purchased for the sum of £5.5M the Rediffusion retail and satellite businesses in Jersey, Guernsey and Alderney as well as its security company Automatic Alarms Limited totalling 135 staff. Interestingly, the original merger had been proposed by Rediffusion some time earlier but talks between the companies had resulted in stalemate. However, when BET finally decided to concentrate on its core support services business, Channel was able to conduct a reverse takeover funded by a rights issue of 770,000 new shares at £3.90 and borrowings of £2.5M. The Rediffusion turnover for the previous year had been £4.3M of which £2.27M had been for television rentals and generating overall profit of £650k. After the acquisition the merged company had 95% of the television rental market in Jersey.

Rediffusion had been a part of island life for almost 40 years. It had sponsored many events including international golf and local football. Over the years it had also supplied the Public Address system for many major events around the island, including the Battle of Flowers. Many of the original shareholders and directors in the company had been local businessmen and it had enjoyed a good reputation for supplying first class training and employment to local workers in its varied activities.

After the buyout of the Rediffusion network, Jersey Cable continued to offer the same services. CTV increased its shareholding from 2% to 11% through the purchase of Rediffusion Channel Islands Limited retail division as a wholly owned subsidiary, but overall it was business as usual. Peter Funk, was the company chairman an American national who extensive experience in the media business on both sides of the Atlantic. At takeover the network served approximately 5,000 subscribers of which 1,450 also took its CableVision service.

During 1988 the development of the Sky television satellite system brought more competition to the television market, although initially the cost of connecting to satellite television was high and installing a receiving dish was complicated by outdated planning regulations. But gradually the satellite companies expanded their offering, first Sky opened its Astra satellite service with four channels and in 1990 British Satellite Broadcasting would launch five channels in competition. As the UK uptake grew prices for equipment came down. It was clear that in order to retain its market share Jersey Cable would need to expand its offer beyond the two additional channels not offered off air.

In late 1989 Jersey Cable began planning an upgrade to its network to increase the number of channels on offer. Already at this time in the UK cable networks were expanding and the availability of media was growing. However, at this time CTV was busy bidding to renew its Channel Islands ITV franchise and decided to dispose of is shareholding in the company. Nevertheless, Peter Funk stated that while this was disappointing it would not affect the plans for network expansion. The company pressed ahead with a request to the Telecommunications Board to amend its licence to permit expansion from 6 to 30 channels. The company had continued to operate from the former Rediffusion shop in Beresford Street which also served as the Channel Rentals outlet after the takeover, but when CTV withdrew from the consortium Jersey Cable had to find new premises. For a while it operated only from the cable distribution centre at La Pouquelaye but in 1991 it opened a retail outlet at 3a Colomberie.

The company pressed ahead with a request to the Telecommunications Board to amend its licence to permit expansion from 6 to 30 channels. In order to increase the number of channels offered it would be necessary to upgrade the distribution network from twisted pair to co-axial or fibre-optic cable in order to enable the additional bandwidth required for the transmission of all the additional channels. This would require a considerable investment in infrastructure. The purchase of the Rediffusion network therefore amounted to no more than the value of its existing subscriber base.

In the spring of 1990 the Telecommunications Board still had not responded to the request. Funk was becoming frustrated with the lack of progress and so went to the press to express his dissatisfaction. The Director of Telecommunications, Tom Ayton, responded to this publicity by blaming the lack of clear regulation for cable television in Jersey. However, Enid Quenault, president of the Broadcasting Committee speaking in support of Jersey Cable, suggested a temporary licence.

The existing network served much of the social housing developments in and around St Helier, and the Housing Committee expressed a preference for cable over a proliferation of television aerials and satellite dishes. Funk also believed that the Telecommunications Board was dragging its feet because it feared that the new proposed fibre-optic system could be used in competition for telephony, but he denied that this was the intention and underlined that Jersey Cable was primarily a media supplier. Under pressure from States' members the Telecommunications Board sought advice from the Attorney General on the legalities of any potential licence as when the

Telecommunications Law had been drafted there was no consideration of cable television systems.

In November 1990 Peter Funk was elected chairman of the UK Cable Television Association, the industry body that at that time represented more than 50 members with some 125 cable television franchises across the UK. The cable system was then at its peak serving up to 14M homes, IVS itself had around 250,000 subscribers in Andover, Oxford, Salisbury and Stafford.

The Telecommunications Board finally announced in May 1991 that it would issue a licence up to the end of 1993. This was issued on the understanding that suitable legislation for the control of cable television would be adopted when the UK Broadcasting Act came into force which, it was hoped, would be in 1991. A further stipulation for the building of the network was that it should be installed underground. This would of course greatly increase both the time and cost of deployment. However, Funk felt satisfied with the new licence and announced that work would continue the upgrade work that it had already begun to pass up to 7,500 homes to connect its existing 5,000 subscribers to the new network.

In May 1993, the Telecommunications Board proposed a new 10 year licence. This was to be tightly controlled in the absence of a specific Cable Television Law which had been expected since 1990, although other priorities had prevented the drafting of such legislation. The States approved the issuing of the licence the following month. The licence included conditions such as the need to install cable solely underground and also limiting the ability of the company to use its network for anything other than the delivery of television and associated audio.

Jersey Cable had a much lower profile than Rediffusion and played a less active role in island affairs. It got on quietly with the upgrade of its network adding extra channels to its service as they became available. In November 1993 Steve Wells retired as managing director and was replaced by ex-Rediffusion colleague Brian Hamilton. Hamilton had joined Rediffusion in 1976 and had transferred to Jersey Cable on takeover. He was formerly its technical director and had managed the upgrade programme. By the end of that year the company offered a choice of 22 channels in addition to the

Image by the author

Street cabinet and cable duct manhole cover installed by Jersey Cable during fibre upgrade.

four terrestrial broadcast services.

In 1993 the separation from CTV was completed with the relocation of the Jersey Cable technical department to Springside Industrial Estate. The old transmission network was now fully shut down as the upgrade of the network was now complete. Although all its customers were now connected via the new underground fibre-optic trunk network, the old twisted pair overhead network was still largely in place and was abandoned as it stood. The original plan for the network had, however, been considerably truncated and effectively was limited to the main built-up areas around St Helier, encompassing much of the States' social housing developments and larger private estates.

While Jersey Cable relied on satellite channels for its network in the UK many cable franchises were able to share content over relatively low cost circuits leased from BT or other network operators. The cost of individual satellite systems was still high and thus cable was an economic alternative for many Jersey viewers. But by the mid 1990s satellite television was beginning to have an impact on the profitability of cable television in the UK. Thus Peter Funk decided in 1995 to concentrate on Jersey Cable. The IVS business was sold to KPN of the Netherlands and the 59% of Jersey Cable shares held by IVS was sold to Carveth Limited one of the businesses in which Funk had a controlling interest.

By 1997 the number of subscribers had reduced to around 1,300 although the number of channels on offer had increased to over 50. Satellite receiver systems had reduced in price and Sky had become more aggressive in its marketing. However, when Channel 5 Television began broadcasting in the UK, Jersey Cable ran a series of adverts that proclaimed that Channel 5 would only be available to cable subscribers in the island. Channel 5 would not be relayed on Jersey terrestrial television broadcasts from Frémont Point as there was no available frequency for the additional channel. The island's geographic location close to France had a limited the ability of the island to utilise the same broadcasting plan as the UK. Thus Channel 5 would remain as a cable and, eventually, a satellite station until the introduction of digital terrestrial broadcasting scheduled for 2010. In

October of that year the Racing Channel was added to the network in an attempt to increase subscriptions.

Finally, in November 1997 Jersey Cable was absorbed by the recently formed Newtel Holdings Limited. Newtel comprised a number of investors including the Jersey Electricity Company and Carveth Limited although a number of other smaller investors also had holdings. The newly structured company was to shift the business focus towards telecommunications, effectively relegating the cable television business to the back burner.

As part of the revamp of the business, the Jersey Cable shop on the corner of Colomberie and Snow Hill, St Helier, was fully refurbished and at the same time extended to incorporate the telecommunications business. Newtel continues to supply cable television services to a diminishing subscriber base, the majority of its customers being within social housing developments. Cable television in the UK and particularly in Jersey faces increasingly stiff competition from satellite services, both subscription and free-to-air. Newtel was licensed as a telecommunications operator in January 2003 by the Jersey Competition Regulatory Authority under the Telecommunications (Jersey) Law 2002. The business continues to provide both telecommunication and cable television services from its new offices in David Place, St Helier.

Although it is now nearly 30 years since the replacement of the original Rediffusion overhead distribution network with an underground fibre distribution system, there is still considerable evidence of the former cables in and around St Helier.

Chapter 25 - Local Broadcasting Services

After the First World War, the military had realised the importance of wireless as a means not only of communication but also as a tool that could be used for the dissemination of propaganda. Therefore, the British Government came under pressure to introduce some level of regulation. By the early 1920's there were thousands of amateur wireless operators in the UK and thus the Government introduced a system of licensing, to be collected by the GPO. In February 1920 the Marconi Company began public broadcasting from Chelmsford. However, under the pretext that it was interfering with military communications broadcasting was suspended by the order of the Government. Two years later in February 1922 the Marconi Company was issued a licence, in October with the help of other British and American companies it formed the original British Broadcasting Company.

At this time most domestic radio receivers were quite crude and most belonged to enthusiasts who were prepared to make their own sets. However, as broadcasting started to develop the major electrical manufacturers began producing commercial receivers.

The British Government was still wary of radio and thus it stipulated strict conditions in its licence to the company. At the same time it introduced the Wireless Licence, set at 10/-, which again was to be collected by the GPO and a portion was handed to the company. The requirement for a licence to listen to broadcasts was therefore to extended to Jersey under the Wireless

Telegraphy Act, although it received a fairly poor service from the few transmitters around the UK. The Conservative Government decided to control all broadcasting and thus the British Broadcasting Corporation (BBC) was formed in December 1926.

During the 1920's many continental stations started broadcasting, including stations based in Paris, Toulouse, Berlin and Hilversum. In 1933 a new station very close to the Channel Islands began an English service broadcasting music after midnight. This was Radio Normandie[180] which had a 25kW transmitter located at Fécamp on the French coast east of Le Harvre and broadcast on a wavelength of 269.5 metres. Many islanders joined *l'Association des Auditers de Radio Normandie.* Radio Normandie became the platform for entrepreneur Captain Leonard Plugge who saw an opportunity to outflank the British authorities and established the International Broadcasting Company in 1931. He bought air time from many stations but Normandie (anglicized to Normandy in publicity literature) was closest to the UK and therefore became the company's main focus. Islanders were regular listeners to its recorded music broadcasts, so rare on the BBC, with such presenters as Roy Plomley, later famous for his BBC programme *Desert Island Disks.* Unfortunately, the station did not survive the war; the occupying German forces used the equipment to broadcast propaganda to the UK but the transmitter towers were blown up by the Germans following the D-Day landings. Also in 1933 the famous Radio Luxembourg[181] started transmissions from the Grand Duchy with a transmitter of 150kW on 1,191 metres LW.

The quality of reception of UK stations in the Channel Islands was less than satisfactory, especially with the growing number of continental stations causing interference. Despite this indifferent service, the GPO nevertheless levied the same licence fee for local listeners. Although as early as November 1934 there were calls for a local transmission service for the Channel Islands, it was not until after the Second World War that any

---

180 The history of broadcasting in the United Kingdom, Asa Briggs, Oxford University Press, 1995
ISBN 0192129309, 9780192129307
181 ibid

move was made to improve the quality of local reception, despite the requirement to pay the full Wireless Licence fee to operate a receiver set.

There was much enthusiasm for wireless in the island which spawned many radio dealers selling the new commercially made super-heterodyne radio receivers from manufacturers such as Philips, that were more reliable and easier to operate than earlier devices. The radio dealers also provided services for recharging batteries, often using their own dynamos or generators since there was no widely available commercial electricity supply before July 1925 and even after the opening of the Jersey Electricity Company, many properties in St Helier continued to use gas as the main source of energy.

Broadcasting of a sort was carried out, albeit on a very small scale, by the Jersey Evening Post in conjunction with the States Telephone Department, which used to relay the bi-annual Muratti cup football match. Speakers were set up outside the newspaper offices by a local radio dealer and crowds could follow the progress of the game as it happened.

A limited music and drama broadcast service was set up over the States Telephone network during the occupation. This was as a substitute for the confiscation of wireless receivers by the occupation forces. Special patching arrangements were made at the local switchboards to enable subscribers to listen to the service using their telephone receivers.

After the occupation broadcasting services in the UK gradually began to return to normal. Television was restarted in June 1946 and thereafter a number of enthusiastic local radio dealers reported intermittent reception of programmes from Alexandra Palace. Towards the end of the 1940s the Post Office re-designated frequencies for the expanding television service, opening up Band 1 VHF frequencies for the new service. This meant that reception in the Channel Islands became more difficult as the higher frequencies did not transmit as far as the former lower frequency ones.

In 1948, however, the BBC West of England Home Service began broadcasting highlights from the the annual local inter-island football match, the Muratti Cup, thus for the first time local events received wireless coverage. Wireless reception was still on the long and medium wavebands and the level of interference from both foreign stations and local electrical sources made for very difficult listening. The arrival of Rediffusion in 1949 and the opening of its wired relay network in late 1950 improved the quality of reception for many town dwellers.

Meanwhile, the BBC broadcast network was quickly being developed in the UK and there was hope that a Channel Islands transmitter would be built in the near future. Rediffusion opened a TV relay service in time for the Coronation of Elizabeth II in May 1953 but because of an administrative error in amending its GPO licence, the service was again closed until the following April.

Local transmissions of broadcast services did not progress in Jersey until 1955 when the BBC announced its intention in April of that year to open a television repeater transmitter at Les Platons to serve all the islands. The Les Platons site was selected because of its height above sea level, being practically the highest point in the Channel Islands. The contract for the building of the transmitter hall for the new relay transmitter was awarded to local builders, Messrs J A Parr, in April 1955 and the station commenced test transmissions in September.

The station opened initially broadcasting the BBC 405 line VHF television service on Channel 4 horizontally polarized (61.75 MHz video, 58.25 MHz audio) with a 1kW transmitter. Originally signals were taken off-air from the Wenvoe Band I Channel 5 vertically polarized (66.75 MHz video, 63.25 MHz audio) transmitter and rebroadcast as the BBC West of England Service. Band I (40 MHz – 70 MHz) transmissions had a significantly better propagation pattern than later Band III (174–240 MHz) VHF frequencies and thus this method of receiving and rebroadcasting the off-air signals at Les Platons worked reasonably well. Signals were vertically or horizontally polarized in order to enable a better usage of frequencies when planning the overall network. Different polarizations affected the way in which the receiving aerial was mounted. In the Channel Islands 405 line receiving aerials were mounted in the horizontal plane.

The quality of the retransmitted signal was subject to weather conditions and often during the equinox periods and in warm summer weather there would be significant co-channel interference from French channels and 'ghosting' caused by multi-path reception. Because of the difference in technology of the video signal (the French system used 819 lines) the interference resulted in two half pictures superimposed on the BBC picture, there was less interference on the sound than the vision since the French system used a different 'offset' frequency spacing for the sound carrier. There would often be the need to display notices to the effect that the reception was subject to interference and that there was no need for the viewer to adjust their set. The BBC changed its off-air reception from Wenvoe to North Hessary Tor when that transmitter opened for service in 1956 using Channel 2. This improved the stability of the received signal, although it occasionally reverted to Wenvoe when prevailing weather conditions caused interference or poor signal quality from North Hessary Tor.

Rediffusion had meanwhile been experimenting with reception of the new Independent Television broadcasts from London. It set up a relay station in Alderney and modified its network to enable the transmission of a second channel. Reception from Croydon on Channel 9 was not good, however, and no London service was ever relayed. It did briefly relay a French station from January 1958 until the opening of Southern Television in September that year when it switched its service to the IOW Chillerton Down transmitter which it relayed from its receiving station on Alderney over a GPO microwave link.

FM broadcasting (the Light Programme, the Home Service and the Third Programme) started in 1961 with the upgrading of the transmitter station at Les Platons. A new 47m (155 foot) transmitter tower was constructed to bear the additional aerial load. Stereo FM was introduced in 1984 but Radio 1 was not brought onto FM until spring 1992.

Independent Television

Independent Television for the Channel Islands was first mooted in 1959. Two separate companies were registered in the Royal Court: CI TV Limited and CI Communications TV Limited in September 1959. In January 1960 the share nominations of 84,000 10/- (50p) shares and 80,000 2/- (10p) shares respectively was registered. This was in preparation for the forthcoming applications for the ITV franchise for the Channel Islands. There were three bids entered from CITV, Channel Islands Communications (Television) Limited and Westward Television Limited. The winning bid from Channel Islands Communications (Television) Limited was announced in March by the Independent Television Authority.

Work on the new station began immediately. The Independent Television Authority (ITA), which was responsible for the operation and regulation of the industry, started preparatory engineering work to test the routing of the UK signal to the Channel Islands. A test site was installed on Alderney and by February 1961 plans for a relay station were being drawn up. A new transmitter at Stockland Hill near Sidmouth, Devon was opened for the new south-west station operated by Westward Television. This would be used as the relay transmitter for the Channel Islands. In March the provisions of the Television Act (1954) were extended to the Channel Islands by Order in Council at the request of the States of Jersey and Guernsey, permitting the broadcasting of ITV in the islands. Westward Television opened on 29 April and Rediffusion immediately substituted its programmes for the Southern service that it had been relaying.

The founders and investors of the company were drawn from local buinesses. Founder members included: Chairman - Senator George Troy (a long-time campaigner for local broadcasting), Harold Fielding, Arthur Harrison, managing editor of the Jersey Evening Post, Gervaise Peek, managing director of the Guernsey Evening Press, Senator Wilfred Krichefski, Mr Bertram Bartlett, Mr C. Forbes Cockell, Sir Reginald Biddle, Mr W. N. Rumboll, Mr I. McCormick, and Advocate K. Hooper Valpy. In all 181 shareholders were recorded.

It was not until the following year that a site for the new studios was found at Rouge Boullion near Robin Hood, St Helier. In September the ITA finally got the agreement of the Alderney States to build an off-air receiver and relay station and construction and testing was started using signals from the new Stockland Hill transmitter. Ken Killip was appointed the first general manager of the company in August. He had formerly been with the ITV franchise holder ABC Television in the Midlands. Channel Television opened its UK London office at 195 Knightsbridge early in 1962. This was a necessary requirement because of the rules applied by the ITA and also for attracting national advertising. In January the results of a competition for the new company logo was announced and at the same time building work on the new studio commenced. Channel TV also announced plans to publish its own ITV programme guide to be known as the Channel Viewer which would retail at 1/- (5p).

Meanwhile, planning permission was granted to the ITA for a new transmitter at Frémont Point. This was a controversial decision. The Committée de Beautés Naturelles had railed against the grant of planning permission by the Island Development Committee (IDC) at a site of some environmental sensitivity. A full debate over the decision was held in the States, but in the end it was agreed that the IDC had acted lawfully and therefore nothing could be done other than censure the committee. Construction of the 139m (456 feet) transmitter mast, which had been formerly operational at Lichfield (and before that Croydon), was begun in February by the contractors British Insulated Calendar Cables Limited and was quickly completed by the third week of March. Meanwhile construction of the transmitter hall continued. The relay station in Alderney was completed by the end of July and testing began between it and the now completed Frémont Point site using a diversified microwave link provided by the GPO. A cable link was constructed between the transmission station and the CTV studios via the GPO repeater station at Trinity Gardens, St Helier. The public enthusiasm for the new local station resulted in the August share issue being greatly oversubscribed. The ITA commenced test broadcasts

Image by the author

Fremont point transmitter tower showing the microwave dishes for the Guernsey links.

around the same time to allow the trade to set-up television sets for the new channel. Transmission parameters were 405 lines monochrome on VHF channel 9. A 10kW transmitter with horizontal polarization for both vision (194.75 MHz) and sound (191.25 MHz), in Band III was installed at the Frémont site. Thus for existing Band I BBC viewers it was necessary to install a second television aerial of smaller dimensions in order to obtain satisfactory reception. This resulted in something of a bonanza for local radio retailers.

Channel TV started broadcasts on 1 September 1962 with the official opening at 5.00pm presided over by the Lieutenant Governor General Sir George Erskine, Irvine Kilpatrick of the ITA and various dignitaries from the various islands. It began with some local programme content including local news, magazine programmes, a farming spot, local weather and a late evening news review in French. The main programme content was of course delivered from the ITV network and this was expected to bolster viewing figures. Early on it transpired that the transmitter coverage was not as good as expected and various adjustments were made over the next few months to improve coverage across Jersey and the other islands. Surprisingly, in the early days, surveys showed that locally made programmes were the most popular, and the national 'soaps' did not appear to inspire audiences. Naturally, this reflected in advertising revenues and after two years of operation the company was trading with an accumulated deficit of £70,000 (equivalent to more than £1M at the time of writing). This would have been a very large sum to recover from advertising revenues although it was subsidized from good revenues from the Channel Viewer. Negotiations with the ITA reduced the charges for transmitter rental which went some way to relieve the immediate problems.

Despite this deficit, the company's franchise was renewed in the 1964 bidding round, largely because tit was unopposed. The company sought a firmer financial footing and thus negotiated a sale of 51% of its stock to Associated British Picture Corporation (part owner of the Midlands region franchisee ABC Television). The same year the ban on cigarette advertising on television seriously affected the company's advertising revenues, the then buoyant local tourist industry relying on cheap tobacco products for a large proportion of its income. Channel was again awarded the franchise in 1967 once again being unopposed. In 1969 ITV joined BBC in introducing colour broadcasting using new 625 line UHF transmissions. For various technical and economic reasons, colour would not arrive in the Channel Islands until 1976, thus Channel was the only ITV station stranded in the black and white era.

Channel began to look for other options to boost its revenues and set up a subsidiary publishing company Channel Sun Publications. In 1971 it negotiated a contract with the States Telephone Department to publish the telephone directory in association with QB Printers of Colchester. The contract was for 40,000 copies per year for 5 years. The same year it opened its own studio link to Guernsey over a private microwave circuit from Frémont Point to the Guernsey Market building in St Peter Port. During the early 1970s Channel also made several abortive attempts to run a local radio station. In 1972 the ITA was renamed the Independent Broadcasting Authority (IBA) following the passing of the UK Sound Broadcasting Act 1972 and further consolidated by the Independent Broadcasting Authority Acts of 1973 and 1974. This gave the former independent television regulator further powers with regard to sound as well as vision broadcasting.

At the end of 1975 Channel further diversified its operations by entering the television rental market through a new subsidiary Channel Rentals Limited. This was strategically placed just ahead of the opening of Colour TV broadcasting which would begin at the end of July 1976. Colour receivers were at that time still complex and expensive, and consumers were much drawn to the rental market as a consequence of the relatively low initial cost and the inclusive maintenance support.

Planning for colour television had begun in 1974 with the IBA beginning tests in Alderney. In order to have colour television services in the Channel Islands an improved off-air receiving and relay system would be required. Colour television was allocated spectrum in the Ultra High Frequency (UHF) band (470 to 854 MHz) for use with an improved quality 625 line system using the PAL colour encoding system. These higher frequencies have a shorter range than the lower 405 line system frequencies and thus a more powerful relay station would be required on Alderney to ensure an acceptable picture quality. Alternative means of delivering the signal, such as via submarine cable, were, at that time, infeasible because of the required bandwidth and the capacity of available submarine cable technology. Direct microwave links to the nearest point in the UK were also infeasible because of the distance and the propagation path across the sea[182], which has detrimental effects on microwave signals. Consequently a new off-air receiver and relay station was commissioned for installation in Alderney. A revolutionary type of aerial called SABRE (Steerable Adaptive Broadcast Reception Equipment) was being jointly developed by engineers at the BBC and ITA. The new receiver used circuitry to enhance the response of the array in order to obtain the best signal from Stockland Hill whilst at the same time excluding interference from transmitters in Europe that vary with changing propagation conditions.

The plans to bring colour to the Channel Islands meant that some rethinking of the existing arrangements was necessary. Up to now the BBC and ITV had operated separate transmitters form different points in Jersey that were supposed to cover all the islands. This worked to some extent using 405 line transmissions (although Alderney often received a poor signal because of weather and other propagation reasons), but when colour was introduced using 625 lines the coverage would need enhancing. The BBC had improved its reception system in 1968 by moving the off-air receiver to Guernsey and relaying the signal to Les Platons over a microwave link. However, with the advent of colour it became apparent that sharing technology with ITV and utilizing its higher transmitter mast and improved off-air relay would be of great advantage.

Work on a new transmitter hall for the 625 equipment began in 1974 and at first the BBC, uncertain about funding, refused to give a commitment regarding whether it would join the project. However, in January it confirmed that it would start to deploy resources to install its section of the transmitter. The Frémont mast was increased in height by 5m to

---

[182] However, a microwave telephone circuit was constructed from the IOW to Alderney later. See above.

accommodate the 625 aerial array while permitting continued 405 operation.

From the beginning of May 1976 ITA engineers started tests in colour from Frémont initially broadcasting a locally generated test card, although occasional live transmissions were relayed as testing of the Alderney link was undertaken. At the same time Bill Kidd, local ITA engineer manager, announced that further relay stations would shortly be opened in Guernsey and later in Sark and Alderney to improve coverage of the new higher frequency transmission. Initially the retransmissions relied on temporary off-air receivers in Alderney until the new computer controlled SABRE system was commissioned. Meanwhile the BBC was working on its side of the new service and issued a statement hoping to start colour 'by February'.

The 625 line service broadcasting on Channel 41, horizontally polarized, was officially opened on 26 July 1976 by Lady Plowden, chair of the ITA. Also present were other ITA officials including Howard Steele and W Collingwood. This was followed by the opening of the relay station at Les Toulliers in Guernsey on Channel 54, vertically polarized. At the start of the ITV colour service CTV remained in monochrome as its budget had not planned for the early opening. The local television equipment manufacturer, RCA (Jersey) Limited, in seasonal good will, loaned CTV colour equipment for use over the Christmas period 1976. In the event the BBC began test transmissions in November and opened a full service on 16 December 1976, initially retransmitting an off-air signal received from Guernsey but this was changed in March 1977 to the new SABRE receiver collocated with the ITA in Alderney. BBC 2 was finally brought to the Channel Islands on Channel 44 with BBC 1 on Channel 51. The 405 line transmitter at Les Platons continued broadcasting BBC 1 only. Although the new SABRE receiver was the most advanced available and far superior to the earlier system, it did not provide a perfect solution and the quality of the television signals in the Channel Islands remained dependent on the vagaries of the weather.

In early January 1977 CTV demonstrated the new ITV Oracle service with a view to deploying locally when suitable computer equipment had been installed. At the time it was estimated that it would increase costs of television receivers by as much as £800! That same month ITV engineers adjusted the directionality of the Frémont point transmitter aerials to reduce interference on the nearby French coast.

In October 1978 a relay transmitter was being installed on the JEC power station chimney at La Collette, St Helier. The IBA engineers confirmed it would be commissioned in time for Christmas to improve reception along the south coast of Jersey. The BBC also shared the transmitter but work on its section was carried out separately. The vertically polarized Band C/D transmitter would provide BBC1 on Channel 55, BBC 2 on Channel 67 and ITV on Channel 59.

Throughout the summer of 1979, ITV technicians staged a strike and work to rule. This had a knock-on effect on CTV which had to reduce its programme schedule and fill missing national programming from locally derived material. The company estimated that the industrial action was costing it £7,000 per week in lost revenue. However, the company struggled on and was still in a financially fit position to bid for its franchise renewal in 1979. This time its application also came under the scrutiny of the States Broadcasting Committee. However, at a public meeting held in St Helier's Town Hall the company received the full support of the public and was commended by the ITA as a good example of a small franchise and also for pioneering new electronic news gathering technology. However, in February 1980 the ITA announced increased fees that would hike the annual bill for CTV by twelve-fold to £60,000 per year. This caused the directors of the small station to put a hold on its franchise application. There was worse to come. Fresh industrial action threatened the very existence of the company. It entered negotiations with local technicians eventually offering a 9% increase on 1 April 1980. Despite these difficulties the company made its successful franchise bid which was renewed unopposed. However, it continued to negotiate with the ITA over the increased fees, £50,000 of which was required to fund the newly proposed Channel 4.

At the end of 1981 CTV introduced the ITA Oracle service which included local content from September 1982. BBC had already started broadcasting its Ceefax service some time earlier after a joint technical standard had been agreed between it and the ITA. On 2 November 1982 Channel 4 started broadcasts from the upgraded Frémont point transmitter. The signal was again relayed over the SABRE system. The costs of Channel 4 through the ITA levy continued to be a bone of contention with CTV, however later in the month the ITA agreed a reduction to £10,000 per year. Killip, CTV's Managing Director, claimed that none of the promised Channel 4 advertising revenues had found its way to the company as a result of ongoing industrial action. Nevertheless, at the end of December CTV announced an 8% increase in profits. In October it finally opened a local studio in Guernsey following many administrative delays.

In spring 1984 a second television relay transmitter was opened at Gorey. The aerial was located at Mont Orgeil Castle, mounted on it outside wall. Transmitting BBC 1 on channel 54, BBC 2 channel 26, ITV channel 23 and Channel 4 on channel 29 all vertically polarized.

In July 1984 CTV announced a rights issue to raise more working capital. It increased its issued shares from £150,000 to £1M. By the end of the year colour TV licences outnumbered monochrome by two to one. At the end of 1985 CTV announced that it would switch mainland network provider to Television South. The same year the board of CTV became unnerved by proposed advertising on the BBC that was part of the Peacock Committee review of broadcasting initiated by the Thatcher Conservative government. This would, they asserted, surely spell the end of local television in the Channel Islands. However, in the end the idea was abandoned.

For some time CTV had sought to enlarge its studios in order to expand its business, particularly in related areas operated by its subsidiary companies. Its plans to expand its Rouge Boullion site had been rejected by the local planning authorities but in July 1987 it announced a wholesale move to a new site. The former Reditronics factory at La Pouquelaye had become vacant after the Rediffusion parent company had pulled out of cable relay services in 1985 and sought to sell off its business interests. The new owners of the factory operation had encountered financial difficulties and decided to close the business. CTV bought the site for £1.8M and later sold its existing studios to Yew Holdings for £1.5M. In August its long-serving managing director Ken Killip announced his retirement. He was replaced by the internal promotion of John Henwood.

The Rediffusion cable network was also sold the following year and CTV was part of a consortium that took it over, renaming the operation Jersey Cable. Later in 1988 CTV bought the remainder of the Rediffusion retail division that operated a sales and rental business in the Channel Islands for a sum of £5.5M, This purchase fitted in well with the CTV Channel Rentals business and created a near monopoly in the television rental sector. The purchase was funded by the release of 770,000 new ordinary shares and a £2.5M loan. The move to the new studios was completed in November and officially opened in April 1989.

Meanwhile the UK broadcasting review had resulted in the new Broadcasting Act which sought to shake up the way that radio and television was regulated. John Henwood express the view that this could result in the disappearance of small local stations. The CTV franchise was shortly due for renewal and a rival bidder was waiting in the wings headed by former CTV employee John Rothwell and local politician Senator Dick Shenton. In 1990 the company sold its interest in Jersey Cable in order to concentrate on its franchise bid. The proposed changes turned more political in May 1990 when the States Greffier made a plea to protect the Channel Islands area in an article in the IBA in-house magazine, Airwaves. However, George Russell, chairman designate of the new proposed Independent Television Commission (ITC) that would take over from the IBA under the new legislation, confirmed that local representation was important and that mergers permitted under the proposed Channel 3 franchise that would replace the current system should not be done without consideration of local needs. The same year the IBA transmitter network was privatized and renamed National Transcommunications Limited (NTL).

In September the changes under the new UK Broadcasting Act 1990 started to become clearer. Channel 4 would become independent and the current revenue and cost sharing arrangement with ITV would end. This encouraged CTV in its franchise bid as future costs were likely to be lower. The new tenders would be placed in January 1991 for services to begin under the new organization in 1993. In the event the threat posed by Rothwell did not materialize, however another bidder, CI3 entered the fray. This was headed up by Guy de Faye, another ex-CTV employee.

CTV had long produced a weekly television guide originally called the Channel Viewer and later changed to Channel TV Times, in order to take advantage of national advertising. However, changes under the Broadcasting Act required it to publish not only CTV programmes but also those of rivals. This additional cost was thought to make the publication unprofitable and thus it ceased publication on 15 October 1991. A further blow to its profits came on 5 October when it ceased tobacco advertising. However, its fortunes turned when it was awarded the new franchise on 15 November with a bid of just £1,000 per year. Its rival, CI3, had offered a bid of £102,000 but the ITC had taken into account CTV's long-standing performance and renewed its franchise. Henwood confirmed that it would continue its relationship with the Southern franchise for national programming. However, not all was good news. The company announced in January 1992 that it was in the red. At the end of the year it sold its shares in Independent Television News.

In December 1992 it was announced that the soon to be independent Channel 4 had awarded its network distribution services to BT. As a consequence a negotiation between BT and the Channel Islands telecommunications authorities resulted in the setting up of a direct connection for Channel 4 programmes to be transmitted from the UK to the Frémont point transmitter. This service would pass over a dedicated fibre in the CI No7 cable to Guernsey and then on to Jersey via the inter-island microwave system. This would greatly improve the reliability of the signal received in the islands. The Guernsey to Jersey microwave link was replaced in 1994 by a new inter-island submarine cable.

CTV returned to profit in 1996 although it again dipped into the red the following year. In June 2000 the company was sold for £16.3M to Media Holdings Limited, led by two former Harlech Television (HTV) executives, Huw Davies (chief executive) and David Jenkins (finance director). Under the ownership of Media Holdings an attempt was made to take over HTV which its then owner, Granada, was instructed to sell by the ITC. This was later abandoned and HTV was sold to the Carlton Group. CTV was again sold to the Yattendon Investment Trust plc 2001.

In 2004 NTL sold its broadcasting unit to the Australian owned Macquarie Communications Infrastructure Group. This was subsequently renamed Aquiva and later the Macquarie group divested its interests.

Finally, in 2011, CTV was sold by the Yattendon Group to ITV plc on 18 October 2011, after clearance by the Jersey Competition Regulatory Authority, thus ending 50 years of independence.

Chapter 26 - Local Sound Radio

Local commercial radio was first proposed in Jersey as early as 1945, when it was suggested that redundant German military transmitters could be used for a service. However, nothing came from the suggestion and, thereafter, other than frequent commentaries on the quality of BBC reception and the desirability of a local transmitter, local radio was not a matter of public concern before the late 1960s. Radio listening had become the main source of family entertainment since its beginnings in the 1920s but with the advent of television in the 1950s the BBC moved many of its stars over to the new medium, making radio the poor cousin. Television thus became mainly the preserve of the older adult population.

Younger people started to adopt radio as theirs, particularly with the advent of cheaper transistor radios and the increasing popularity of the car radio. Radio thus became a more mobile medium with the ability to take receivers out of the living room and onto the beach and into the workplace. The BBC did not immediately recognize this and young listeners that wanted to keep current with latest music and record releases turned to the only other source of English speaking radio – Radio Luxembourg.

Radio Luxembourg had been around since before WWII and had gone through many iterations, but in the late 50s had become a platform for exposure of record companies new records through many sponsorship deals with the main labels. Thus Luxembourg became a shop window with the record companies cramming as many new releases as possible into their sponsored slot, playing only tantalizing snippets of the recordings. This demand for popular music was quickly seen as a business opportunity by entrepreneurs and a search began to obtain an entry into the broadcasting market.

The British government was reluctant to address this matter and thus was born the offshore Pirate Radio stations. The phenomena was not new; there had been offshore broadcasting around Europe since the late 1950s but the British took to the idea like ducks to water and soon there was almost a dozen stations around the British coast. The pirates provided what Tony Benn[183], the Labour PMG, came to call 'audible wallpaper'. The stations were extremely popular, broadcasting almost non-stop music every day, tailored to suit different tastes. Radio Scotland's ship, the MV Comet, was fitted out in St Peter Port during December 1965 and shortly after that the JEP reported that a Jersey group of businessmen had made an application to the PMG for a radio frequency and if not forthcoming would consider a pirate ship. Publicity leaflets were said to be being distributed. This was the first time that radio in (or perhaps from Jersey – since this initiative may have been aimed at a wider audience) had been mentioned.

Despite their distance, reception of the offshore pirate stations was reasonably good in the Channel Islands, particularly during daylight hours. This was because of the 'skip distance' signal, which is the signal that is reflected back to earth by the ionosphere. Often such signals are stronger than the direct line signal, especially in built-up areas (it was often reported that the Thames estuary based pirates could not be clearly heard in the city of London which was the target audience!). Such was the case in Jersey, as the signals, particularly those of Radio Caroline South, Radio London and Radio 390 were readily received on car and portable radios.

While the pirate stations ruled the waves there was little more interest in the proposed local station, the threatened local pirate ship never materialized, although several local interests made (unsuccessful) applications to the PMG for frequency allocation during the pirate years. In April 1967 the States debated the order for the extension of the Marine

---

183 Benn, Tony (1988). *Out of the Wilderness: Diaries 1963-7*. Arrow

Broadcasting (Offences) Act 1967[184] that would, if agreed, extend the Law to Jersey. There was little opposition in the chamber of the States and the Law was duly introduced into Jersey concurrently with the UK on 14 August 1967.

In the UK the government recognized that the demise of the pirates would need to be addressed since they had created an expectation among the voting public. The BBC was consequently reorganized to provide a pop station. The new format for BBC radio was a range of stations named quite unimaginatively Radio 1, 2, 3 and 4. Apart from Radio 1, which portrayed a quasi-pirate image (and, indeed, employed many of the former pirate station presenters) but with a very much limited 'needle time' allowance, the remaining stations were very much in the traditional BBC mode. The pirates had generated a different approach to broadcasting. In contrast to the very formal paternalistic BBC presentation, the offshore presenters had created a more intimate relationship with the listener, similar to the small station format in the USA. This approach had created a cosier image for radio which in itself sparked off further debate over local radio within the UK. If the mass audience pirates could touch the local soul, then surely there was a place for local or community radio?

The reorganization of the BBC as a result of the closure of the pirate stations involved reallocation of the radio frequencies. Radio 2 was removed from its 1214 kHz (247m) MW slot but continued on its 200 kHz (1500m) LW frequency. Initially Radio 1 shared some of its output time with Radio 2, thus some programmes were simultaneously broadcast on both stations. The 247m wavelength was broadcast from a number of synchronized transmitters across the UK, thus in Jersey where reception relied on the skip distance signal, reception was less than satisfactory. The simultaneous reception of two or more skip signals caused a number of unpleasant effects including phasing, fading and distortion for the listeners. This was only mitigated by the occasions when it was in tandem with Radio 2 which was also available on the much clearer FM frequency, although at that time the FM band was not included on budget portable or car radios.

---

184 1967/1275 The Marine, etc, Broadcasting (Offences) (Jersey) Order

The scene was set and the following February the States began talks with the PMG for the allocation of a MW frequency for local radio. The JEP reported in its 28 February 1968 edition that the States had conducted discussions with the radio manufacturer Pye Limited of Cambridge as early as 1964. Pye had been involved with the setting up of Manx Radio that year and had sought to extend its business to another offshore jurisdiction. These talks were not disclosed at the time and nothing further developed from this meeting. At this time, however, the UK government was wont to allocate spectrum without a general European agreement for frequency coordination. Although this had been attempted several times since the war, many countries simply flouted agreed rules. As a consequence the airwaves were jammed with coexisting stations competing for space. In particular the medium waveband, which was at that time the preferred spot on the dial because the propagation patterns of these frequencies was particularly good for wireless broadcasting, was crowded with channels. Thus interference, unless very close to the transmitter, was a given particularly after dark. This reluctance to allocate frequencies had not seemed to be a problem for Manx radio, which started broadcasting in June 1964, initially on FM only. It received its 188m MW frequency in October that year. (This may have been because it was further away from the continent than the Channel Islands.) In 1968, the Labour government was also somewhat lukewarm on any commercial broadcasting ventures and instead commissioned the BBC to set up experimental community stations in various UK towns and cities.

The States of Jersey Sound Radio Special Committee was formed in May 1968 and initially comprised Senators J Le Marquand (President) and White, Deputies Ellis and Riley, and the Constable of St Lawrence. Deputy Bob Smale later replaced Le Marquand as President and he and vice president Deputy John Ellis went on a fact-finding visit to BBC Radio Bristol in October that year. At this time the Labour Government had permitted the BBC to set up a few experimental stations to sound out the demand for local radio broadcasting. For reasons which are not entirely clear, the BBC considered the Channel Islands an important strategic market for their local radio plans and made considerable effort to impress the Committee including the promise of outline plans for a local station. This is despite of the fact that the CI would represent, by a very large margin, the smallest regional station and potential audience proposed by the BBC.

At around the same time as the setting up of the Jersey committee, the Guernsey States also sanctioned a committee (the Post Office Board) to evaluate local radio. During the next year or so the committees from both islands met and compared ideas. Meanwhile, the pirates reappeared in the winter of 1970 when Radio Nordsee International (RNI)[185] started broadcasting first off the coast of Holland and then moving to the Thames estuary in March in time for the UK General Election, restyling itself as Radio Caroline during the run-up to voting day. Using the largest MW transmitter installed on a ship, its signal was clearly audible in the Channel Islands. Eventually in June 1970 the long awaited report, which was prepared by both the Jersey and Guernsey committees, was lodged *au Greffe* for discussion. It would appear that at this time the committees had come to a general consensus as to what a local station should be like. The emerging view from the report was that of largely talk and information based programming rather than the familiar image (with few exceptions) of commercial and BBC local radio today. Great emphasis was placed on the importance to the tourist and agriculture industries, women's magazine style programmes, sporting coverage, schools programmes and public affairs, the imparting of weather and traffic information and as an information service in the event of emergencies. No mention was made of music programming, whether recorded or live.

It is not clear how the committee arrived at this format. Perhaps it was by way of comparison to the Channel Television local programme format, which was a mix of news and cosy living room style magazine and discussion shows, albeit for just a few hours per week, or it may have been as a result of its discussions with the BBC. The BBC local radio agenda had been set with typical establishment concern for the arts and education with not a little steer from the Labour government, but it seems unlikely that such a prescriptive style would attract any commercial sponsorship. However,

---

[185] Broadcasting from the High Seas (Edinburgh, Paul Harris Publishing, 1976), ISBN 0-904505-07-3

consideration was given in the report to a States funded station as well as commercial or BBC licence fee based options. Channel Television was also considered as a potential operator, even though under the ITA rules television operators were barred from holding interest in other media, the ITA had indicated that there could perhaps be an exception to permit CTV from operating through a subsidiary company. CTV for its part recognized that if it were to operate the radio service, it would be prepared to work under any rules imposed by the insular authorities. In conclusion, the report recommended that CTV should be offered the contract given that the BBC's stated position was that it would not address any Channel Island station 'for at least 4 to 5 years'. The report also recognized the importance of a MW frequency for local listeners. At that time FM receivers were largely confined to Hi-Fi units and few car or portable radios had FM bands.

Needless to say the JEP immediately railed against this report. It first of all challenged the actual demand for a radio station. This was an obvious weakness in the report since both the islands' committees had acted in isolation, making their investigations and report without canvassing public opinion. There was a presumed requirement for a local radio station based, seemingly, on the gap left by the popular pirate broadcasts, since it was only after their demise that the issue had surfaced. Of course, the JEP also sought to protect its revenues. It concluded that a radio station would, it asserted, suck advertising away and result in redundancies and reduction of its public services. (It is now known that radio advertising has very little impact on newspapers since the types of advert that each medium carries is somewhat different[186]). The JEP was also miffed that it had been excluded from the report as a potential operator. Indeed, at that time the UK newspaper owners and the press unions were campaigning strongly to be included in any future commercial radio operations. Of course the JEP had reacted in a similar fashion to the opening of CTV itself, but at that time there had been no committee investigation into the provision of a local ITV station; it had simply been waved through the States with the extension of the Television Act 1954 to the island. The JEP editorial claiming that any additional advertising medium would seriously affect its revenues set off an

---

186 UK Competition Commission: Newspapers markets and the effects of the transfers

immediate spat between it and Kenneth Killip, the Managing director of CTV. Clearly each company considered that it had much to gain or lose following the outcome of the States' deliberations.

The JEP published in September 1970 a report from the Newspaper Society which was a partisan assessment of the perceived need and viability of commercial radio. The newspapers were clearly rattled by the prospect of competition for local advertising. The report's assessment was that commercial radio was not viable for populations less than 200,000 to 300,000. The report concluded that the local press should be majority shareholders in any radio ventures and that it could otherwise be harmed in terms of content or quality otherwise. In other words, it was a highly protectionist document designed to prevent competition. This was followed by a strongly worded editorial on the eve of the States debate declaring that all the community news and information services proposed by the Radio committee were already being provided by the press (ignoring the instant availability aspect of radio and the fact that the JEP publishes late in the day) thus questioning the very need of radio. The JEP was clearly also riled that CTV should be favoured instead of it (despite it then being a shareholder in the station).

The debate centred around the three proposals in the report: (a) to agree the principle of a radio station and to request allocation of a MW frequency from the UK authorities; (b) to agree the principle that a subsidiary company of CTV should operate the station and (c) to agree that the committee should be empowered to act on behalf of the States in preparing the appropriate legislation. The debate opened with Deputy Smale, the committee president, outlining the committee's conclusions with its Guernsey counterpart, that a 'community' radio station should be established jointly with Guernsey. Much reliance on the experience of the only such station in the UK, Manx Radio, was expounded. He referred to the immediacy of radio and that in the Isle of Man it was a much liked medium along the style of the service provided by the pirate stations. In the view of the committee this type of service could only be provided by radio. However, he emphasized that the importance of this debate was not who should operate the station but whether there should be one.

The debate centred on the commercial viability of a station. Estimates by the BBC to the committee had suggested an annual running cost of £25,000 with a further £20,000 in salaries. Annual CI spend on advertising was estimated at £2M and in the view of Deputy Smale, the gross loser would be CTV as its revenue would be split. (There was a general assumption that there was a fixed pot of advertising revenue that had to be shared. Subsequent experience shows that radio advertising only overlaps slightly with other media[187]). However the deputy did point to the fact that the IOM had no separate TV station and that local IOM TV advertising on the receivable ITV channels was practically non-existent. Senator Vibert, with considerable foresight, expressed concern that the Performing Rights Society would limit 'needle time' (the restriction on the playing of recorded music had been a severe limiting factor to the BBC for many years[188]) and thus the station would be filled with parish pump talk and thus become a 'dreary mediocrity' as, in his opinion, the best interview and discussion programmes would gravitate towards television. Recently elected Deputy Averty wanted the report referred back for further study claiming that it lacked depth and substance, which to some extent was true. Towards the end of the debate it was revealed that the Guernsey States had voted in favour of the proposition but had baulked at the idea of CTV operating the station. They would rather a Guernsey registered company, underlining the historic rivalry between the islands. At the division, however, the house voted 29 to 19 in favour of a local station and empowering the committee to request radio spectrum from UK authorities. However the proposition that CTV should run it was lost by 23 to 25. Averty's proposition to refer the matter back for further consideration was passed by 32 to 12. Thus the only outcome of substance was the decision to seek a suitable MW frequency allocation.

Meanwhile, in the UK, the recently elected Heath Conservative

---

187 Independent News & Media PLC and Trinity Mirror plc: A report on the proposed newspaper merger. Appendix 4.2 The newspaper industry in the UK.

188 The Performing Rights Society and the Musicians Union exercised a virtual stranglehold over the broadcasting of recorded music in the UK until The Monopolies and Mergers Commission's Investigation of the U.K. Music Market was conducted in 1988

government was pressing forward with plans for local commercial radio. The new Minister of Posts and Telecommunications, Christopher Chattaway, announced plans for a Sound Broadcasting Bill and a White Paper was published in March 1971. In May the Sound Radio Committee attended talks in London which were described as 'useful'. The UK Sound Broadcasting Act was given Royal Assent in July, however by August 1972 the committee reported that still no MW wavelength had been allotted to Jersey and that consultations with Guernsey on the matter were to take place. In the meantime, reorganization of BBC radio frequencies resulted in a degradation of its West of England service to the CI in both reception quality and news content.

This was followed by a long radio silence until May 1974, during which time the first independent local stations in the UK started, beginning with the London Broadcasting Company. The pot was further stirred with the reappearance of the pirate station, Radio Caroline on 23 February. At this time the committee president, Bob Smale, proposed the forming of a general Broadcasting Committee to oversee all aspects of local broadcasting. This was conditionally welcomed by CTV, as it still hoped to run the radio station, notwithstanding the new rules under the recently formed Independent Broadcasting Authority (IBA) ,which incorporated the regulatory functions of the ITA, that required a 'large measure of independence between radio and television contractors in the same area'. Indeed, maximum limits on shareholding in radio companies were specified for existing television companies and newspapers that held monopoly positions in any given area. The States approved the setting up of a permanent committee at its sitting on 12 June but despite heavy lobbying by CTV no decision was made regarding how a station should be set up or by whom. This was followed by more talks in Guernsey at the end of November although no further decisions were made. Frustration at the lack of progress surfaced in the summer of 1975 when a local businessman was quoted in the JEP to be considering setting up a local pirate station. The Broadcasting committee, however, was still undecided. It was still pondering the question of whether indeed a radio station was needed. Talks in London during July revealed that the BBC had no immediate plans for local radio in the islands since it was now awaiting the results of Lord Annan's inquiry into UK broadcasting set up by the new Labour government. However, the committee had woken up to the fact that perhaps a Channel Islands radio station was not achievable, and that a purely Jersey station may be the answer. This view was underpinned by the Defence Committee's view that a local Jersey station would be a great advantage in case of civil emergency. However, still no MW frequency was forthcoming and this would not be discussed until the next international spectrum coordination meeting in 1976. Radio Lions opened for hospital broadcasting in December, although not a wireless station and with a very limited audience, it was nevertheless the first local content service in Jersey.

Following the elections at the end of 1975 a new Broadcasting committee was formed consisting of the president, colourful local politician Senator Dick Shenton, with Deputies Carter, Troy, Thomas and Perkins. In the spring of 1976 they visited BBC Radio Solent and independent Radio Victory in Portsmouth and planned further visits to BBC and commercial local stations later. Wavelengths had finally been allocated to the Channel Islands in January and Shenton was altogether more upbeat about a local station claiming 'lots of interest'. Further talks with the Guernsey committee followed that month but without any firm conclusions. In July the committee visited Manx Radio and this was followed by visits to BBC Radio Carlisle and Border TV. In the meantime Shenton railed against the level of the TV licence payments in Jersey, claiming that the payment was the same as the rest of the UK but without the same level of service. He planned a revolt that would see the States withdrawing from full payment. However a meeting with the Home Office placated the Senator as promises of future services and a possible local BBC radio station were offered. Following this meeting the committee promised a new report by the end of the year.

In October, however, a series of public meetings to campaign for a local commercial radio station were held. Frustration at the glacial progress of the various committees over the years had prompted local interests to form a pressure group in an attempt to hasten progress. The promised report was delayed and the committee held further talks with the BBC in January 1977 after which Shenton bullishly announced that local radio would be on the air by November 1978, although at this stage it was still not clear in what form. Further talks were held in Guernsey with CTV and the National Union of Journalists which was campaigning for a role in local broadcasting. Shenton was nevertheless adamant that the island governments would retain control over local broadcasting.

The second report on local radio was presented to the States in April 1977. This was almost 10 years after the original committee had been set up and thus far very little progress had been made. It was claimed that this report had been delayed by the Guernsey committee. The report laid before the States was in essence not dissimilar from that of the earlier committee's. After investigating the alternatives it concluded that any station should be a community station, disseminating information and a mix of talk and music programming, though it should not be a 'wireless jukebox' (a recognition of the value of music not contained in the earlier report). The recommendation was that the States should follow the Guernsey States' 1970 decision that the station should be a self-funding commercially operated venture; that the Broadcast committee should be responsible for setting standards of advertising and hours of broadcasting; for the adjudication of any tenders for the operation; and that an independent Channel Islands Community Radio Advisory Council should be established. The report also revealed that talks with the BBC had shown that it was still enthusiastic about opening a CI station and that this would be unaffected by the eventual Annan inquiry report. The States debate was scheduled for the autumn.

The Broadcasting committee was quietly confident that its proposals would be adopted by the States. In its view the impact of a local radio station could only be good. Evidence from the UK had suggested that local commercial radio had had little impact on advertising revenue for other media. However, the committee was not prepared for the trump card played by the JEP. The managing director and future politician and subsequently Jersey's first First Minister, Frank Walker, commissioned an 'independent' report on local radio in Jersey[189]. The report appeared to be protective of the JEP's advertising and other revenues from a perceived threat from commercial radio[190]. It claimed that its research showed that there was no demand for any local station, contrary to the opinion of the Broadcasting Committee. It further sought to demonstrate that irreversible damage could be inflicted on the JEP [profits] if a commercial station were allowed to compete for what it claimed to be a finite pool of advertising. (The JEP failed to mention that since it had entered the local media market it had seen off two rivals). It issued a veiled threat that such damage would result in the closure of the paper, thus removing a pillar of the local establishment. It therefore concluded that if any station were to be sanctioned by the States it should be operated by the BBC thus funded from the TV licence fee. This report was published in the JEP and circulated in full to all States members just one week before the debate. Naturally, Shenton was outraged by this tactic but this was rebuffed by Walker as 'a predictable and emotive response'.

Prior to the debate Senator Ralph Vibert tabled an amendment that would require the full States to approve the station operator. Shenton dismissed this and also the amendment proposed for the Guernsey States' debate that the BBC should not be excluded from applying. Shenton declared that any local operator awarded a franchise by the committee would not be the BBC. At the same time the Guernsey politicians were incensed that Jersey should presume to itself appoint any future pan-island operator. Michael Barton of the BBC also stepped into the melee by issuing a statement to the effect that the BBC would be prepared to operate a station on either island if the other rejected them.

The debate was set for 28 September 1977; the opposition to the Broadcasting Committee's report was led by Senator John Averty, the long-time critic of any local radio. He claimed that there was no evidence that there was a need for a local station, this point was rebuffed by Shenton who pointed to the fact that the States had already voted in favour of a station

---

[189] Jersey Evening Post 19 September 1977
[190] The Advertising Association's Advertising Statistics Yearbook consistently shows that commercial radio only attracts about 4% of the total yearly spend on advertising.

in 1974. Averty was clearly of the view that existing businesses should be protected, assuming, without convincing evidence, that the advertising pool was finite. He went on at length at the possible damage to existing media in the island and expounded the view that if any station were needed then only CTV should be allowed to operate the station in order to prevent the total collapse of the local media industry. Other States members, including Senator Cyril Le Marquand, alluded to the lack of evidence that people wanted a station. (It is worth noting that in the 10 years since the formation of the first committee no canvassing of public opinion had ever been carried out). The discussion ping-ponged back and forth between both sides of the argument; clearly the JEP report and the Averty backed lobby had had the desired effect on the debate. Deputy Stan Le Ruez said that it was regrettable that the JEP report had been circulated in the way that it had been and that he believed that a local radio station would be of immense benefit to island life. There was also much feeling among the States' members present that the island was already well served by existing media and that radio would not add anything. Eventually after four hours a vote was called and the motion was narrowly defeated by 23 votes to 21. Immediately Shenton resigned from the presidency of the committee.

Following Shenton's resignation the States voted to appoint Averty as president of the Broadcasting Committee. Clearly the feeling within the house was that commercial radio was now a non-starter, despite the close vote. Initially Averty announced his intention to wind-up the committee claiming the matter was dead. However, shortly afterwards the BBC announced another reorganization of radio frequencies (as part of a European agreement to rationalize frequencies) which would impact heavily on the reception of services in the island. Consequently there was now still at least a lobbying position for the Broadcasting Committee and thus Averty followed up with a survey of public opinion early in November in order 'to establish once and for all whether there is a demand [for local radio]'. His new committee, however, was expressly against any station other than one run by the BBC. The survey was conducted by the Economic Advisor's Office which approached people in the street. The questions included: the quality of current radio reception; whether radio services could be improved;

whether they would prefer better reception on existing stations or the extension of stereo to the island or the introduction of local radio; if opting for local radio whether they would prefer it to be run (and thus funded) by the island authorities or the BBC; whether they would be prepared to contribute to the running costs; and general questions about broadcasting hours and programme content. Hardly open questions on the issues debated by the States since no option for commercial radio was included.

The results of the survey predictably favoured better reception over a local station. But of those that did want a local station the majority were against the BBC but only 12% would be prepared to contribute to a local station finances (strangely, this was never an option debated in the States or considered by previous committees). The survey, which was heavily biased towards general radio reception questions rather than a simple choice, concluded that (unsurprisingly) Radio 2 was the most listened to station and this would be the station most seriously affected by the BBC's frequency changes (Radio 2 being shifted from the Droitwich Long Wave transmitter in favour of Radio 4). From these results the committee somehow concluded that a local station should not be pursued, especially considered along with the Housing Committee's concern that such a station might require a number of addition 'essentiall employed' immigrants (which would swell the population by 2 or 3 possibly!). The committee also planned talks with the BBC which had already expressed continuing interest in the operation of a local station in the islands.

However, on November 29 a private sponsor published a separate survey in the JEP. Although this was somewhat akin to shutting the stable door after the horse had bolted, it did pose a more open set of questions:

1. Do you listen to ANY radio services during the day? Yes/No
2. Of all the radio services available in Jersey during the day, to which do you listen most? ………

3. Are you satisfied with the BBC's coverage of island affairs? Yes/No

4. Would you welcome LOCAL radio services? Yes/No

5. If answer to Q4 was Yes, briefly explain why you would welcome LOCAL radio. …….

6. If answer to Q4 was No, briefly explain why you would not welcome LOCAL radio. …….

7. Do you think the islands needs another news reporting service? Yes/No

This was followed a few days later by the setting up of a petition set up by the Action for Local Radio group. The petition attracted almost 6,000 signatures by 12 December, the day before Senator Averty's proposed debate on the abandonment of a local station. Written responses to the published survey numbered 181 of which 105 were for and 73 against a local station. The petition was presented to the States by Deputy Pierre Horsfall and in the debate Averty was accused by Shenton of bias and of having vested interests. Shenton also quoted from a letter sent from Frank Walker to the chairman of CTV (in which he was, by ownership of the Guiton Group, a shareholder) in which it was stated that he would not support any application for the BBC to get a wavelength. This was an affront to free speech, Shenton claimed. In the division the Averty proposal was defeated and the Broadcasting Committee was effectively told to get on with its job and investigate further with the BBC. Cracks in the all-island community station widened when the Guernsey Broadcasting Committee president, Colin McCathie, suggested that each island should perhaps follow separate roads.

In the New Year the action group started petitioning for the reappointment of Shenton as president of the Broadcasting Committee. The group spokesman, Vince Bobim, said that the group had no confidence in Averty. A campaign to increase membership of the group was launched and it vowed to pursue its agenda to the fullest. John Rothwell, later to become a States member, joined the group in February. Shenton asked questions in the States regarding the links between the JEP and CTV, requesting that an enquiry into this relationship should be undertaken. In answer to the questions posed, Averty confirmed that Guiton related companies and their board members held more than 30% of the company. This would have been contrary to the Annan Report's recommendations had that part of the Broadcasting Act applied in Jersey, although under recommendation from the States Law Officers, those parts had been removed from the Order in Council. However, Averty said that Frank Walker would be prepared to stand down from the board of CTV if it were to compromise the interest of either company. In addition Averty then concluded that an inquiry would serve no useful purpose.

Meanwhile at the end of January 1978 the Home Office approved plans for the BBC to seek a site for a MW transmitter in the islands. This would provide a better signal when Radio 2 was moved from the LW and could, the BBC asserted, also be used for a local station. It was by now becoming increasingly clear that the original all-island station was a non-starter. A meeting in May between the Guernsey and Jersey committees agreed a strategy to approach the BBC with a view to setting up a Channel Islands station, but with separate facilities for each island's news. Rothwell had earlier said that the campaign group did not want a BBC operation as it had already declared that it would only provide limited local programming daily. But Averty had effectively railroaded the process albeit having failed at his attempt to scupper the whole idea with the winding up of the committee. A meeting was then held in London at the end of June between the islands' committees and top BBC officials including its Director for Local Radio, Aubery Singer. Following the meeting Averty confidently announced that the new station would be operating 'within two years'. It was a strange coincidence that two weeks later the Guiton Group, owners of the JEP, announced that it had disposed of its CTV shares following an attractive offer from 'a local resident'. However CTV policy was to offer any shares to existing shareholders and its employees first and this offer was taken up. Nevertheless, it made no difference; the Guiton Group completed the disposal stating its view that impending legislation changes informed its decision.

By September talks between the BBC and the two island's committees were progressing and the BBC began the process of looking for suitable studio sites. This was prior to any States approval but nevertheless the BBC felt confident enough that it would be providing the service. Talks continued between the three parties and Averty announced that a proposition would soon be brought to the States. In the meantime the BBC located a site at Rue de Brabant, Trinity where it could erect a MW transmitter in order to improve local reception of Radio 2. Averty was campaigning for reelection to the States during this period and was making his version of local radio an issue on the basis that anything other than the BBC would not be good for Jersey. His stance was roundly criticized by Rothwell and others via letters to the JEP. Nevertheless, the issue was not a deal breaker and Averty was duly reelected and again appointed to the Broadcasting Committee as President. The committee justified its pro-BBC stance on a number of bases including the shifting of all costs to the BBC (through the licence fee), the BBC's track record for news gathering and the backing of its national network.

While negotiations with the BBC continued, the JEP again commissioned an independent report into the viability of commercial radio in the Channel Islands which it published on 9 March 1979. The Economist Intelligence Unit (EIU) produced a report on behalf of the paper that showed that commercial radio would appear to be unable to be self-supporting through advertising. The audience was estimated at a maximum of 110,000 being less than half of the audience of the IBA's current smallest station, Plymouth Sound. The total advertising market for the Channel Islands was estimated at £3M and the cost of running a station was estimated at £200,000 per annum (considerably more than the BBC's earlier estimate). In the UK the average share of advertising revenue for commercial radio was 3%[191]. Thus, the report concluded, that the station would need to attract at least twice the average share of advertising, naturally affecting drastically the profitibality

---

[191] The Advertising Association's Advertising Statistics Yearbook consistently shows that commercial radio attracts about 4% of the total yearly spend on advertising.

Image by the author

The BBC FM transmitter at Les Platons now operated by Arqiva.

of the JEP (and CTV). Its estimate of the loss of advertising for the JEP was set at a startling £60,000. No mention was made of the fact that the JEP was a virtual monopoly in its own sector.

This was followed by the publication of the Broadcasting Committee's proposals which set out 18 key points. In essence this set out that the BBC would fund the station entirely out of licence funds; that the States had the power to revoke the licence if standards were not met; that the BBC committed to 4 hours of local output per day for each island; that each island would set up a Radio Advisory Committee with certain mandated meeting requirements; that the BBC would supplement broadcasting with services from its national or other local services (both the two preceeding conditions typical of a BBC local station); that a person from the Channel Islands would be appointed to the general advisory council of the BBC and a number of ancilliary administrative requirements. In other words a typical BBC local station as elsewhere in the UK but with the addition of a States 'licence'. The BBC promised to bring the station on-air 'within two years'.

The local unions were split on the proposal; the TGWU supported a BBC station while the local NUJ branch expressed concerns on the 'independence of national news collection' used by the BBC. Shenton initially tabled a 'no-confidence' motion but withdrew it on the day of the States debate on 8 May 1979. The main opposition to the Committee's proposals was led by the newly elected Deputy John Rothwell, a long-time opponent of the BBC option and former spokesman for the Jersey Association for Local Radio. He argued in a lengthy speech that the independence of the islands would be compromised by the use of the national broadcaster for the local service. But in the end the house was persuaded to go with the Committee's proposals and accepted the motion by 29 votes to 20. The fact that the BBC had been engaged at length in discussions and that there was no obvious independent alternative provider was also a major factor. The Committee, having rejected a commercial player (including CTV) had foreclosed the matter, leaving only the choice of local radio or not.

Having achieved his aim, Averty resigned from the Broadcasting Committee at the end of the month. The States elected Deputy Arthur

Carter to replace him in June. Rothwell's bid failed as did Deputy Jane Sandeman's – both being opposed to the BBC option. Averty expressed concern that had Sandeman been elected that all his work could be undone. The process of adopting the BBC was complete when in July the Guernsey States also voted to accept the joint committee's proposals.

Local Radio at Last

After the formal invitation from the islands to set up local stations the BBC began its work. Tom Beesley of the BBC Local Radio Development Group arrived in Jersey in September 1978 to look for suitable premises for a studio. (This was more than 6 months prior to the States approving the BBC as the provider). Concurrently the BBC was also seeking planning permission for a MW transmitter site at Trinity, largely for Radio 2 but which would be shared with its local station. Despite the States decision there was still condiderable public concern over the BBC appointment but its selection was constantly asserted as the best option by Averty.

After the States vote the new committee visited BBC Carslile to view the operation of a local station even though this station had been visited by a previous committee. In September 1979 Peter Redhouse of the BBC visited Jersey again seeking a studio site. In the meantime the newly elected Thatcher government was looking closely at the BBC's licence fees and doubt was being expressed that the BBC would be able to procede with its local radio plans according to its original plans. This was likely to affect the expected opening date of the island station. It was now also coming to the public's attention that the local station would be far from local. Experience elsewhere showed how much the BBC relied on its networked programmes to support its community radio stations. The BBC tried to play down this aspect, emphasizing the importance (in its view) of local speeched-based radio.

The BBC finally admitted in February 1980 that its finance review would affect its planned roll-out of the local station. Carter tried to put on a brave face but it was clear that the original planned timescale would not be met. The budget restrictions also reduced Channel Island news coverage on the BBC southwest regional Radio 4 broadcasts. In May the BBC announced that it now intended to provide separate reporting for Jersey and Guernsey and admitted that it would also rely on Radio 2 relay to back-up its service, it also admitted that it would also be unlikely to be able to provide the four hours per day of local content originally negotiatied with Averty's committee. This was seized upon by Rothwell who criticized the previous Broadcasting Committee's naivety. It was now clear that the BBC had no idea exactly when the new station would open.

The Broadcasting Committee arranged a meeting with the BBC to get assurances on the number of hours of local broadcasts. The BBC were clearly suffering from the UK government imposed cash restrictions and would only guarantee an hour a day, which was described by Carter, the committee President, as 'a waste of time'. Some relief came with the opening of the BBC local Radio 2 MW transmitter in August 1980 on 330 metres (909 KHz), improving the MW reception on the island which had been poor since the frequency rearrangement in November 1978.

The talks between the committee and the BBC were stalled during this period which, the BBC asserted, was due to an ongoing dispute with the musicians union. However, this was dismissed as 'rubbish' by Rothwell who claimed it was just delaying tactics and that the wrong choice of local radio provider had been made. Eventually Carter, in a more upbeat statement, announced that the BBC was now offering more than one hour per day, although the exact amount was not clear. His committee was closely questioned in the States and it was clear that some members were now becoming disillusioned by the lack of progress. Carter's response was that the talks were progressing.

In January 1981 the BBC finally produced an offer in which they promised a station by the end of the year. The BBC, however, was still only offering one hour of local content per day, relying heavily on its networked programmes for the remainder. Many States members were disappointed with the state of play. Deputy Don Fillieu proposed that the offer be rejected and that the committee should start over. But Carter stuck to his guns claiming that he was merely carrying out the brief voted for by the States. The BBC offer was debated on 27 January. The BBC offer was roundly critized by Filieu and others. Carter defended the BBC position saying that a staff of four could not be expected to produce more than one hour per day of local content. Fillieu countered saying that that was derisory given the annual licence revenue collected in Jersey (around £3M at the time). The critics were, however in the minority and it was generally accepted by the States that the BBC should procede. Rothwell claimed that the output of the station was not what the public wanted. Carter countered in his summing up that a community station should not cater for either the young or old and that the young had a choice of stations to tune into (hardly true, given the poor reception of the BBC Radio 1 in the island and the absence at that time of any daytime commercial or even pirate stations). Clearly, the position had changed substantially from 'local' radio when the original committee was formed to 'community' radio with the decision to go with the BBC. At the division the BBC proposal was accepted by 29 votes to 19. Two days later Guernsey fell in line.

Having attained certainty, the BBC began in earnest setting up the station. After a spat with the Housing Committee the BBC announced that it would only employ three staff locally. This limited staff quota caused the local branch of the NUJ to criticize the station as having inadequate local reporters for news coverage. In April the Jersey Radio Council was set up to monitor the station with Bernard Dubras appointed as first chairman. By June the BBC announced that it would now not open until the spring of the following year and the station was estimated to require about £70,000 per year to operate. It considered a site at the former school in La Motte Street, St Helier. This was quickly rejected and it then next proposed using a portable studio provided by the BBC outside broadcast unit.

It appointed a station manager, Mike Warr, a BBC stalwart with 22 year's service, most recently with BBC Radio Solent. Two further staff were then proposed, both at salaries of approximately £8,500 per year. By October it had employed two reporting staff, Peter Gore also ex- Radio Solent and local journalist Mike Vibert. The BBC at last found a permanent studio site in Rouge Boullion, St Helier on the site of Somerland, a local textile company. The office building was swiftly renamed 'Broadcasting House' even though the BBC only occupied part of the total area.

BBC Radio Jersey came on air at 7.05 on 15 March 1982 broadcasting now on 1026 kHz (292m) and thus displacing Radio 2 from the local transmitter, albeit for just over 1 hour per day. The format for the station output was essentially talk with a limited amount of recorded music at weekends when it tied up with Radio Lions for a local hospital request programme. The BBC had also suggested that it could broadcast States' sittings. The station was opened officially by the Bailiff, Sir Frank Eraut, with the BBC represented by George Howard, Aubrey Singer and his successor Dick Francis. The BBC chose the same day to announce the extension of stereo FM to the island which it hoped to implement by Christmas 1983 for Radio 2, 3 and 4. In fact the stereo service began on 14 January 1984. By the year end the local station claimed to be broadcasting nearly 15 hours per week of local content, with just over 1 hour during each weekday and the remainder at the weekend. Initially its output relied heavily on local freelance contributors to fill its airtime. At this time the BBC local stations still had severe limitations on their recorded music output as a result of its negotiations with the PRS. This 'needle time' had to be shared across all its networks and thus left little over for local radio, while UK independent stations each had 9 hours or more per day available[192].

The station continued with this format for its first few years. At the end of March 1984 the station also had access to FM stereo on 88.8 MHz when the new transmitter was commissioned at Les Platons. By this time FM was becoming more widely available on budget receivers and car radios, and improvements in receiver technology ensured better reception of FM signals. There was a further disagreement with the Housing Committee over staff accommodation. The committee refused permission for the BBC to buy a house at St Mary and sublet it to its station manager. The BBC countered that it was unreasonable for its manager to continue to live in a boarding house. Eventually the matter was settled in the courts when the BBC's appeal was granted. Despite its defeat and a critical judgement, the Housing Committee stood firm claiming that the BBC had misled it over the matter. The committee President Hendric Vandervliet even threatened that the BBC's licence could be withdrawn unless it appointed a locally qualified

---

[192] Independent Radio: The first 30 years. Meg Carter. Published by the Radio Authority, Holbrook House, 14 Great Queen Street, London WC2B 5DG (Now Ofcom)

manager. The BBC's Director General, Alistair Milne, responded that the BBC would prefer to lose its licence rather than appoint an unqualified manager.

The BBC reported later that year that audience research indicated that its CI stations were amongst the most listen to of its local network. At the same time a report from the Channel Islands Broadcasting Advisory Council, one of the bodies overseeing local broadcasting, recommended more resources for the local stations, given the lack of BBC televison coverage locally. This would, it suggested, help to improve the balance and independence of local news. The BBC responded with extra funding. The local station would receive £25,000 extra per year and thus the station was able to advertise for two more part-time reporters and a permanent receptionist. The stations output had by now risen to an average of almost three hours per day of local content including a hour at lunchtime.

However, not all local listeners were content with the BBC station. In March 1985 rumours were abounding about a new 'pirate' station that could be set up to broadcast from France. The company involved was Normandie Sound Limited which had been recently formed by two local entrepreneurs: Nicholas Thorne and Peter McClinton (a local club disc jockey). The French government had the previous year liberalized the allocation of spectrum for local broadcasting and the company had applied for a local broadcasting licence in Normandy. It proposed a music station that would broadcast from Normandy to the Channel Islands with an English service aimed at the 18 – 40 agegroup. Naturally, with this new commercial threat, both the JEP and CTV expressed alarm and concern.

The new company met with the States Broadcasting Committee to outline its plans and also discussed the possibility of a local station. However, the States had agreed only to a BBC run station and any changes to that policy would need a fresh debate. At this time in the UK, the Conservative government was also starting a review of broadcasting and Leon Britain, the Home Secretary, commented in the press that deregulation of broadcasting would be considered and the broadening of its scope with the addition of community radio licences for very localized radio. It could be argued that because of its small size Jersey already had community radio. But the futher definition of community radio would be even smaller than the population of each of the islands. In a written reply to the House of Commons on the 11th July 1985, the Secretary of State for the Home Department the Rt. Hon. Leon Brittan, announced that, 'I have now decided to establish an experiment to test the viability of and scope for a range of different types of community radio, set up and financed in different ways in different locations. I hope that frequencies will be available for about 20 experimental stations which could begin broadcasting early next year'. On this basis Normandie Sound made an application to the Home Office for a licence. Unfortunately this application was not included in the stations that eventually received licences under the scheme.

In the meanwhile, the BBC made a further request to broadcast the States sittings. Radio Guernsey already carried government debates and the Jersey States debated the BBC's request in September. However the assembly decided to defer the decision pending a report from the Broadcasting Committee. The report was ready by April 1986 and the committee urged the States to accept the principle of broadcasting. The proposition was debated at the early June sitting and the principle was agreed. The first experimental airing of a States sitting was on 30 September. The house later agreed on 25 November to continue with the broadcasts.

At the end of 1986 the BBC licence was due for renewal and the then president of the Broadcasting Committee, Deputy Enid Quenault, in a report to the States recommending the renewal, praise was heaped upon the station's operation. The States subsequently agreed early in 1987 to the renewal of the BBC Jersey licence. Indeed the station did have a substantial following, according to BBC surveys it was the most listened to of its local stations with up to 39% of the population tuning-in during the average week. Given the lack of realistic choice and the sharing of its MW transmitter with Radio 2 in Jersey, this was hardly surprising. BBC local stations elsewhere did not generally have a large following, especially where confronted by commercial alternatives. However, the station now employed eight full time and 10 part-time staff and had increased its daily output to an average of six hours per day and was beginning to wean itself off its reliance on relayed programmes from the main networks.

Moves were also afoot in the UK to ease the restrictions on BBC needle time. The Monopoly and Mergers Commision began an investigation into the Performing Rights Society's management of copyrights. Around the same time the government Green Paper on the future of broadcasting was released with proposals for up to three national and more local commercial stations, and to ease regulation of the licencees. The Broadcasting Committee doubted that it would affect Jersey as there were no available frequencies, the BBC holding the only MW allocation. The JEP noted that national commercial stations would offer no serious challenge to its advertising.

The Great Storm during the night of 9/10 October 1987 took the BBC MW transmitter off the air for several days before rigging crews could be despatched to make the repair. That month also saw the opening of the English service on the French based station Radio Force 7 which broadcast from St Malo on 99.1 MHz FM. Initially three local club DJs, Stever Ross, Andy J and Biko broadcast in English for about 2 hours each day between 8-9AM and 1-2PM Later Jackie Monkman (ex-BBC Jersey) joined as a news reader. The station was fundamentally a music-based service and had been broadcasting in French since 1982 during which time it had gained popularity among the younger radio listeners in Jersey. The new English service quickly attracted a number of local advertisers.

In the spring of 1988 changes to the broadcasting regulations mean that the BBC local stations were at last granted more freedom to broadcast recorded music. The BBC started to rearrange its local stations into regional units. This meant that there were proposals to simulcast programmes across a number of local stations in the south-west region. These stations included BBC stations in both Jersey and Guernsey. A number of other tentative proposals, including the merging of the stations management, led the States of both islands' to believe that individual identities may be lost. This was in contrast to the intial idea of a Channel Islands station pursued by both islands broadcasting committees for almost 10 years. However, in the event, the only real change was a shared lunchtime programme and during the evenings when instead of retransmitting national network programmes, the BBC reverted to shared local station content broadcasting up to 19 hours per day.

In the summer of 1988 the French Commission Nationale de la Communication granted a licence to a station calling itself Contact 94 and the station announced that it would begin programmes on 5 September. Initial tests were transmitted from the stations studio at Cambernon near Lessay, Normandy on 94.4 MHz FM which was located above the pub/disco *La Campagnette* owned by one of its financial backers Frenchman Alain Tardiff. Other backers included directors Colin Nixon, Stephen Wallace Clipp, sales, and Chris Kirby also sales and marketing. Clipp later pulled out and was replaced by Michael Voison and John Billington. On 14 September the French authorites requested that the station moved to 96.3 MHz to reduce its interference on nearby staions. It moved frequency again in mid December to 97.7 MHz for similar reasons. The station rapidly became popular in the island broadcasting 24 hours per day of non-stop music with local and national advertising. Its music mix was based on pop from the previous 30 years and targeted the 20 to 40 age groups. Many of the stations DJs had previous radio experience, particularly with the offshore 'pirate' stations: Kevin Turner (aka Kevin Stewart, later to become a States of Guernsey Deputy), formerly with Radio Caroline and Laser 558; Mark Mathews, also ex-Caroline; John Tyler ex Caroline; Steve Ryan ex Capitol Radio Dublin; Neil McLeod and Liam Mayclem.

However, in the New Year the French authority for broadcasting changed to the Consiel Superior de l'Audio/Visual CSA) and it started a process of reviewing licenses issues under the former authority. As a consequence the licence was revoked and Contact 94 went off air in February while the company appealed the CSA's decision. However, the CSA did not interfere with Force 7 since it was only partially English it was declared to be of minority interest. It did, however, reduce the stations maximum output power, thus reducing its footprint in Jersey. The problems with Contact 94 soon spawned rumors of financial difficulties, which were strongly rebuffed, and also promped questions in the States. Senator Shenton asked the President of the Broadcasting Committee whether it had suggested to the French authorities that the station should shut down, but this was denied by the president of the Broadcasting Committee, Enid Quenault. She did, however, admit that the Broadcasting Committee had contacted the Home Office about the station, but denied that this had led to the French action. Quenault referred to the statement from the Home Office supplied to the company Radio Normandie Limited in its 1985 application:

`.. the United Kingdom would expect the French regulatory authorities to co-ordinate any such proposal with them and it would be open to the United Kingdom to make representations to the French authorities should it appear that the proposals entailed deliberate coverage of the Channel Islands or are otherwise in contravention of the ITU Regulations. Number 2666 of the Radio Regulations provides that, outside the high frequency broadcasting bands (within which overseas broadcasting is permitted), broadcasting should only use sufficient power to provide an effective national service. The United Kingdom could, and would, insist, therefore, that a station was not set up in France with the intention of providing a service to the Channel Islands.'

Given this position, it was clear that the Home Office could if it wished pursue this with the French authorities, but this would seem unlikely without the States intervention unless the signal interfered with the UK mainland. She denied that the States had instigated any action against the station. Further, despite the earlier position declared by the Home Office, the station had nevertheless been permitted to broadcast without, it would appear, any formal representation from the UK government. However, in her statement Quenault reported that the BBC had protested about Contact 94 broadcasting from French territory It is interesting to note that in answer to a further question from Shenton as to whether vested interests had protested (here Shenton was presumably alluding to both the JEP and CTV which had both previously had concerns about commercial radio). Quenault noted in her reply that both the JEP and CTV, in its TV programme guide, published the programme details of both Contact 94 and Force 7.

The French authorities prosecuted Contact 94 for broadcasting without a licence and for contravening the terms of any such licence by broadcasting on the wrong frequency and thus interfering with Radio Force 7. A court hearing at Coutances on 25 July 1989 nevertheless found them not guilty of broadcasting without a licence but fined them the equivalent of about £500 for the interference with £700 costs. The court also ordered the release of only some of the consficated transmitter and studio equipment. Steven Clipp, the company managing director, immediately lodged an appeal for the full return of assets.

The station returned to the air on 3 August on 94.6 MHz with reduced power using its secondary transmitter based at Cambernon pending the outcome of the appeal and the release of the main equipment. Normal programmes were resumed much to the relief of its listeners. A double page spread in the Jersey Evening Post followed in October describing both the station and the troubles with the French courts. The article noted that its original frequency had also interfered with the BBC Radio 3 signal under certain weather conditions but the new frequency was now more compatible. Danielle Berrou, the station's only French DJ, joined the team at around the same time. The station's format remained the same with local broadcasting from 6AM to 11PM with Sky channel Supergold relayed for the overnight period. It claimed to gather local news from a number of freelance sources in the islands and tried to keep it well balanced as it was broadcasting to a bi-lingual audience. It was not until the following April that the French courts released the main equipment and the station was able to return to its full output power from the transmitter at Lessay which was connected to the studio over a local microwave link.

Also in April 1990 the BBC appointed Robert (Bob) Bufton as manager of the Channel Islands stations, replacing Mike Warr in Jersey and Reg Brookes in Guernsey. This completed the reorganization of the Channel Islands network as a result of the cost cutting measures introduced by the BBC under pressure from the UK government.

In October the IBA announced its intention to make available franchises and associated wireless spectrum for commercial stations in the Channel Islands under the Broadcasting Act which would receive Royal Assent in November. The new stations would be licenced by the new Radio Authority and proposals to extend the legislation to Jersey was brought before the States by the Broadcasting Committee on 23 October. The Committee was completely behind the idea of introducing an additional commercial station, a dramatic change of position from its previousl stance. The debate in the house was relatively brief. Enid Quenault, the Broadcasting Committee president, stated that BBC Radio Jersey had no objection to the new station as it perceived it to be addressing a different audience. CTV, however, wanted the decision deferred in order to consult with local media, but Quenault dismissed this as the Home Office was pressing for a decision before the Royal Assent. This view was supported by Senator Derek Carter who did not believe that either CTV or the JEP would be adversely affected by commercial radio. Senators Brookes and Jeune and Deputy Maltwood all backed the opposite view, proposing that the Broadcasting Committee should have the final decision. Senators Jeune and Le Maitre also proposed that any new station might run contrary to the presiding zero growth policy of the States. However, this was dismissed by Quenault saying that the current debate was about the principle and that such matters should be properly addressed by the Economics Committee and the Regulations of Undertakings Law. However, she did say that the Broadcasting Committee would hold a veto over the Radio Authority's decision. At the division the proposal was passed by a large majority.

When the proposal was debated in Guernsey, the States waived the Broadcasting Act rule of a maximum stake of 20% for existing media companies in any radio franchise. Quenalut expressed surprise that such an action should be taken and that her committee had no such plans. However, both CTV and the JEP seized upon this differential and John Henwood, CTV's managing director, began lobbying for Jersey to follow suit. Henwood claimed that it would be a lost opportunity for the island to forgo the expertise already gained by the local broadcaster. He further claimed that CTV's local Oracle adverting service could be irrecovably damaged by the introduction of commercial radio. John Averty of the JEP backed Henwood, claiming that the island's advertising revenue cake was finite, and that inevitably it would have to be divied more ways, although he failed to confirm whether the newspapers parent company, the Guiton Group, would be making a bid for a local radio franchise.

Following the States approval Broadcasting Committee started the process of extending the Broadcasting Act to Jersey under Order in Council, which came into effect in July 1991. In the meantime local companies began jockeying for position ready for the eventual opening of the franchise invitation. In March Chris Kirby, commercial director of Contact 94, which continued broadcasting from France, announced its intention to enter the fray. There was a degree of confidence that its experience in the local market would stand it in good stead. Normandie Sound, headed by Spencer Pryor, who had lobbied unsuccessfully for a licence in 1985, also entered the race. Normandie Sound had opened a radio service, Eclipse FM, on the Sky satellite service. Another early contender was the newly formed company Channel Radio Limited headed by Richard Whatmore with Deputy Robin Rumbold.and local publisher Malcolm Hall. This consortium was later joined by Mike Warr, the ex-station manager of BBC Radio Jersey as a non-executive director. The Broadcasting Committee stated that it would invite the Radio Authority to oversee the award but retained the right to veto its decision and would prefer a locally based applicant.

In July Contact 94 received a boost when the French authorities agreed a five year extension to its licence. However, its frequency was once again changed, this time to 93.1 MHz. The change came into immediate effect but the stations loyal listeners were not easily confused. The station continued its broadcasting while in the meantime it planned its bid for a Jersey licence. By October when the Radio Authority, spokesman David Vick, announced that the 25 page franchise application paper would be posted soon after the Guernsey Order in Council had been approved, probably within the next few weeks, four potential bidders were in the frame. The successful bidder would receive an eight year licence and although the Broadcasting Committee would be consulted, the decision would be made by the Radio

Authority based on the commercial offer and viability of the franchisee. In preparation for the bid, Contact 94 announced that it would cease broadcasting on 29 November. This was no doubt so that its questionable legal status would not be a consideration in its bid.

In January 1992 the Jersey Evening Post announced that it too was entering a consortium including Channel Television (CTV) to make a bid for the commercial radio franchise. The consortium was chaired by the States Greffier, Edward Potter and other directors included Peter Picher, president of the Chanber of Commerce; Jurat Michael Bonn; Philip Mallet de Carteret; Senator Corrie Stein; John Henwood MD of CTV and John Averty MD of the Guiton Group, the JEP's owner. The consortium called itself Jersey FM and planned only 9 hours per day of local content, the remainder being outsourced. The station planned to use a studio at the CTV centre at La Pouquelaye. The consortium confirmed that both CTV and the JEP would hold the maximum 20% shares permitted under the Broadcasting Act.

However, shortly after the announcement, John Averty began lobbying States members to permit both CTV and the JEP to have a greater stake in the consortium. Averty argued that: *'As owners of the Jersey Evening Post Limited and Michael Stephen Publishers Limited we have a potential commercial interest in the matter, but our most important priority is to ensure that the people of Jersey get the best possible service in terms of news coverage and information.'* It should be recalled that this was the very same John Averty who had lobbied successfully in the late 1970's to prevent the introduction of commercial radio in the island. He further argued that in the prevailing recession the advertising revenues would be restricted and that sharing of facilities and expertise would be 'highly desirable'. However, Senator Dick Shenton, under whose direction the Broadcasting Committee had suffered the narrowest of defeats to introduce commercial radio in 1977, perhaps influenced by Averty's lobbying, proposed an amendment to the Broadcasting Act that would limit local media companies to just 5% holdings in the interest of preventing the continuance of the monopoly on news provision. Averty denied that it would be in the consortium's interest to fix advertising rates to disadvantage business and was confident that at worst the States would compromise at a holding of 12.5%.

On this occasion Averty's lobbying failed to convince the States and Senator Shenton had his revenge, if not sweet, satisfying, in that his amendment was easily passed in the chamber. As a consequence of this decision the Guiton Group decided to immediately withdraw from the consortium, although CTV remained in contention. Averty tried to put a brave face on the matter by declaring that the JEP would be willing to develop a good working relationship with the eventual winner. However, it later transpired that this 5% limit had not made its way into the law.

In February one of the potential bidders, Normandie Sound Limited, was declared '*en désastre*' (bankcrupt under Jersey law), effectively finishing its hopes of entering the competition. But it was replaced by the entry of a fifth bidder in March when the Wave 101 FM consortium entered a bid just hours before the deadline.

The final lineup before the Radio Authority were:

Jersey 1st FM headed by local politician Senator John Rothwell and whose lineup included locally based singer-songwriter Gilbert O'Sullivan. The group planned a 24 hour service with overnight service taken from Chiltern FM. It planned a varied format but its target audience was the 25 - 44 agegroup. It did not have a studio but planned a town centre location.

Channel Radio which had invoked the experience of Mike Warr to bolster its bid. It planned a 24 hour service of music aimed at the 25 – 55 agegroup with hourly news and overnight service from a satellite station from Midnight to 6AM. The company proposed to use a studio in Tunnell Street, St Helier.

Jersey FM was backed by the CTV group and proposed to use studio facilities on the CTV site. Its format was to be varied with a mixture of talk and music. It planned an afternoon and nightime tie-up with Capital Gold. Its 24 hour format would include lengthy news programmes at lunch and tea times and hourly bulletins.

Jersey Music Radio (the former Contact 94 team) chaired by David Overland with Chris Kirby and Jon Myers among the other investors. Its target audience was the 18 – 45 agegroup in a 24 hour format with 75% music of which 60% of the output would be dedicated to 60s, 70s and 80s pop. It planned only a headline news format and planned a linkup with a satellite station for overnight while also promising evening programmes dedicated to new releases. It stated that it would prefer an out of town studio location.

Wave 101 FM was chaired by Terry Lavery and also included John Dingle, John Dee and Chris Bee with Spencer Pryer and Peter Maclinton (who were both formerly with Normandie Sound); John Uphoff a radio presenter from Liverpool and Keith Skues former pirate radio presenter and latterly programme director with UK station Radio Hallam. It promised to use staff and expertise from Normandie Sound if successful. A studio located at 17 Bond Street, St Helier was proposed with local programming from 0600 to 2200 daily. The target audience was the 18 – 44 agegroup and the company promised to provide news coverage and programming for minority interest groups such as the local Portugese community. It also proposed to set-up a listeners panel to advise on content.

The full bids were published in the JEP shortly after the deadline for submissions to the Radio Authority.

The results of the bid were published on 13 May 1992. The winning bidder was Channel Radio. While the winners were naturally delighted the other contenders all expressed disappointment and gave explanations of their failure to win. The CTV backed consortium in Guernsey won the franchise with its Island FM bid. However, having failed to win the franchise in Jersey with its Jersey FM offer, John Henwood, the CTV MD, expressed doubt whether it would continue in the consortium.

Immediately after Channel Radio's success, Richard Johnson, one of BBC Jersey's most popular presenters jumped ship to take over as its station manager. He announced that the station would be different, not a competitor of Radio 1. He also said the name the station would broadcast under would depend on the final allocation of a frequency. The Radio

Authority allocated 103.7 MHz as the frequency for the new station in July and Channel Radio announced that it would begin broadcasting as Channel 103 from 10AM on 25 October 1992. The first programme was presented by Richard Johnson. Other presenters included John Upoff, Matt Howells, Marc Curtis and Caroline Leonard in the newsroom. The station started with a 24 hour format with overnight music programming from Chiltern FM via a satellite link. A 2KW transmitter was supplied by the IBA at the Fremont Point station.

The station quickly established itself and by August 1993 audience figures showed a weekly audience of 32,000 although BBC Jersey still claimed an audience of 53,000. Nevertheless, this audience share was in line with its business plan of 40% of the population and its chairman, Deputy Rumbold express satisfaction with the data.

At the BBC in October the management changed. Bob Button moved back to the UK at BBC Radio Devon and would be replaced by Bob Lloyd-Smith from BBC Radio Wiltshire in February 1994 concurrent with a move to a new building in Parade Gardens, St Helier. The new £300,000 studio complex provided more space for the BBC operations on Jersey. The new building was opened by the Lt Governor Air Marshall Sir John Sutton and other guests included the veteran broadcaster Alan Wicker, a resident of Jersey.

In July the Broadcasting Committee announced that is was in discussions with its Guernsey equivalent over a possible deregulation of the airwaves. Although the CI had limited options as the spectrum was already crowded as a consequence of the proximity to France. A grant of £80,000 was expected over the next 10 years from the new Channel 3 licensing arrangement. Deputy Crespel the committee president did, however, rule out the use of this money for subsidizing TV licenses for old age pensioners.

The radio market changed very little over the next few years save the gradual increase in the popularity of the commercial station over the BBC, by 1997 statistics showed that it now had the larger share of the local audience. The same year Classic FM decided not to start broadcasts from Jersey leaving the allocated 101.3 MHz frequency unused. In 1998 Channel 103 was acquired by the UK based Tindle Group which operated a number of small stations in the south-east. In September 1999 the local commercial radio franchise came up for renewal. Channel TV again attempted to enter into the local radio sector. A consortium calling itself Sunshine Radio Limited, which intended to broadcast as Sunny FM, was formed and consisted of CTV, local businessman Tom Scott and a number of CTV shareholders. It also placed a bid before the Radio Authority. However, in January 2000 Channel 103 had its eight year franchise renewed which would begin in October.

Alphabetical Index

3
30 channel PCM.................299, 300

A
AC1 signalling............368, 370, 371
AC11 Signalling.................269, 299
AC9 Signalling.........................299
Access Deficit Charge................320
ADSL.........332, 336, 339, 340, 342
Alcatel.......293, 296, 315, 326, 327
Assmann...........................246, 269
Asynchronous Digital Subscriber Line...332
Asynchronous Transfer Mode.....332
AT&T........................160, 293, 296
ATM..................................245, 332
Automatic Line Insulation Tester
....262
Automatic Telephone Manufacturing Company...........157
ayton..10
Ayton................293, 308, 348, 415

B
BABT..293
Bennett. 14, 48, 55, 85, 86, 87, 97, 98, 99, 100, 101, 102, 147, 149, 151, 157, 158, 159, 160, 165, 234, 348, 362
Breguet.....................................352
Bridgeman Committee..............113
Bright................................18, 34
British Approvals Board for Telecommunications..................293
BT and Securicor......................301
BT standard Telecommunications Equipment Practice 1E..............298
Bullivant and Allen................40, 46

C
Cable and Wireless....41, 297, 328, 345, 346
Cable Charge....................113, 129
CB10 133, 160, 164, 175, 177, 196, 197, 202, 203, 205, 209, 368, 369, 389
Cellnet.....301, 302, 313, 315, 320, 335

Centrex..........................327, 341
Chadwick............................39, 43
Channel Islands Electric Grid.....343
Channel Television...223, 344, 345, 375, 381, 405, 406, 410, 413, 427, 443, 444, 471
Chapter 23 - Civil Defence........385
CIEG.................................343
Cinergy........................330, 337
CIT-Alcatel....................293, 296
Civil Defence.................385, 386
Class Licence.....................338
Compagnie Générale de Construction Téléphonique........248
Coppock..213, 221, 242, 247, 256, 262, 263, 348
CPE...........293, 295, 301, 308, 314
Crampton............................11
CS Alert.....41, 115, 205, 291, 371, 373, 375, 381
CS Ariel............................142
CS Discovery...................316, 319
CS Electra.........................117
CS Faraday...............109, 132, 142
CS Iris............................316
CS Lady Carmichael 40, 41, 71, 381
CS Libation........................288
CS Monarch. 24, 26, 27, 28, 62, 63, 110, 115, 121, 125, 167, 276, 282, 291, 377
CS Poolster........................377
CS Resolute..................24, 34, 38
Customer Premises Equipment. 293

D
De Quetteville...........14, 29, 30, 45
Department of Trade and Industry 301
Digital Electronics Corporation..293
Directory Assistance System.....303
DTMF..................268, 269, 275
Dual Tone Multi-Frequency........268

E
Eady..................82, 85, 91, 95, 348
Ericsson...196, 230, 231, 248, 256, 258, 277, 279, 293, 296, 326
Ericsson of Beeston...196, 230, 231
Eurofone Limited.......................333
Exchange Telegraph Company. .256

F
Fabbrica Apparecchiature per Comunicazioni Elettriche...........248
FACE..........248, 257, 262, 263, 296
Freephone 304, 317, 318, 319, 320, 328, 333, 335, 344

G
GEC. 127, 132, 133, 202, 203, 205, 208, 209, 210, 211, 213, 214, 217, 231, 232, 238, 239, 243, 245, 248, 256, 257, 258, 259, 293, 296, 298, 299, 304, 371
Gerhardi..35, 36, 37, 38, 39, 50, 51
Gfeller........................237
Glass, Elliot and Company.....24, 46
GPT.........................304, 306
Graves.................18, 23, 25, 27, 28
Grimstone.................14, 15, 44
GSM.315, 320, 321, 323, 324, 327, 328, 329, 331, 335, 338, 339, 340,

343

Guernsey.net..............................331

H

Henley...18, 34, 45, 46, 47, 48, 49, 50, 51, 52, 55, 60, 62, 63, 109, 213, 351, 362

I

IBA...412, 430, 433, 435, 447, 455, 469

India Rubber Gutta Percha and Telegraph Co.............................361

Integrated Switched Digital Network......................................299

Interactive Communications.....327

International Subscriber Dialling 260

International Telegraph and Telephone..................................248

Internet Service Provider...........327

ISDN 299, 310, 315, 323, 327, 331, 340

ISP.............327, 329, 330, 331, 337

ITA...284, 376, 381, 405, 408, 426, 427, 429, 430, 431, 432, 433, 434, 444, 447

ITT........................248, 258, 296

J

JCRA 338, 339, 340, 341, 342, 343, 344, 346

JEC...185, 290, 326, 328, 332, 333, 336, 339, 403, 404, 405, 409, 433

Jersey Competition Regulatory Authority...................338, 419, 437

Jersey Eastern Railway Company Limited......................................361

Jersey Electricity Telecom..........326

Jersey Railways and Tramways Company Limited......................361

JT Rapid.............................339, 340

L

Le Feuvre 36, 45, 47, 48, 49, 50, 51, 55, 59

Lemon..........................81, 82, 348

Linkline......................................304

LocalDial............327, 329, 331, 344

Long Distance International Limited 333

Luxon......185, 189, 190, 191, 192, 193, 196, 205, 348, 367, 369, 370

M

May...........264, 281, 293, 348, 382

McNutt..............................341, 342

MCVF.................278, 366, 373, 377

Mercury Communications 297, 314, 318, 319, 320

microwave 282, 283, 284, 287, 291, 295, 299, 301, 303, 305, 308, 314, 321, 324, 339, 340, 376, 379, 381, 405, 425, 427, 430, 431, 436, 468

Mobile Non-Director automatic Exchange..................................263

Mobile Switching Centre............321

Morse. .11, 19, 54, 66, 68, 93, 108, 111, 126, 183

Motorola....................................281

MSC...321

N

National Telephone Company....73, 75, 77, 78, 91, 210, 355

Newall............14, 15, 16, 17, 18, 24

Newtel.....332, 333, 336, 338, 339, 341, 342, 343, 344, 345, 346, 419

No. 5 Electronic Switching System 296

O

Ofcom..346

Office of Utility Regulation.........344

Oftel 304, 309, 312, 316, 319, 320, 321, 328, 333, 335

OUR....................................344, 345

P

Packet Switchstream..................297

Passive Optical Network............311

Pentaconta......248, 261, 264, 273, 296, 297, 300, 312

Philips 45, 258, 266, 278, 292, 293, 296, 408, 412, 423

Philips en AT&T Telecommunicatie BV..296

Plessey...............246, 248, 258, 304

Preece...23, 24, 25, 26, 27, 28, 29, 45, 50, 52, 56, 135

Premium Rate Services.............303

Prestel................................287, 288

Printed Meter Check..................276

PRS..303, 462

PRX..258, 259, 260, 261, 263, 264, 266, 268, 270, 272, 273, 275, 276, 278, 280, 284, 285, 286, 292, 297, 300, 312, 322

PSINet..331

Pulse Code Modulation..............295

Pye TMC. .227, 244, 248, 258, 259, 263, 264, 266, 270, 275, 280, 281, 377

R

Racal-Vodafone.........................304

Rack Mounted Power Supply.....300

RCU.............................298, 300, 303

Rediffusion......202, 213, 223, 395, 396, 397, 399, 402, 403, 404, 405, 406, 409, 410, 412, 413, 414, 415, 416, 424, 425, 426, 434, 435

RMPS...300

Rumfitt.............................168, 177

SDH...........319, 324, 326, 331, 339

Siemens....19, 121, 141, 142, 143, 157, 160, 210, 367

sleeve control............................368

Sleeve Control...................208, 368

Société Anonyme de Télécommunications, Radioélectrique et Téléphonique 295

Standard Telephones and Cables 131, 210, 235, 244, 248, 258, 283, 293, 375, 377

Stanhope.123, 131, 133, 159, 166, 168, 170, 184, 189, 191, 348

Subscriber Automatic Line Tester 231, 262

Subscriber Carrier.....................261

subscriber private meter...........284

Subscriber Private Meter...........276

SuperNet............................329, 331

Synchronous Digital Hierarchy. .324

System X. 296, 298, 299, 300, 302, 303, 304, 305, 306, 308, 309, 311, 312, 313, 314, 318, 322, 326

Syvret......191, 196, 205, 207, 209, 213, 215, 217, 221, 239, 348, 370, 404

T

Telephone Engineering Centre 266, 305

Telephone Manufacturing Company 237

telex..................................276, 291

Telex 278, 280, 284, 286, 295, 297, 306, 314, 372, 373, 377, 380

TEP 1E...298

Terrestrial Trunked Radio...........390

Thorn-Ericsson...........................256

Total Access Communications System......................................301

V

Vanderhof...................................308

Varley...27

Vodafone....................301, 302, 304

W

Wave Telecom...........................345

Westward Television..........406, 426

Wheatstone. 10, 54, 63, 66, 68, 93, 108, 117, 352

www.ingramcontent.com/pod-product-compliance
Lightning Source LLC
Chambersburg PA
CBHW080538230426
43663CB00015B/2633